T0178135

Huyên Pham

# Continuous-time Stochastic Control and Optimization with Financial Applications

 Springer

Huyên Pham
Université Paris 7 - Denis Diderot
UFR Mathématiques
Site Chevaleret, Case 7012
75202 Paris Cedex 13
France
pham@math.jussieu.fr

*Managing Editors*

Boris Rozovskiĭ
Division of Applied Mathematics
Brown University
182 George St
Providence, RI 02912
USA
rozovsky@dam.brown.edu

Geoffrey Grimmett
Centre for Mathematical Sciences
University of Cambridge
Wilberforce Road
Cambridge CB3 0WB
UK
g.r.grimmett@statslab.cam.ac.uk

ISSN 0172-4568
ISBN 978-3-642-10044-4          e-ISBN 978-3-540-89500-8
DOI 10.1007/978-3-540-89500-8
Springer Dordrecht Heidelberg London New York

Mathematics Subject Classification (2000): 93E20, 91B28, 49L20, 49L25, 60H30

*Cover design*: deblik

Printed on acid-free paper

Springer is part of Springer Science+Business Media (www.springer.com)

To Châu, Hugo and Antoine

# Preface

Dynamic stochastic optimization is the study of dynamical systems subject to random perturbations, and which can be controlled in order to optimize some performance criterion. It arises in decision-making problems under uncertainty, and finds numerous and various applications in economics, management and finance.

Historically handled with Bellman's and Pontryagin's optimality principles, the research on control theory has considerably developed over recent years, inspired in particular by problems emerging from mathematical finance. The dynamic programming principle (DPP) to a stochastic control problem for Markov processes in continuous-time leads to a nonlinear partial differential equation (PDE), called the Hamilton-Jacobi-Bellman (HJB) equation, satisfied by the value function. The global approach for studying stochastic control problems by the Bellman DPP has a suitable framework in viscosity solutions, which have become popular in mathematical finance: this allows us to go beyond the classical verification approach by relaxing the lack of regularity of the value function, and by dealing with degenerate singular controls problems arising typically in finance. The stochastic maximum principle found a modern presentation with the concept of backward stochastic differential equations (BSDEs), which led to a very active research area with interesting applications in stochastic analysis, PDE theory and mathematical finance. On the other hand, and motivated by portfolio optimization problems in finance, another approach, called the convex duality martingale method, developed and generated an important literature. It relies on recent results in stochastic analysis and on classical methods in convex analysis and optimization. There exist several monographs dealing with either the dynamic programming approach for stochastic control problems ([FR75], [BL78], [Kry80], [FSo93], [YZ00], [T04]) or backward stochastic differential equations ([ElkM97], [MY00]). They mainly focus on the theoretical aspects, and are technically of advanced level, and usually difficult to read for a nonexpert in the topic. Moreover, although there are many papers about utility maximization by duality methods, this approach is rarely addressed in graduate and research books, with the exception of the forthcoming one [FBS09].

The purpose of this book is to fill in this gap, and to provide a systematic treatment of the different aspects in the resolution of stochastic optimization problems in continuous time with a view towards financial applications. We included recent developments and original results on this field, which appear in monograph form for the first time. We paid attention to the presentation of an accessible version of the theory for those who are not necessarily experts on stochastic control. Although the results are stated in a rather general framework, useful for the various applications, with complete and detailed proofs, we have outlined the intuition behind some advanced mathematical concepts. We also take care to illustrate each of the resolution methodologies using several examples in finance. This monograph is directed towards graduate students and researchers in mathematical finance. It will also appeal to applied mathematicians interested in financial applications and practitioners wishing to know more about the use of stochastic optimization methods in finance.

The book is organized as follows. Since it is intended to be self-contained, we start by recalling in Chapter 1 some prerequesites in stochastic calculus. We essentially collect notions and results in stochastic analysis that will be used in the following chapters and may also serve as a quick reference for knowledgeable readers. In Chapter 2, we formulate in general terms the structure of a stochastic optimization problem, and outline several examples of real applications in economics and finance. Analysis and solutions to these examples will be detailed in the subsequent chapters by different approaches. We also briefly discuss other control models than the one studied in this book. Chapter 3 presents the dynamic programming method for controlled diffusion processes. The classical approach based on a verification theorem for the HJB equation when the value function is smooth, is detailed and illustrated in various examples, including the standard Merton portfolio selection problem. In Chapter 4, we adopt the viscosity solutions approach for dynamic programming equations to stochastic control problems. This avoids the a priori assumption of smoothness of the value function, which is desirable as it is often not smooth. Some original proofs are detailed in a unifying framework embedding both regular and singular control problems. A section is devoted to comparison principles, which are key properties in viscosity solutions theory, as they provide unique characterization of the value function. Illustrative examples coming from finance complete this chapter. In Chapter 5, we consider optimal stopping and switching control problems, which constitute a classical and important class of stochastic control problems. These problems have attracted an increasingly renewed interest due to their various applications in finance. We revisit their treatment by means of viscosity solutions to the associated dynamic programming free boundary problems. We give explicit solutions to several examples arising from the real options literature. As mentioned above, the Pontryagin maximum principle leads naturally to the notion of backward stochastic differential equations. Chapter 6 is an introduction to this theory, insisting especially on the applications of BSDEs to stochastic control, and to its relation with nonlinear PDEs through Feynman-Kac type formulae. We also consider reflected BSDEs, which are related to optimal stopping problems and variational inequalities. Two applications in option hedging problems are solved by the BSDE method. In Chapter 7, we present the convex duality martingale approach that originates from portfolio optimization problem. The starting point of this method is a dual representation for the superreplication cost of options relying on powerful decomposition theorems in stochastic analysis. We then state a general existence and characterization result for the utility maximization problem by duality methods, and illustrate in some particular examples how it leads to explicit solutions. We also consider the popular mean-variance hedging problem that we study by a duality approach.

This book is based mainly on my research studies, and also on lecture notes for graduate courses in the Master's programs of mathematical finance at Universities Paris 6 and Paris 7. Part of it was also used as material for an optional course at ENSAE in Malakoff. This edition is an updated and expanded version of my book published in French by Springer in the collection *Mathématiques et Applications* of the SMAI. The text is widely reworked to take into account the rapid evolution of some of the subjects treated. A new Chapter 5 on optimal switching problems has been added. Chapter 4 on the viscosity solutions approach has been largely rewritten, with a detailed treatment of the terminal condition, and of comparison principles. We also included in Chapter 6 a new section on reflected BSDEs, which are related to optimal stopping problems, and generate a very active research area.

I wish to thank Nicole El Karoui, who substantially reviewed several chapters, and made helpful comments. Her seminal works on stochastic control and mathematical finance provided a rich source for this book. Several experts and friends have shown their interest and support: Bruno Bouchard, Rama Cont, Monique Jeanblanc, Damien Lamberton, Philip Protter, Denis Talay and Nizar Touzi. I am grateful to Monique Pontier, who reviewed the French edition of this book for MathSciNet, and pointed out several misprints with useful remarks. Last but not least, I would like to thank Châu, Hugo and Antoine for all their love.

Paris, December 2008                                              *Huyên PHAM*

# Contents

# Notation

## I. General notation

For any real numbers $x$ and $y$:

$x \wedge y = \min(x, y), \qquad x \vee y = \max(x, y)$

$x^+ = \max(x, 0), \qquad x^- = \max(-x, 0).$

For any nonnegative and nondecreasing sequence $(x_n)_{n \geq 1}$, its nondecreasing limit in $[0, \infty]$ is denoted by $\lim_{n \to +\infty} \uparrow x_n$.

For any sequence $(x_n)_{n \geq 1}$, $y_n \in \text{conv}(x_k, k \geq n)$ means that $y_n = \sum_{k=n}^{N_n} \lambda_k x_k$ where $\lambda_k \in [0, 1]$, $n \leq k \leq N_n < \infty$ and $\sum_{k=n}^{N_n} \lambda_k = 1$.

## II. Sets

$\mathbb{N}$ is the set of nonnegative integers, $\mathbb{N}^* = \mathbb{N} \setminus \{0\}$.

$\mathbb{R}^d$ denotes the $d$-dimensional Euclidian space. $\mathbb{R} = \mathbb{R}^1$, $\mathbb{R}_+$ is the set of nonnegative real numbers, $\mathbb{R}_+^* = \mathbb{R}_+ \setminus \{0\}$ and $\bar{\mathbb{R}} = \mathbb{R} \cup \{-\infty, +\infty\}$. For all $x = (x^1, \ldots, x^d)$, $y = (y^1, \ldots, y^d)$ in $\mathbb{R}^d$, we denote by $.$ the scalar product and by $|.|$ the Euclidian norm:

$$x.y = \sum_{i=1}^{d} x_i y_i \quad \text{and} \quad |x| = \sqrt{x.x}.$$

$\mathbb{R}^{n \times d}$ is the set of real-valued $n \times d$ matrices ($\mathbb{R}^{n \times 1} = \mathbb{R}^n$). $I_n$ is the identity $n \times n$ matrix. For all $\sigma = (\sigma^{ij})_{1 \leq i \leq n, 1 \leq j \leq d} \in \mathbb{R}^{n \times d}$, we denote by $\sigma' = (\sigma^{ji})_{1 \leq j \leq d, 1 \leq i \leq n}$ the transpose matrix in $\mathbb{R}^{d \times n}$. We set $\text{tr}(A) = \sum_{i=1}^{n} a^{ii}$ the trace of a $n \times n$ matrix $A = (a^{ij})_{1 \leq i, j \leq n} \in \mathbb{R}^{n \times n}$. We choose as matricial norm on $\mathbb{R}^{n \times d}$

$$|\sigma| = (\text{tr}(\sigma \sigma'))^{\frac{1}{2}}.$$

$\mathcal{S}_n$ is the set of symmetric $n \times n$ matrices and $\mathcal{S}_n^+$ is the set of nonnegative definite $A$ in $\mathcal{S}_n$. We define the order on $\mathcal{S}_n$ as

$$A \leq B \iff B - A \in \mathcal{S}_n^+.$$

The interior, the closure and the boundary of a set $\mathcal{O}$ in $\mathbb{R}^d$ are denoted respectively by $\text{int}(\mathcal{O})$, $\bar{\mathcal{O}}$ and $\partial \mathcal{O}$. We denote by $B(x, r)$ (resp. $\bar{B}(x, r)$) the open (resp. closed) ball of center $x \in \mathbb{R}^d$, and radius $r > 0$.

### III. Functions and functional spaces

For any set $A$, the indicator of $A$ is denoted by

$$1_A(x) = \begin{cases} 1, \ x \in A, \\ 0, \ x \notin A. \end{cases}$$

$C^k(\mathcal{O})$ is the space of all real-valued continuous functions $f$ on $\mathcal{O}$ with continuous derivatives up to order $k$. Here $\mathcal{O}$ is an open set of $\mathbb{R}^n$.

$C^0(\mathbb{T} \times \mathcal{O})$ is the space of all real-valued continuous functions $f$ on $\mathbb{T} \times \mathcal{O}$. Here, $\mathbb{T} = [0, T]$, with $0 < T < \infty$, or $\mathbb{T} = [0, \infty)$.

$C^{1,2}([0, T) \times \mathcal{O})$ is the space of real-valued functions $f$ on $[0, T) \times \mathcal{O}$ whose partial derivatives $\dfrac{\partial f}{\partial t}, \dfrac{\partial f}{\partial x_i}, \dfrac{\partial^2 f}{\partial x_i \partial x_j}$, $1 \leq i, j \leq n$, exist and are continuous on $[0, T)$ ($T$ may take the value $\infty$). If these partial derivatives of $f \in C^{1,2}([0, T) \times \mathcal{O})$ can be extended by continuity on $[0, T] \times \mathcal{O}$ (in the case $T < \infty$), we write $f \in C^{1,2}([0, T] \times \mathcal{O})$. We define similarly for $k \geq 3$ the space $C^{1,k}([0, T] \times \mathcal{O})$.

Given a function $f \in C^2(\mathcal{O})$, we denote by $Df$ the gradient vector in $\mathbb{R}^n$ with components $\dfrac{\partial f}{\partial x_i}$, $1 \leq i \leq n$, and $D^2 f$ the Hessian matrix in $\mathcal{S}_n$ with components $\dfrac{\partial^2 f}{\partial x_i \partial x_j}$, $1 \leq i, j \leq n$. These are sometimes denoted by $f_x$ and $f_{xx}$. When $\mathcal{O}$ is an open set in $\mathbb{R}$, we simply write $f'$ and $f''$. The gradient vector and the Hessian matrix of a function $x \to f(t, x) \in C^2(\mathcal{O})$ are denoted by $D_x f$ and $D_x^2 f$.

For a function $f$ on $\mathbb{R}^n$, and $p \geq 0$, the notation $f(x) = o(|x|^p)$ means that $f(x)/|x|^p$ goes to zero as $|x|$ goes to zero.

### IV. Integration and probability

$(\Omega, \mathcal{F}, P)$: probability space.

$P$ a.s. denotes "almost surely for the probability measure $P$" (we often omit the reference to $P$ when there is no ambiguity). $\mu$ a.e. denotes "almost everywhere for the measure $\mu$".

$\mathcal{B}(U)$: Borelian $\sigma$-field generated by the open subsets of the topological space $U$.

$\sigma(\mathcal{G})$: the smallest $\sigma$-field containing $\mathcal{G}$, collection of subsets of $\Omega$.

$Q \ll P$: the measure $Q$ is absolutely continuous with respect to the measure $P$.

$Q \sim P$: the measure $Q$ is equivalent to $P$, i.e. $Q \ll P$ and $P \ll Q$.

$\dfrac{dQ}{dP}$: Radon-Nikodym density of $Q \ll P$.

$E^Q(X)$ is the expectation under $Q$ of the random variable $X$.

$E(X)$ is the expectation of the random variable $X$ with respect to a probability $P$ initially fixed. $E[X|\mathcal{G}]$ is the conditional expectation of $X$ given $\mathcal{G}$. $\mathrm{Var}(X) = E[(X - E(X))(X - E(X))']$ is the variance of $X$.

$L^0_+(\Omega, \mathcal{F}, P)$ is the space of $\mathcal{F}$-measurable random variables, which are nonnegative a.s.

$L^p(\Omega, \mathcal{F}, P; \mathbb{R}^n)$ is the space of random variables $X$, valued in $\mathbb{R}^n$, $\mathcal{F}$-measurable and such that $E|X|^p < \infty$, for $p \in [1, \infty)$. We sometimes omit some arguments and write $L^p(P)$ or $L^p$ when there is no ambiguity.

$L^\infty(\Omega, \mathcal{F}, P; \mathbb{R}^n)$ is the space of random variables, valued in $\mathbb{R}^n$, bounded, $\mathcal{F}$-measurable. We sometimes write $L^\infty$.

## V. Abbreviations

ODE: ordinary differential equation

SDE: stochastic differential equation

BSDE: backward stochastic differential equation

PDE: partial differential equation

HJB: Hamilton-Jacobi-Bellman

DPP: dynamic programming principle

# 1

# Some elements of stochastic analysis

In this chapter, we present some useful concepts and results of stochastic analysis. There are many books focusing on the classical theory presented in this chapter. We mention among others Dellacherie and Meyer [DM80], Jacod [Jac79], Karatzas and Shreve [KaSh88], Protter [Pro90] or Revuz and Yor [ReY91], from which are quoted most of the results recalled here without proof. The reader is supposed to be familiar with the elementary notion of the theory of integration and probabilities (see e.g. Revuz [Rcv94], [Rev97]). In the sequel, $(\Omega, \mathcal{F}, P)$ denotes a probability space. For $p \in [1, \infty)$, we denote by $L^p = L^p(\Omega, \mathcal{F}, P)$ the set of random variables $\xi$ (valued in $\mathbb{R}^d$) such that $|\xi|^p$ is integrable, i.e. $E|\xi|^p < +\infty$.

## 1.1 Stochastic processes

### 1.1.1 Filtration and processes

A stochastic process is a family $X = (X_t)_{t \in \mathbb{T}}$ of random variables valued in a measurable space $\mathcal{X}$ and indexed by time $t$. In this chapter and for the aim of this book, we take $\mathcal{X} = \mathbb{R}^d$ equipped with its Borel $\sigma$-field. The time parameter $t$ varying in $\mathbb{T}$ may be discrete or continuous. In this book, we consider continuous-time stochastic processes, and the time interval $\mathbb{T}$ is either finite $\mathbb{T} = [0, T]$, $0 < T < \infty$, or infinite $\mathbb{T} = [0, \infty)$. We often write process for stochastic process. For each $\omega \in \Omega$, the mapping $X(\omega) : t \in \mathbb{T} \rightarrow X_t(\omega)$ is called the path of the process for the event $\omega$. The stochastic process $X$ is said to be càd-làg (resp. continuous) if for each $\omega \in \Omega$, the path $X(\omega)$ is right-continuous and admits a left-limit (resp. is continuous). Given a stochastic process $Y = (Y_t)_{t \in \mathbb{T}}$, we say that $Y$ is a modification of $X$ if for all $t \in \mathbb{T}$, we have $X_t = Y_t$ a.s., i.e. $P[X_t = Y_t] = 1$. We say that $Y$ is indistinguishable from $X$ if their paths coincide a.s.: $P[X_t = Y_t, \forall t \in \mathbb{T}] = 1$. Obviously, the notion of indistinguishability is stronger than the one of modification, but if the two processes $X$ and $Y$ are càd-làg, and if $Y$ is a modification of $X$, then $X$ and $Y$ are indistinguishable.

The interpretation of the time parameter $t$ involves a dynamic aspect: for modeling the fact that uncertainty on the events of $\Omega$ becomes less and less uncertain when time elapses, i.e. one gets more and more information, one introduces the notion of filtration.

H. Pham, *Continuous-time Stochastic Control and Optimization with Financial Applications*, Stochastic Modelling and Applied Probability 61, DOI 10.1007/978-3-540-89500-8_1, © Springer-Verlag Berlin Heidelberg 2009

**Definition 1.1.1** *(Filtration)*
*A filtration on* $(\Omega, \mathcal{F}, P)$ *is an increasing family* $\mathbb{F} = (\mathcal{F}_t)_{t \in \mathbb{T}}$ *of* $\sigma$-*fields of* $\mathcal{F}$: $\mathcal{F}_s \subset \mathcal{F}_t$
$\subset \mathcal{F}$ *for all* $0 \leq s \leq t$ *in* $\mathbb{T}$.

$\mathcal{F}_t$ is interpreted as the information known at time $t$, and increases as time elapses. We set $\mathcal{F}_{\bar{T}} = \sigma(\cup_{t \in \mathbb{T}} \mathcal{F}_t)$, the smallest $\sigma$-field containing all $\mathcal{F}_t$, $t \in \mathbb{T}$. The quadruple $(\Omega, \mathcal{F}, \mathbb{F} = (\mathcal{F}_t)_{t \in \mathbb{T}}, P)$ is called filtered probability space. The canonical example of filtration is the following: if $X = (X_t)_{t \in \mathbb{T}}$ is a stochastic process, the natural filtration (or canonical) of $X$ is

$$\mathcal{F}_t^X = \sigma(X_s, 0 \leq s \leq t), \quad t \in \mathbb{T},$$

the smallest $\sigma$-field under which $X_s$ is measurable for all $0 \leq s \leq t$. $\mathcal{F}_t^X$ is interpreted as the whole information, which can be extracted from the observation of the paths of $X$ between 0 and $t$.

We say that a filtration $\mathbb{F} = (\mathcal{F}_t)_{t \in \mathbb{T}}$ satisfies the *usual conditions* if it is right-continuous, i.e.

$$\mathcal{F}_{t+} := \cap_{s \geq t} \mathcal{F}_s = \mathcal{F}_t, \quad \forall t \in \mathbb{T},$$

and if it is complete, i.e. $\mathcal{F}_0$ contains the negligible sets of $\mathcal{F}_{\bar{T}}$. We then say that the filtered probability space $(\Omega, \mathcal{F}, \mathbb{F} = (\mathcal{F}_t)_{t \in \mathbb{T}}, P)$ satisfies the usual conditions. The right-continuity of $\mathcal{F}_t$ means intuitively that by observing all the available information up to time $t$ inclusive, one learns nothing more by an infinitesimal observation in the future. The completion of the filtration means that if an event is impossible, this impossibility is already known at time 0. Starting from an arbitrary filtration $(\mathcal{F}_t)_{t \in \mathbb{T}}$, one constructs a filtration satisfying the usual conditions, by considering for any $t \in \mathbb{T}$ the $\sigma$-field $\mathcal{F}_{t+}$ to which one adds the class of negligible sets of $\mathcal{F}_{\bar{T}}$. This constructed filtration is called the *augmentation* of $(\mathcal{F}_t)_{t \in \mathbb{T}}$.

In the sequel, we are given a filtration $\mathbb{F} = (\mathcal{F}_t)_{t \in \mathbb{T}}$ on $(\Omega, \mathcal{F}, P)$.

**Definition 1.1.2** *(Adapted process)*
*A process* $(X_t)_{t \in \mathbb{T}}$ *is adapted (with respect to* $\mathbb{F}$*) if for all* $t \in \mathbb{T}$, $X_t$ *is* $\mathcal{F}_t$-*measurable.*

When one wants to be precise with respect to which filtration the process is adapted, we write $\mathbb{F}$-adapted. Thus, an adapted process is a process whose value at any time $t$ is revealed by the information $\mathcal{F}_t$. We say sometimes that the process is nonanticipative. It is clear that any process $X$ is adapted with respect to $\mathbb{F}^X = (\mathcal{F}_t^X)_{t \in \mathbb{T}}$.

Until now, the stochastic process $X$ has been viewed either as a mapping of time $t$ for fixed $\omega$ (when we consider path) or as a mapping of $\omega$ for fixed $t$ (when we consider the random variable as in Definition 1.1.2). One can consider the two aspects by looking at the process as a mapping on $\mathbb{T} \times \Omega$. This leads to the following definitions:

**Definition 1.1.3** *(Progressively measurable, optional, predictable process)*
*(1) A process* $(X_t)_{t \in \mathbb{T}}$ *is progressively measurable (with respect to* $\mathbb{F}$*) if for any* $t \in \mathbb{T}$, *the mapping* $(s, \omega) \to X_s(\omega)$ *is measurable on* $[0, t] \times \Omega$ *equipped with the product* $\sigma$-*field* $\mathcal{B}([0, t]) \otimes \mathcal{F}_t$.

*(2) A process $(X_t)_{t \in \mathbb{T}}$ is optional (with respect to $\mathbb{F}$) if the mapping $(t, \omega) \to X_t(\omega)$ is measurable on $\mathbb{T} \times \Omega$ equipped with the $\sigma$-field generated by the $\mathbb{F}$-adapted and càd-làg processes.*

*(3) A process $(X_t)_{t \in \mathbb{T}}$ is predictable (with respect to $\mathbb{F}$) if the mapping $(t, \omega) \to X_t(\omega)$ is measurable on $\mathbb{T} \times \Omega$ equipped with the $\sigma$-field generated by the $\mathbb{F}$-adapted and continuous processes.*

When we want to specify the filtration, we write $\mathbb{F}$-progressively measurable (optional or predictable). Obviously, any progressively measurable process is adapted and measurable on $\mathbb{T} \times \Omega$ equipped with the product $\sigma$-field $\mathcal{B}(\mathbb{T}) \otimes \mathcal{F}$. It is also clear by definition that any càd-làg and adapted process is optional (the converse is not true). Similarly, any continuous and adapted process $X$ is predictable (the converse is not true): since in this case, $X_t = \lim_{s \nearrow t} X_s$, this means that the value of $X_t$ is announced by its previous values. Since a continuous process is càd-làg, it is clear that any predictable process is optional. The following result gives the relation between optional and progressively measurable process.

**Proposition 1.1.1** *If the process $X$ is optional, it is progressively measurable. In particular, if it is càd-làg and adapted, it is progressively measurable.*

By misuse of language, one often writes in the literature adapted process for progressively measurable process.

### 1.1.2 Stopping times

Having in mind the interpretation of $\mathcal{F}_t$ as the available information up to time $t$, we want to know if an event characterized by its first arrival time $\tau(\omega)$, occurred or not before time $t$ given the observation in $\mathcal{F}_t$. This leads to the notion of stopping time.

**Definition 1.1.4** *(Stopping time)*
*(1) A random variable $\tau : \Omega \to [0, \infty]$, i.e. a random time, is a stopping time (with respect to the filtration $\mathbb{F} = (\mathcal{F}_t)_{t \in \mathbb{T}}$) if for all $t \in \mathbb{T}$*

$$\{\tau \le t\} := \{\omega \in \Omega : \tau(\omega) \le t\} \in \mathcal{F}_t.$$

*(2) A stopping time $\tau$ is predictable if there exists a sequence of stopping times $(\tau_n)_{n \ge 1}$ such that we have almost surely:*
*(i) $\lim_n \tau_n = \tau$*
*(ii) $\tau_n < \tau$ for all $n$ on $\{\tau > 0\}$.*
*We say that $(\tau_n)_{n \ge 1}$ announces $\tau$.*

We easily check that any random time equal to a positive constant $t$ is a stopping time. We also notice that if $\tau$ and $\sigma$ are two stopping times, then $\tau \wedge \sigma$, $\tau \vee \sigma$ and $\tau + \sigma$ are stopping times.

Given a stopping time $\tau$, we measure the infomation cumulated until $\tau$ by

$$\mathcal{F}_\tau = \{B \in \mathcal{F}_{\bar{T}} : B \cap \{\tau \le t\} \in \mathcal{F}_t, \ \forall t \in \mathbb{T}\},$$

which is a $\sigma$-field of $\mathcal{F}$. It is clear that $\tau$ is $\mathcal{F}_\tau$-measurable. We immediately see that if $\tau = t$ then $\mathcal{F}_\tau = \mathcal{F}_t$. We state some elementary and useful properties on stopping times (see e.g. the proofs in Ch. I, Sec. 1.2 of Karatzas and Shreve [KaSh88]).

**Proposition 1.1.2** *Let $\sigma$ and $\tau$ be two stopping times, and $\xi$ a random variable.*
*(1) For all $B \in \mathcal{F}_\sigma$, we have $B \cap \{\sigma \leq \tau\} \in \mathcal{F}_\tau$. In partcular, if $\sigma \leq \tau$ then $\mathcal{F}_\sigma \subset \mathcal{F}_\tau$.*
*(2) The events*

$$\{\sigma < \tau\}, \ \{\sigma \leq \tau\}, \ \{\sigma = \tau\}$$

*belong to $\mathcal{F}_{\sigma \wedge \tau} = \mathcal{F}_\sigma \cap \mathcal{F}_\tau$.*
*(3) $\xi$ is $\mathcal{F}_\tau$-measurable if and only if for all $t \in \mathbb{T}$, $\xi 1_{\tau \leq t}$ is $\mathcal{F}_t$-measurable.*

Given a process $(X_t)_{t \in \mathbb{T}}$ and a stopping time $\tau$, we define the random variable $X_\tau$ on $\{\tau \in \mathbb{T}\}$ by

$$X_\tau(\omega) = X_{\tau(\omega)}(\omega).$$

We check that if $X$ is measurable then $X_\tau$ is a random variable on $\{\tau \in \mathbb{T}\}$. We then introduce the stopped process (at $\tau$) $X^\tau$ defined by

$$X_t^\tau = X_{\tau \wedge t}, \quad t \in \mathbb{T}.$$

**Proposition 1.1.3** *Let $(X_t)_{t \in \mathbb{T}}$ be a progressively measurable process, and $\tau$ a stopping time. Then $X_\tau 1_{\tau \in \mathbb{T}}$ is $\mathcal{F}_\tau$-measurable and the stopped process $X^\tau$ is progressively measurable.*

The next result provides an important class of stopping times.

**Proposition 1.1.4** *Let $X$ be a càd-làg, adapted process, and $\Gamma$ an open subset of $\mathcal{X} = \mathbb{R}^d$.*
*(1) If the filtration $\mathbb{F}$ satisfies the usual conditions, then the hitting time of $\Gamma$ defined by*

$$\sigma_\Gamma = \inf \{t \geq 0 : X_t \in \Gamma\}$$

*(with the convention $\inf \emptyset = \infty$) is a stopping time.*
*(2) If $X$ is continuous, then the exit time of $\Gamma$ defined by*

$$\tau_\Gamma = \inf \{t \geq 0 : X_t \notin \Gamma\}$$

*is a predictable stopping time.*

We end this section with the important section theorem, proved in Dellacherie and Meyer [DM75] p. 220.

**Theorem 1.1.1** *(Section theorem)*
*Let $(X_t)_{t \in \mathbb{T}}$ and $(Y_t)_{t \in \mathbb{T}}$ be two optional processes. Suppose that for any stopping time $\tau$, we have: $X_\tau = Y_\tau$ a.s. on $\{\tau < \infty\}$. Then, the two processes $X$ and $Y$ are indistinguishable.*

### 1.1.3 Brownian motion

The basic example of a process is Brownian motion, a name given by the botanist Robert Brown in 1827 to describe the irregular motion of pollen particles in a fluid. The context of applications of Brownian motion goes far beyond the study of microscopical particles, and is now largely used in finance for modelling stock prices, historically since Bachelier in 1900.

**Definition 1.1.5** *(Standard Brownian motion)*
*A standard d-dimensional Brownian motion on* $\mathbb{T}$ *is a continuous process valued in* $\mathbb{R}^d$,
$(W_t)_{t \in \mathbb{T}} = (W_t^1, \ldots, W_t^d)_{t \in \mathbb{T}}$ *such that:*
*(i)* $W_0 = 0$.
*(ii) For all* $0 \leq s < t$ *in* $\mathbb{T}$, *the increment* $W_t - W_s$ *is independent of* $\sigma(W_u, u \leq s)$ *and follows a centered Gaussian distribution with variance-covariance matrix* $(t - s)I_d$.

As an immediate consequence of the definition, the coordinates $(W_t^i)_{t \in \mathbb{T}}$, $i = 1, \ldots, d$, of a $d$-dimensional standard Brownian motion, are real-valued standard Brownian motion, and independent. Conversely, real-valued independent Brownian motion generates a vectorial Brownian motion. In the definition of a standard Brownian motion, the independence of the increments is with respect to the natural filtration $\mathcal{F}_s^W = \sigma(W_u, u \leq s)$ of $W$. The natural filtration of $W$ is sometimes called Brownian filtration. It is often interesting to work with a larger filtration than the natural filtration. This leads to the following more general definition.

**Definition 1.1.6** *(Brownian motion with respect to a filtration)*
*A vectorial (d-dimensional) Brownian motion on* $\mathbb{T}$ *with respect to a filtration* $\mathbb{F} = (\mathcal{F}_t)_{t \in \mathbb{T}}$ *is a continuous* $\mathbb{F}$-*adapted process, valued in* $\mathbb{R}^d$, $(W_t)_{t \in \mathbb{T}} = (W_t^1, \ldots, W_t^d)_{t \in \mathbb{T}}$ *such that:*
*(i)* $W_0 = 0$.
*(ii) For all* $0 \leq s < t$ *in* $\mathbb{T}$, *the increment* $W_t - W_s$ *is independent of* $\mathcal{F}_s$ *and follows a centered Gaussian distribution with variance-covariance matrix* $(t - s)I_d$.

Of course, a standard Brownian motion is a Brownian motion with respect to its natural filtration.

A major problem concerns the existence, construction and simulation of a Brownian motion. We do not discuss this problem here, and refer to the multiple textbooks on this topic (see e.g. Hida [Hi80], Karatzas and Shreve [KaSh88], Le Gall [LeG89] or Revuz and Yor [ReY91]). We only state a classical property of Brownian motion.

**Proposition 1.1.5** *Let* $(W_t)_{t \in \mathbb{T}}$ *be a Brownian motion with respect to* $(\mathcal{F}_t)_{t \in \mathbb{T}}$.
*(1) Symmetry:* $(-W_t)_{t \in \mathbb{T}}$ *is also a Brownian motion.*
*(2) Scaling: for all* $\lambda > 0$, *the process* $((1/\lambda)W_{\lambda^2 t})_{t \in \mathbb{T}}$ *is also a Brownian motion.*
*(3) Invariance by translation: for all* $s > 0$, *the process* $(W_{t+s} - W_s)_{t \in \mathbb{T}}$ *is a standard Brownian motion independent of* $\mathcal{F}_s$.

We also recall that the augmentation of the natural filtration $(\mathcal{F}_t^W)_t$ of a Brownian motion $W$ is $(\sigma(\mathcal{F}_t^W \cup \mathcal{N}))_t$ where $\mathcal{N}$ is the set of negligible events of $(\Omega, \mathcal{F}_{\bar{T}}, P)$. Moreover,

$W$ remains a Brownian motion with respect to its augmented filtration. By misuse of language, the augmentation of the natural filtration of $W$ is called again natural filtration of Brownian filtration.

### 1.1.4 Martingales, semimartingales

In this section, we consider real-valued processes. The proofs of results stated here can be found for instance in Dellacherie and Meyer [DM80].

**Definition 1.1.7** *(Martingale)*
*An adapted process $(X_t)_{t \in \mathbb{T}}$ is a supermartingale if $E[X_t^-] < \infty$ for all $t \in \mathbb{T}$ and*

$$E[X_t | \mathcal{F}_s] \leq X_s, \quad a.s. \quad for\ all\ 0 \leq s \leq t,\ s, t \in \mathbb{T}. \tag{1.1}$$

*$X$ is a submartingale if $-X$ is a supermartingale. We say that $X$ is a martingale if it is both a supermartingale and a submartingale.*

The definition of a super(sub)-martingale depends crucially on the probability $P$ and on the filtration $\mathbb{F} = (\mathcal{F}_t)_{t \in \mathbb{T}}$ specified on the measurable space $(\Omega, \mathcal{F})$. In this book, the filtration will be fixed, and if it is not specified, the super(sub)-martingale property will always refer to this filtration. However, we shall lead to consider different probability measures $Q$ on $(\Omega, \mathcal{F})$, and to emphasize this fact, we shall specify $Q$-super(sub)-martingale.

An important example of a martingale is the Brownian motion described in the previous section. On the other hand, a typical construction of martingale is achieved as follows: we are given an integrable random variable $\xi$ on $(\Omega, \mathcal{F})$: $E|\xi| < \infty$. Then, the process defined by

$$X_t = E[\xi | \mathcal{F}_t], \quad t \in \mathbb{T},$$

is clearly a martingale. We say that $X$ is closed on the right by $\xi$. Conversely, when $\mathbb{T} = [0, T]$, $T < \infty$, any martingale $(X_t)_{t \in [0,T]}$ is closed on the right by $\xi = X_T$. When $\mathbb{T} = [0, \infty)$, the closedness on the right of a martingale is derived from the following convergence result:

**Theorem 1.1.2** *(Convergence of martingales)*
*(1) Let $X = (X_t)_{t \geq 0}$ be a submartingale, càd-làg, and bounded in $L^1$ (in particular if it is nonnegative). Then, $X_t$ converges a.s. when $t \to \infty$.*

*(2) Let $X = (X_t)_{t \geq 0}$ be a càd-làg martingale. Then $(X_t)_{t \geq 0}$ is uniformly integrable if and only if $X_t$ converges a.s. and in $L^1$ when $t \to \infty$ towards a random variable $X_\infty$. In this case, $X_\infty$ closes $X$ on the right, i.e. $X_t = E[X_\infty | \mathcal{F}_t]$ for all $t \geq 0$.*

In the sequel, we denote by $\bar{\mathbb{T}}$ the interval equal to $[0, T]$ if $\mathbb{T} = [0, T]$, and equal to $[0, \infty]$ if $\mathbb{T} = [0, \infty)$. We also denote by $\bar{T}$ the right boundary of $\mathbb{T}$. With this convention, if $(X_t)_{t \in \mathbb{T}}$ is a càd-làg uniformly integrable martingale, then $X_{\bar{T}}$ is the limit a.s. and in $L^1$ of $X_t$ when $t$ goes to $\bar{T}$. Moreover, $X_{\bar{T}}$ closes $X$ on the right: $X_t = E[X_{\bar{T}} | \mathcal{F}_t]$ for all $t \in \mathbb{T}$.

The next result is a very important property of martingales: it extends the relation (1.1) for dates $t$ and $s$ replaced by stopping times.

**Theorem 1.1.3** *(Optional sampling theorem)*
*Let $M = (M_t)_{t \in \mathbb{T}}$ be a martingale càd-làg and $\sigma$, $\tau$ two bounded stopping times valued in $\mathbb{T}$, and such that $\sigma \leq \tau$. Then,*

$$E[M_\tau | \mathcal{F}_\sigma] = M_\sigma, \quad a.s.$$

A useful application of the optional sampling theorem is given by the following corollary:

**Corollary 1.1.1** *Let $X = (X_t)_{t \in \mathbb{T}}$ be a càd-làg adapted process.*

*(1) $X$ is a martingale if and only if for any bounded stopping time $\tau$ valued in $\mathbb{T}$, we have $X_\tau \in L^1$ and*

$$E[X_\tau] = X_0.$$

*(2) If $X$ is a martingale and $\tau$ is a stopping time, then the stopped process $X^\tau$ is a martingale.*

We state a first fundamental inequality for martingales.

**Theorem 1.1.4** *(Doob's inequality)*
*Let $X = (X_t)_{t \in \mathbb{T}}$ be a nonnegative submartingale or a martingale, càd-làg. Then, for any stopping time $\tau$ valued in $\mathbb{T}$, we have:*

$$P\left[ \sup_{0 \leq t \leq \tau} |X_t| \geq \lambda \right] \leq \frac{E|X_\tau|}{\lambda}, \quad \forall \lambda > 0,$$

$$E\left[ \sup_{0 \leq t \leq \tau} |X_t| \right]^p \leq \left( \frac{p}{p-1} \right)^p E\left[ |X_\tau|^p \right], \quad \forall p > 1.$$

Notice that the first above inequality and the theorem of convergence for martingales imply that if $(X_t)_{t \in \mathbb{T}}$ is a càd-làg uniformly integrable martingale, then $\sup_{t \in \mathbb{T}} |X_t| < \infty$ a.s.

In the sequel, we fix a filtered probability space $(\Omega, \mathcal{F}, \mathbb{F} = (\mathcal{F}_t)_{t \in \mathbb{T}}, P)$ satisfying the usual conditions.

In the theory of processes, the concept of localization is very useful. Generally speaking, we say that a progressively measurable process $X$ is locally "truc" (or has the "truc" local property) if there exists an increasing sequence of stopping times $(\tau_n)_{n \geq 1}$ (called localizing sequence) such that $\tau_n$ goes a.s. to infinity and for all $n$, the stopped process $X^{\tau_n}$ is "truc" (or has the "truc" property). We introduce in particular the notion of locally bounded process, and we see that any continuous adapted process is locally bounded: take as localizing sequence $\tau_n = \inf\{t \geq 0 : |X_t| \geq n\}$ so that $X^{\tau_n}$ is bounded by $n$. Notice that when $X$ is not continuous with unbounded jumps, $X$ is not locally bounded. An important example of localization is given in the following definition.

**Definition 1.1.8** *(Local martingale)*
*Let $X$ be a càd-làg adapted process. We say that $X$ is a local martingale if there exists a sequence of stopping times $(\tau_n)_{n \geq 1}$ such that $\lim_{n \to \infty} \tau_n = \infty$ a.s. and the stopped process $X^{\tau_n}$ is a martingale for all $n$.*

Any càd-làg martingale is a local martingale but the converse property does not hold true: local martingales are more general than martingales, and we shall meet them in particular with stochastic integrals (see Section 1.2). It is interesting to have conditions ensuring that a local martingale is a martingale. The following criterion is useful in practice.

**Proposition 1.1.6** *Let $M = (M_t)_{t \in \mathbb{T}}$ be a local martingale. Suppose that*

$$E\left[\sup_{0 \le s \le t} |M_s|\right] < \infty, \quad \forall t \in \mathbb{T}. \tag{1.2}$$

*Then $M$ is a martingale.*

Actually, we have a necessary and sufficient condition for a local martingale $M$ to be a "true" martingale: it is the so-called condition (DL) stating that the family $(M_\tau)_\tau$ where $\tau$ runs over the set of bounded stopping times in $\mathbb{T}$, is uniformly integrable. The sufficient condition (1.2) is often used in practice for ensuring condition (DL). We also mention the following useful result, which is a direct consequence of Fatou's lemma.

**Proposition 1.1.7** *Let $M$ be a nonnegative local martingale such that $M_0 \in L^1$. Then $M$ is a supermartingale.*

We introduce and summarize some important results on Snell envelopes, which play a key role in optimal stopping problems. For $t \in [0, T]$, $T < \infty$, we denote by $\mathcal{T}_{t,T}$ the set of stopping times valued in $[t, T]$.

**Proposition 1.1.8** *(Snell envelope)*
*Let $H = (H_t)_{0 \le t \le T}$ be a real-valued $\mathbb{F}$-adapted càd-làg process, in the class (DL). The Snell envelope $V$ of $H$ is defined by*

$$V_t = \operatorname*{ess\,sup}_{\tau \in \mathcal{T}_{t,T}} E\big[H_\tau | \mathcal{F}_t\big], \quad 0 \le t \le T,$$

*and it is the smallest supermartingale of class (DL), which dominates $H$: $V_t \ge H_t$, $0 \le t \le T$. Furthermore, if $H$ has only positive jumps, i.e. $H_t - H_{t-} \ge 0$, $0 \le t \le T$, then $V$ is continuous, and for all $t \in [0, T]$, the stopping time*

$$\tau_t = \inf\{s \ge t : V_s = H_s\} \wedge T$$

*is optimal after $t$, i.e.*

$$V_t = E[V_{\tau_t} | \mathcal{F}_t] = E[H_{\tau_t} | \mathcal{F}_t].$$

We now define the important concept of quadratic variation of a (continuous) local martingale. We say that a process $A = (A_t)_{t \in \mathbb{T}}$ has finite variation if every path is càd-làg and has finite variation, i.e. for all $\omega \in \Omega$, $t \in \mathbb{T}$,

$$\sup \sum_{i=1}^{n} \big|A_{t_i}(\omega) - A_{t_{i-1}}(\omega)\big| < \infty, \tag{1.3}$$

where the supremum is taken over all subdivisions $0 = t_0 < t_1 < \ldots < t_n = t$ of $[0, t]$. The process $A$ is nondecreasing if every path is càd-làg and nondecreasing. Any process $A$ with finite variation can be written as $A = A^+ - A^-$ where $A^+$ and $A^-$ are two nondecreasing processes. There is uniqueness of such a decomposition if one requires that the associated positive measures $A^+([0,t]) = A_t^+$ and $A^-([0,t]) = A_t^-$ are supported by disjoint Borelians. We denote by $A$ the signed measure, the difference of the two finite positive measures $A^+$ and $A^-$. The (random) positive measure associated to the increasing process $A^+ + A^-$ is denoted by $|A|$: $|A|([0,t]) = A_t^+ + A_t^-$ equal to (1.3), and is called the variation of $A$. For any process $\alpha$ such that

$$\int_0^t |\alpha_s(\omega)| d|A|_s(\omega) < \infty, \quad \forall t \in \mathbb{T}, \ \forall \omega \in \Omega,$$

the process $\int \alpha dA$ defined by the Stieltjes integral $\int_0^t \alpha_s(\omega) dA_s(\omega)$, for all $t \in \mathbb{T}$ and $\omega \in \Omega$, has finite variation. Moreover, if $A$ is adapted and $\alpha$ is progressively measurable, then $\int \alpha dA$ is adapted. For any process $A$ with finite variation, we define $A^c$:

$$A_t^c = A_t - \sum_{0 \leq s \leq t} \Delta A_s, \quad \text{where} \quad \Delta A_s = A_s - A_{s-} \quad (\Delta A_0 = A_0).$$

$A^c$ is a continuous process with finite variation, and is called the continuous part of $A$.

**Proposition 1.1.9** *Let $M$ be a local martingale, $M_0 = 0$. If $M$ has predictable finite variation, then $M$ is indistinguishable from 0.*

**Theorem 1.1.5** *(Quadratic variation, bracket)*
*(1) Let $M = (M_t)_{t \in \mathbb{T}}$ and $N = (N_t)_{t \in \mathbb{T}}$ be two local martingales, and one is locally bounded (for example continuous). Then, there exists a unique predictable process with finite variation, denoted by $< M, N >$, vanishing in 0, such that $MN - < M, N >$ is a local martingale. This local martingale is continuous if $M$ and $N$ are continuous. Moreover, for all $t \in \mathbb{T}$, if $0 = t_0^n < t_1^n < \ldots t_{k_n}^n = t$ is a subdivision of $[0, t]$ with mesh size going to 0, then we have:*

$$< M, N >_t = \lim_{n \to +\infty} \sum_{i=1}^{k_n} \left( M_{t_i^n} - M_{t_{i-1}^n} \right) \left( N_{t_i^n} - N_{t_{i-1}^n} \right),$$

*for the convergence in probability. The process $< M, N >$ is called the bracket (or cross-variation) of $M$ and $N$. We also say that $M$ and $N$ are orthogonal if $< M, N > = 0$, i.e. the product $MN$ is a local martingale.*

*(2) When $M = N$, the process $< M, M >$, also denoted by $< M >$ and called the quadratic variation of $M$ or increasing process of $M$, is increasing. Moreover, we have the "polarity" relation*

$$< M, N > = \frac{1}{2} \left( < M + N, M + N > - < M, M > - < N, N > \right).$$

**Example**
If $W = (W^1, \ldots, W^d)$ is a $d$-dimensional Brownian motion, we have

$$< W^i, W^j >_t = \delta_{ij} t$$

where $\delta_{ij} = 1$ if $i = j$ and $0$ otherwise. Moreover, the processes $W_t^i W_t^j - \delta_{ij} t$ are martingales.

The following inequality is useful for defining below the notion of a stochastic integral.

**Proposition 1.1.10** *(Kunita-Watanabe inequality)*
*Let $M$ and $N$ two continuous local martingales and $\alpha$, $\beta$ two measurable processes on $\mathbb{T} \times \Omega$ equipped with the product $\sigma$-field $\mathcal{B}(\mathbb{T}) \otimes \mathcal{F}$. Then for all $t \in \mathbb{T}$,*

$$\int_0^t |\alpha_s||\beta_s| |d| < M, N > |_s \le \left( \int_0^t \alpha_s^2 d < M, M >_s \right)^{\frac{1}{2}} \left( \int_0^t \beta_s^2 d < N, N >_s \right)^{\frac{1}{2}} \quad a.s.$$

This shows in particular that almost surely the signed measure $d < M, N >$ is absolutely continuous with respect to the product measure $d < M >$.

The following fundamental inequality for (local) martingales is very useful when we shall focus on local martingales defined by stochastic integrals for which one can often calculate the quadratic variation.

**Theorem 1.1.6** *(Burkholder-Davis-Gundy inequality)*
*For all $p > 0$, there exist positive constants $c_p$ and $C_p$ such that for all continuous local martingales $M = (M_t)_{t \in \mathbb{T}}$ and all stopping times $\tau$ valued in $\bar{\mathbb{T}}$, we have*

$$c_p E[< M >_\tau^{p/2}] \le E\left[ \sup_{0 \le t < \tau} |M_t| \right]^p \le C_p E[< M >_\tau^{p/2}].$$

By combining the Burkholder-Davis-Gundy inequality with the condition (1.2), we see in particular for $p = 1$ that if the continuous local martingale $M$ satisfies $E[\sqrt{< M >_t}]$ $< \infty$ for all $t \in \mathbb{T}$, then $M$ is a martingale.

We say that a càd-làg martingale $M = (M_t)_{t \in \mathbb{T}}$ is square integrable if $E[|M_t|^2] < \infty$ for all $t \in \mathbb{T}$. We introduce the additional distinction (in the case $\mathbb{T} = [0, \infty)$), and say that $M$ is bounded in $L^2$ if $\sup_{t \in \mathbb{T}} E[|M_t|^2] < \infty$. In particular, a bounded martingale in $L^2$ is uniformly integrable and admits a limit a.s. $M_{\bar{T}}$ when $t$ goes to $\bar{T}$. We denote by $\mathbb{H}_c^2$ the set of continuous martingales bounded in $L^2$. The next result is a consequence of Doob and Burkholder-Davis-Gundy inequalities.

**Proposition 1.1.11** *(Square integrable martingale)*
*Let $M = (M_t)_{t \in \mathbb{T}}$ be a continuous local martingale. Then $M$ is a square integrable martingale if and only if $E[< M >_t] < \infty$ for all $t \in \mathbb{T}$. In this case, $M^2 - < M >$ is a continuous martingale and if $M_0 = 0$, we have*

$$E[M_t^2] = E[< M >_t], \quad \forall t \in \mathbb{T}.$$

*Moreover, $M$ is bounded in $L^2$ if and only if $E[< M >_{\bar{T}}] < \infty$, and in this case*

$$E[M_{\bar{T}}^2] = E[< M >_{\bar{T}}].$$

*The space $\mathbb{H}_c^2$ endowed with the scalar product*

$$(M, N)_{\mathbb{H}^2} = E[< M, N >_{\bar{T}}]$$

*is a Hilbert space.*

The next theorem is known as the Doob-Meyer decomposition theorem for supermartingales.

**Theorem 1.1.7** *(Doob-Meyer decomposition)*
*Let $X$ be a càd-làg supermartingale. Then $X$ admits a unique decomposition in the form*

$$X = X_0 + M - A$$

*where $M$ is a càd-làg local martingale null in 0, and $A$ is a predictable process, increasing and null in 0. If $X$ is nonnegative, then $A$ is integrable, i.e. $E[A_{\bar{T}}] < \infty$ where $A_{\bar{T}} = \lim_{t \to \bar{T}} A_t$ a.s.*

We finally introduce a fundamental class of finite quadratic variation processes, extending the (local) super(sub)-martingales, and largely used in financial modeling, especially in the context of this book.

**Definition 1.1.9** *(Semimartingale)*
*A semimartingale is a càd-làg adapted process $X$ having a decomposition in the form:*

$$X = X_0 + M + A \qquad (1.4)$$

*where $M$ is a càd-làg local martingale null in 0, and $A$ is a adapted process with finite variation and null in 0. A continuous semimartingale is a semimartingale such that in the decomposition (1.4), $M$ and $A$ are continuous. Such a decomposition where $M$ and $A$ are continuous is unique.*

We define the bracket of a continuous semimartingale $X = X_0 + M + A$ by: $< X, X > = \; < M, M >$, and the following property holds: for all $t \in \mathbb{T}$, if $0 = t_0^n < t_1^n < \ldots t_{k_n}^n = t$ is a subdivision of $[0, t]$ with mesh size going to 0, we have the convergence in probability

$$< X, X >_t = \lim_{n \to \infty} \sum_{i=1}^{k_n} \left( X_{t_i^n} - X_{t_{i-1}^n} \right)^2.$$

This property is very important since it shows that the bracket does not change under a change of absolute probability measure $Q$ under which $X$ is still a $Q$-semimartingale.

The main theorems stated above for super(sub)-martingales considered càd-làg paths of the processes. The next theorem gives sufficient conditions ensuring this property.

**Theorem 1.1.8** *Let $\mathbb{F} = (\mathcal{F}_t)_{t \in \mathbb{T}}$ be a filtration satisfying the usual conditions, and $X = (X_t)_{t \in \mathbb{T}}$ be a supermartingale. Then $X$ has a càd-làg modification if and only if the mapping $t \in \mathbb{T} \to E[X_t]$ is right-continuous (this is the case in particular if $X$ is a martingale). Moreover, in this case, the càd-làg modification remains a supermartingale with respect to $\mathbb{F}$.*

For a proof of this result, we refer to Karatzas and Shreve [KaSh88], Theorem 3.13 in Ch. 1.

In the sequel, we say that a vectorial process $X = (X^1, \ldots, X^d)$ is a martingale (resp. supermartingale, resp. semimartingale) (local) if each of the real-valued components $X^i$, $i = 1, \ldots, d$ is a martingale (resp. supermartingale, resp. semimartingale) (local). We also define the matricial bracket of a vectorial continuous semimartingale $X = (X^1, \ldots, X^d)$: $< X > = < X, X' >$, by its components $< X^i, X^j >$, $i, j = 1, \ldots, d$.

## 1.2 Stochastic integral and applications

### 1.2.1 Stochastic integral with respect to a continuous semimartingale

In this section, we define the stochastic integral with respect to a continuous semimartingale $X$. We first consider the case where $X$ is unidimensional. With the decomposition (1.4), we define the integral with respect to $X$ as the sum of two integrals, one with respect to the finite variation part $A$ and the other with respect to the continuous local martingale $M$. The integral with respect to $A$ is defined pathwise (for almost all $\omega$) as a Stieljes integral. On the other hand, if the martingale $M$ is not zero, it does not have finite variation, and one cannot define the integral with respect to $M$ pathwise like for Stieljes integrals. The notion of a stochastic integral with respect to $M$ is due to Itô when $M$ is a Brownian motion, and is based on the existence of a quadratic variation $< M >$, which allows us to define the integral as a limit of simple sequences of Riemann type in $L^2$.

A simple (or elementary) process is a process in the form

$$\alpha_t = \sum_{k=1}^{n} \alpha_{(k)} 1_{(t_k, t_{k+1}]}(t), \tag{1.5}$$

where $0 \leq t_0 < t_1 < \ldots < t_n$ is a sequence of stopping times in $\mathbb{T}$ and $\alpha_{(k)}$ is a $\mathcal{F}_{t_k}$-measurable random variable, and bounded for all $k$. We denote by $\mathcal{E}$ the set of simple processes. When $M$ is bounded in $L^2$, i.e. $M \in \mathbb{H}_c^2$, we define $L^2(M)$ as the set of progressively measurable processes $\alpha$ such that $E[\int_0^{\bar{T}} |\alpha_t|^2 d < M >_t] < \infty$. It is a Hilbert space for the scalar product

$$(\alpha, \beta)_{L^2(M)} = E\Big[ \int_0^{\bar{T}} \alpha_t \beta_t d < M >_t \Big],$$

and the set of simple processes is dense in $L^2(M)$. The stochastic integral of a simple process (1.5) is defined by

$$\int_0^t \alpha_s dM_s = \sum_{k=1}^{n} \alpha_{(k)} \cdot (M_{t_{k+1} \wedge t} - M_{t_k \wedge t}), \quad t \in \mathbb{T},$$

and belongs to $\mathbb{H}_c^2$. Moreover, we have the isometry relation

$$E\Big[ \Big( \int_0^{\bar{T}} \alpha_t \, dM_t \Big)^2 \Big] = E\Big[ \int_0^{\bar{T}} |\alpha_t|^2 d < M >_t \Big].$$

By density of $\mathcal{E}$ in $L^2(M)$, we can extend the mapping $\alpha \to \int \alpha dM$ to an isometry from $L^2(M)$ into $\mathbb{H}_c^2$. For any $\alpha \in L^2(M)$, the stochastic integral $\int \alpha dM$ is characterized by the relation

$$< \int \alpha dM, N > = \int \alpha \, d < M, N >, \quad \forall N \in \mathbb{H}_c^2.$$

Moreover, if $\tau$ is a stopping time, we have

$$\int_0^{t \wedge \tau} \alpha_s dM_s = \int_0^t \alpha_s 1_{[0,\tau]} dM_s = \int_0^t \alpha_s dM_{s \wedge \tau}, \quad t \in \mathbb{T}.$$

This identity relation allows us to extend by localization the definition of stochastic integral to a continuous local martingale, and for a larger class of integrands.

Let $M$ be a continuous local martingale. We denote by $L_{loc}^2(M)$ the set of progressively measurable processes $\alpha$ such that for all $t \in \mathbb{T}$

$$\int_0^t |\alpha_s|^2 d < M >_s < \infty, \quad a.s.$$

For any $\alpha \in L_{loc}^2(M)$, there exists a unique continuous local martingale null in 0, called the stochastic integral of $\alpha$ with respect to $M$ and denoted by $\int \alpha dM$, such that

$$< \int \alpha dM, N > = \int \alpha \, d < M, N >$$

for all continuous local martingale $N$. This definition extends the one above when $M \in \mathbb{H}_c^2$ and $\alpha \in L^2(M)$.

When considering stochastic integrals with respect to a vectorial continuous local martingale, a first idea is to take the sum of stochastic integrals with respect to each of the components. However, in order to get "good" representation and martingale decomposition (see Section 1.2.4) properties, we have to construct the stochastic integral by using an isometry relation as in the one-dimensional case to obtain suitable closure properties. We speak about vectorial stochastic integration. Sufficient conditions ensuring that these two notions of vectorial and componentwise stochastic integration are studied in [CS93]. This is for example the case when $M$ is a $d$-dimensional Brownian motion. On the other hand, when one defines the stochastic integral with respect to a semimartingale, while it is natural and sufficient in practical applications to consider in a first stage integrands with respect to the finite variation and martingale parts, this class of integrands is not large enough in theory when studying closure and stability properties for stochastic integrals. The semimartingale topology was introduced by Emery [Em79] and is quite appropriate for the study of stochastic integrals.

The following three paragraphs, marked with an asterisk and concerning vectorial stochastic integration and semimartingale topology, may be omitted in a first reading. We start by defining the integral with respect to a vectorial process with finite variation, and then with respect to a vectorial continuous local martingale. The presentation is inspired by Jacod [Jac79], ch. IV.

**Stieltjes integral with respect to a vectorial process with finite variation***

Let $A = (A^1, \ldots, A^d)$ be a vectorial process whose components are continuous with finite variation. There exists an increasing process $\Gamma$ (for example $d\Gamma = \sum_{i=1}^d d|A^i|$) and a vectorial predictable process $\gamma = (\gamma^1, \ldots, \gamma^d)$ (derived from the Radon-Nikodym theorem) such that

$$A^i = \int \gamma^i d\Gamma, \quad i = 1, \ldots, d. \tag{1.6}$$

We denote by $L_S(A)$ the set of progressively measurable processes $\alpha = (\alpha^1, \ldots, \alpha^d)$ such that

$$\int_0^t \left| \sum_{i=1}^d \alpha_u^i \gamma_u^i \right| d\Gamma_u < \infty, \quad a.s. \ \forall t \in \mathbb{T}.$$

We then set

$$\int \alpha dA = \int \sum_{i=1}^d \alpha^i \gamma^i d\Gamma,$$

with the convention that $\int_0^t \alpha_u(\omega) dA_u(\omega) = 0$ if $\omega$ is in the negligible set where $\int_0^t \left| \sum_{i=1}^d \alpha_u^i \gamma_u^i(\omega) \right| d\Gamma_u(\omega) = \infty$. The process $\int \alpha dA$ does not depend on the choice of $(\gamma, \Gamma)$ satisfying (1.6), and is also continuous with finite variation. Notice that $L_S(A)$ contains (and in general strictly) the set of progressively measurable processes $\alpha = (\alpha^1, \ldots, \alpha^d)$ such that for all $i = 1, \ldots, d$, $\alpha^i \in L_S(A^i)$, i.e. $\int_0^t |\alpha_u^i| d|A^i|_u < \infty$ for all $t$ in $\mathbb{T}$ for which we have: $\int \alpha dA = \sum_{i=1}^d \int \alpha^i dA^i$, often denoted by $\int \alpha' dA$.

**Stochastic integral with respect to a vectorial continuous local martingale***

Let $M = (M^1, \ldots, M^d)$ be a continous local martingale valued in $\mathbb{R}^d$. We denote by $< M >$ the matricial bracket with components $< M^i, M^j >$. There exists a predictable increasing process $C$ such that $d < M^i, M^j >$ is absolutely continuous with respect to the positive measure $dC$: for example $C = \sum_{i=1}^d < M^i >$. By the Radon-Nikodym theorem, there exists a predictable process $c$ valued in $\mathcal{S}_d^+$ such that

$$< M > = \int c \, dC, \quad i.e. \ < M^i, M^j > = \int c^{ij} dC, \ i, j = 1, \ldots, d. \tag{1.7}$$

We define $L_{loc}^2(M)$ as the set of progressively measurable processes $\alpha = (\alpha^1, \ldots, \alpha^d)$ valued in $\mathbb{R}^d$ such that for all $t \in \mathbb{T}$

$$\int_0^t \alpha_u' d < M >_u \alpha_u := \int_0^t \alpha_u' c_u \alpha_u dC_u < \infty, \quad a.s. \tag{1.8}$$

We mention that the expression $\int_0^t \alpha_u' d < M >_u \alpha_u$ does not depend on the choice of $c$ and $C$ satisfying (1.7). For all $\alpha \in L_{loc}^2(M)$, there exists a unique continuous local martingale null in 0, called the stochastic integral of $\alpha$ with respect to $M$, and denoted by $\int \alpha dM$, such that

$$< \int \alpha dM, N > = \int \alpha \, d < M, N >$$

for any continuous local martingale $N$. The integral on the right-hand side of the above equality is the Stieltjes integral with respect to the vectorial process with finite variation $< M, N > = (< M^1, N >, \ldots, < M^d, N >)$. This Stieltjes integral is well-defined by the Kunita-Watanabe inequality. In particular, we have

$$< \int \alpha dM, \int \alpha dM > = \int \alpha' d < M > \alpha.$$

Notice that if for all $i = 1, \ldots, d$, $\alpha^i \in L^2_{loc}(M^i)$, i.e. $\int_0^t |\alpha_u^i|^2 d < M^i >_u < \infty$ for all $t$ in $\mathbb{T}$, then $\alpha = (\alpha^1, \ldots, \alpha^d) \in L^2_{loc}(M)$ and $\int \alpha dM = \sum_{i=1}^d \int \alpha^i dM^i$. The converse does not hold true in general: the set $L^2_{loc}(M)$ is strictly larger than the set $\{\alpha = (\alpha^1, \ldots, \alpha^d) : \alpha^i \in L^2_{loc}(M^i), i = 1, \ldots, d\}$. However, if the processes $M^i$ are pairwise orthogonal, i.e. $< M^i, M^j > = 0$ for $i \neq j$, we have equality of these two sets, and we often write instead of $\int \alpha dM$:

$$\int \alpha . dM := \sum_{i=1}^d \int \alpha^i dM^i.$$

This is typically the case when $M$ is a $d$-dimensional Brownian motion $W = (W^1, \ldots, W^d)$, the condition (1.8) on $L^2_{loc}(W)$ is then written as

$$\int_0^t |\alpha_u|^2 du = \sum_{i=1}^d \int_0^t |\alpha_u^i|^2 du < \infty, \quad a.s., \forall t \in \mathbb{T}.$$

Moreover, $\int \alpha dM$ is a bounded martingale in $L^2$ if and only if $\alpha \in L^2(M)$, defined as the set of progressively measurable processes $\alpha$ such that $E[\int_0^{\bar{T}} \alpha_t' d < M >_t \alpha_t] < \infty$. In this case, we have

$$E\left[\left(\int_0^{\bar{T}} \alpha_t dM_t\right)^2\right] = E\left[\int_0^{\bar{T}} \alpha_t' d < M >_t \alpha_t\right].$$

In the general case, for $\alpha \in L^2_{loc}(M)$, a localizing sequence of stopping times for the local martingale $\int \alpha dM$ is for example

$$\tau_n = \inf\left\{t \in \mathbb{T} : \int_0^t \alpha_u' d < M >_u \alpha_u \geq n\right\},$$

for which the stopped stochastic integral $\int^{\cdot \wedge \tau_n} \alpha dM$ is a bounded martingale in $L^2$ with $E[< \int^{\cdot \wedge \tau_n} \alpha dM >_{\bar{T}}] \leq n$. When $M$ is a $d$-dimensional Brownian motion $W$ and $\alpha$ is a continous process, another example of a localizing sequence for the continuous local martingale $\int \alpha . dW$ is

$$\tau_n = \inf\{t \in \mathbb{T} : |\alpha_t| \geq n\}.$$

In this case, the stopped stochastic integral $\int^{\cdot \wedge \tau_n} \alpha . dW$ is a square integrable martingale with $E[< \int^{\cdot \wedge \tau_n} \alpha . dW >_t] = E[\int_0^{t \wedge \tau_n} |\alpha_u|^2 du] \leq n^2 t$, for all $t \in \mathbb{T}$.

**Stochastic integral with respect to a vectorial continuous semimartingale***

Let $X$ be a vectorial continous semimartingale written in the form $X = M + A$ where $M$ is a vectorial continuous local martingale and $A$ is a vectorial continuous process with finite variation. Naturally, for all $\alpha \in L^2_{loc}(M) \cap L_S(A)$, one can define the stochastic integral of $\alpha$ with respect to $X$ by setting

$$\int \alpha dX = \int \alpha dM + \int \alpha dA.$$

Notice that the locally bounded progressively measurable processes $\alpha$, i.e. $\sup_{0 \leq s \leq t} |\alpha_s| < \infty$ a.s. for all $t$ in $\mathbb{T}$, belong to $L^2_{loc}(M) \cap L_S(A)$. This class of integrands, sufficient in most practical applications, is not large enough in theory when one is looking at closure and stability properties of stochastic integrals. The semimartingale topology is quite suitable for the study of stochastic integrals. This topology is defined by the distance between two semimartingales (here continuous) $X = (X_t)_{t \in \mathbb{T}}$ and $Y = (Y_t)_{t \in \mathbb{T}}$:

$$D_E(X, Y) = \sup_{|\alpha| \leq 1} \left( \sum_{n \geq 1} 2^{-n} E \left[ \left| \int_0^{\bar{T} \wedge n} \alpha_t dX_t - \int_0^{\bar{T} \wedge n} \alpha_t dY_t \right| \wedge 1 \right] \right),$$

where the supremum is taken over all progressively measurable processes $\alpha$ bounded by 1.

**Definition 1.2.10** *Let $X$ be a continuous semimartingale. Let $\alpha$ be a progressively measurable process and $\alpha^{(n)}$ the bounded truncated process $\alpha 1_{|\alpha| \leq n}$. We say that $\alpha$ is integrable with respect to $X$ and we write $\alpha \in L(X)$, if the sequence of semimartingales $\int \alpha^{(n)} dX$ converges for the semimartingale topology to a semimartingale $Y$, and we then set $\int \alpha dX = Y$.*

We have the following properties for this general class of integrands:

- If $X$ is a local martingale, $L^2_{loc}(X) \subset L(X)$.
- If $X$ has finite variation, $L_S(X) \subset L(X)$.
- $L(X) \cap L(Y) \subset L(X + Y)$ and $\int \alpha dX + \int \alpha dY = \int \alpha d(X + Y)$.
- $L(X)$ is a vector space and $\int \alpha dX + \int \beta dX = \int (\alpha + \beta) dX$.

Moreover, the space $\{\int \alpha dX : \alpha \in L(X)\}$ is closed in the space of semimartingales for the semimartingale topology. Finally, since the semimartingale topology is invariant by change of equivalent probability measure, the same holds true for $L(X)$. Warning: if $X$ is a continuous local martingale and $\alpha$ lies in $L(X)$, the stochastic integral $\int \alpha dX$ is not always a local martingale. Actually, $\int \alpha dS$ is a local martingale if and only if $\alpha$ lies in $L^2_{loc}(M)$. We also know that when the process $\int \alpha dX$ is lower-bounded, then it is a local martingale and also a supermartingale.

### 1.2.2 Itô process

In finance, we often use Itô processes as continuous semimartingales for modeling the dynamics of asset prices.

**Definition 1.2.11** *Let* $W = (W^1, \ldots, W^d)$ *be a d-dimensional Brownian motion on a filtered probability space* $(\Omega, \mathcal{F}, \mathbb{F}, P)$. *We define an Itô process as a process* $X = (X^1, \ldots, X^n)$ *valued in* $\mathbb{R}^n$ *such that a.s.*

$$X_t = X_0 + \int_0^t b_s ds + \int_0^t \sigma_s dW_s, \quad t \in \mathbb{T}, \tag{1.9}$$

*i.e.* $\quad X_t^i = X_0^i + \int_0^t b_s^i ds + \sum_{j=1}^d \int_0^t \sigma_s^{ij} dW_s^j, \quad t \in \mathbb{T}, \, 1 \le i \le n,$

*where* $X_0$ *is* $\mathcal{F}_0$-measurable, $b = (b^1, \ldots, b^n)$ *and* $\sigma = (\sigma^1, \ldots, \sigma^n) = (\sigma^{ij})_{1 \le i \le n, 1 \le j \le d}$ *are progressively measurable processes valued respectively in* $\mathbb{R}^n$ *and* $\mathbb{R}^{n \times d}$ *such that* $b^i \in L_S(dt)$ *and* $\sigma^i \in L_{loc}^2(W)$, $i = 1, \ldots, n$, *i.e.*

$$\int_0^t |b_s| ds + \int_0^t |\sigma_s|^2 ds < \infty, \quad a.s., \; \forall t \in \mathbb{T}.$$

We often write (1.9) in the differential form

$$dX_t = b_t dt + \sigma_t dW_t.$$

### 1.2.3 Itô's formula

**1.** Let $X = (X^1, \ldots, X^d)$ be a continuous semimartingale valued in $\mathbb{R}^d$ and $f$ a function of class $C^{1,2}$ on $\mathbb{T} \times \mathbb{R}^d$. Then $(f(t, X_t))_{t \in \mathbb{T}}$ is a semimartingale and we have for all $t \in \mathbb{T}$

$$f(t, X_t) = f(0, X_0) + \int_0^t \frac{\partial f}{\partial t}(u, X_u) du + \sum_{i=1}^d \int_0^t \frac{\partial f}{\partial x_i}(u, X_u) dX_u^i$$

$$+ \frac{1}{2} \sum_{i,j=1}^d \int_0^t \frac{\partial^2 f}{\partial x_i \partial x_j}(u, X_u) d < X^i, X^j >_u .$$

In this expression, the various integrands are continuous, hence locally bounded, and the stochastic integrals or Stieltjes integrals are well-defined.

**2.** In this book, we shall also use in an example (see Section 4.5) Itô's formula for a semimartingale $X = (X^1, \ldots, X^d)$ in the form

$$X^i = M^i + A^i, \quad i = 1, \ldots, d,$$

where $M^i$ is a continuous martingale and $A^i$ is an adapted process with finite variation. We denote by $A^{i,c}$ the continuous part of $A^i$. If $f$ is a function of class $C^{1,2}$ on $\mathbb{T} \times \mathbb{R}^d$, $(f(t, X_t))_{t \in \mathbb{T}}$ is a semimartingale and we have for all $t \in \mathbb{T}$

$$f(t, X_t) = f(0, X_0) + \int_0^t \frac{\partial f}{\partial t}(u, X_u) du + \sum_{i=1}^d \int_0^t \frac{\partial f}{\partial x_i}(u, X_u) dM_u^i$$

$$+ \frac{1}{2} \sum_{i,j=1}^d \int_0^t \frac{\partial^2 f}{\partial x_i \partial x_j}(u, X_u) d < M^i, M^j >_u$$

$$+ \sum_{i=1}^d \int_0^t \frac{\partial f}{\partial x_i}(u, X_u) dA_u^{i,c} + \sum_{0 < s \le t} [f(s, X_s) - f(s, X_{s-})] .$$

### 1.2.4 Martingale representation theorem

In this section, we state some martingale representation theorems by means of stochastic integrals.

The first result is the Brownian martingale representation theorem, also known as the Itô representation theorem.

**Theorem 1.2.9** *(Representation of Brownian martingales)*
*Assume that $\mathbb{F}$ is the natural (augmented) filtration of a standard $d$-dimensional Brownian motion $W = (W^1, \ldots, W^d)$. Let $M = (M_t)_{t \in \mathbb{T}}$ be a càd-làg local martingale. Then there exists $\alpha = (\alpha^1, \ldots, \alpha^d) \in L^2_{loc}(W)$ such that*

$$M_t = M_0 + \int_0^t \alpha_u . dW_u = M_0 + \sum_{i=1}^d \int_0^t \alpha_t^i dW_t^i, \quad t \in \mathbb{T}, \quad a.s.$$

*Moreover, if $M$ is bounded in $L^2$ then $\alpha \in L^2(W)$, i.e. $E[\int_0^{\bar{T}} |\alpha_t|^2 dt] < \infty$.*

For a proof, we may consult ch. 3, sec. 3.4 in Karatzas and Shreve [KaSh88] or ch. V in Revuz and Yor [ReY91]. This result shows in particular that any martingale with respect to a Brownian filtration is continuous (up to an indistinguishable process).

The second result is the projection theorem on the space of stochastic integrals with respect to a continuous local martingale. It originally appeared in Kunita and Watanabe [KW67], and then in Galtchouk [Ga76]. We may find a proof in Jacod [Jac79], ch. IV, sec. 2.

**Theorem 1.2.10** *(Galtchouk-Kunita-Watanabe decomposition)*
*Let $M = (M^1, \ldots, M^d)$ be a continuous local martingale valued in $\mathbb{R}^d$ and $N$ a real-valued càd-làg local martingale. Then there exist $\alpha \in L^2_{loc}(M)$ and $R$, a càd-làg local martingale orthogonal to $M$ ($< R, M^i > = 0$, $i = 1, \ldots, d$) null in $0$ such that*

$$N_t = N_0 + \int_0^t \alpha_u dM_u + R_t, \quad t \in \mathbb{T}, \quad a.s.$$

*Furthermore, if $N$ is bounded in $L^2$ then $\alpha \in L^2(M)$ and $R$ is also bounded in $L^2$.*

The next result, due to Yor [Yo78], considers the problem of the limit in $L^1$ of integrands with respect to a martingale.

**Theorem 1.2.11** *Let $M$ be a continuous local martingale and $(\alpha^{(n)})_{n \geq 1}$ a sequence of processes in $L(M)$ such that for all $n$, $\int \alpha^{(n)} dM$ is uniformly integrable and the sequence $(\int_0^{\bar{T}} \alpha_t^{(n)} dM_t)_{n \geq 1}$ converges in $L^1$ to a random variable $\xi \in L^1$. Then there exists $\alpha \in L(M)$ such that $\int \alpha dM$ is a uniformly integrable martingale and $\xi = \int_0^{\bar{T}} \alpha_t dM_t$.*

### 1.2.5 Girsanov's theorem

In this section, we focus on the effect of a change of probability measure on the notions of semimartingales and martingales. The presentation is inspired by ch. 4, sec. 3.5 in

Karatzas and Shreve [KaSh88] and ch. VIII in Revuz and Yor [ReY91]. We are given a filtered probability space $(\Omega, \mathcal{F}, \mathbb{F} = (\mathcal{F}_t)_{t\in\mathbb{T}}, P)$ satisfying the usual conditions. To simplify the notation, we suppose that $\mathcal{F} = \mathcal{F}_{\bar{T}}$.

Let $Q$ be a probability measure on $(\Omega, \mathcal{F})$ such that $Q$ is absolutely continuous with respect to $P$, denoted by $Q \ll P$. Let $Z_{\bar{T}}$ be the associated Radon-Nikodym density, often denoted by

$$Z_{\bar{T}} = \frac{dQ}{dP} \quad \text{or} \quad Q = Z_{\bar{T}}.P.$$

For all $t \in \mathbb{T}$, the restriction of $Q$ to $\mathcal{F}_t$ is absolutely continuous with respect to the restriction of $P$ to $\mathcal{F}_t$. Let $Z_t$ be the corresponding Radon-Nikodym density, often denoted by

$$Z_t = \left.\frac{dQ}{dP}\right|_{\mathcal{F}_t}.$$

Then the process $(Z_t)_{t\in\mathbb{T}}$ is a positive martingale (under $P$ with respect to $(\mathcal{F}_t)_{t\in\mathbb{T}}$), closed on the right by $Z_{\bar{T}}$:

$$Z_t = E[Z_{\bar{T}}|\mathcal{F}_t], \quad a.s. \ \forall\, t \in \mathbb{T}.$$

Up to modification, we may suppose that the paths of $Z$ are càd-làg. $Z$ is also called the martingale density process of $Q$ (with respect to $P$). $Z$ is positive, and actually we have

$$Z_t > 0, \quad \forall\, t \in \mathbb{T}, \ Q \ a.s,$$

which means that $Q[\tau < \infty] = 0$, where $\tau = \inf\{t : Z_t = 0 \text{ or } Z_{t-} = 0\}$: this follows from the fact that the martingale $Z$ vanishes on $[\tau, \infty)$. When $Q$ is equivalent to $P$, denoted by $Q \sim P$, we write without ambiguity $Z_t > 0$, for all $t$ in $\mathbb{T}$, a.s. In the sequel, we denote by $E^Q$ the expectation operator under $Q$. When a property is relative to $Q$, we shall specify the reference to $Q$. When it is not specified, the property is implicitly relative to $P$, the initial probability measure on $(\Omega, \mathcal{F})$.

**Proposition 1.2.12** *(Bayes formula)*
*Let $Q \ll P$ and $Z$ its martingale density process. For all stopping times $\sigma \le \tau$ valued in $\mathbb{T}$, $\xi$ $\mathcal{F}_\tau$-measurable random variable such that $E^Q[|\xi|] < \infty$, we have*

$$E^Q[\xi|\mathcal{F}_\sigma] = \frac{E[Z_\tau\xi|\mathcal{F}_\sigma]}{Z_\sigma}, \quad Q \quad a.s.$$

The Bayes formula shows in particular that a process $X$ is a $Q$-(super)martingale (local) if and only if $ZX$ is a (super)martingale (local).

**Theorem 1.2.12** *(Girsanov)*
*Let $Q \ll P$ and $Z$ its martingale density process. We suppose that $Z$ is continuous. Let $M$ be a continuous local martingale. Then the process*

$$M^Q = M - \int \frac{1}{Z} d<M, Z>$$

*is a continuous local martingale under $Q$. Moreover, if $N$ is a continuous local martingale, we have*

$$< M^Q, N^Q > = < M, N > .$$

*If $Q \sim P$, i.e. $Z$ is strictly positive a.s., then there exists a unique continuous local martingale $L$ null in $t = 0$, such that*

$$Z_t = \exp\left( L_t - \frac{1}{2} < L, L >_t \right) =: \mathcal{E}_t(L), \quad t \in \mathbb{T}, \text{ a.s.}$$

*and $L$ is given by*

$$L_t = \int_0^t \frac{1}{Z_s} dZ_s, \quad t \in \mathbb{T}, \text{ a.s.}$$

*The $Q$-local martingale $M^Q$ is then written also as:*

$$M^Q = M - < M, L > .$$

In the case of a Brownian motion, we have the following important result.

**Theorem 1.2.13** *(Cameron-Martin)*
*Let $W$ be a Brownian motion. Let $Q \sim P$ with martingale density process*

$$\left. \frac{dQ}{dP} \right|_{\mathcal{F}_t} = \mathcal{E}_t(L),$$

*where $L$ is a continuous local martingale. Then the process*

$$W^Q = W - < W, L >$$

*is a $Q$-Brownian motion.*

In the applications of the Girsanov-Cameron-Martin theorem, we start with a continuous local martingale $L$ null on 0 and we construct $Q \sim P$ by setting $Q = \mathcal{E}_t(L).P$ on $\mathcal{F}_t$. This obviously requires the process $\mathcal{E}(L)$ to be a martingale. In general, we know that $\mathcal{E}(L)$ is a local martingale, called the Doléans-Dade exponential. It is also a (strictly) positive process and so a supermartingale, which ensures the existence of the a.s. limit $\mathcal{E}_{\bar{T}}(L)$ of $\mathcal{E}_t(L)$ when $t$ goes to $\bar{T}$. The Doléans-Dade exponential $\mathcal{E}(L)$ is then a martingale if and only if $E[\mathcal{E}_{\bar{T}}(L)] = 1$. In this case, we can define a probability measure $Q \sim P$ with Radon-Nikodym density $\mathcal{E}_{\bar{T}}(L) = dQ/dP$, and with martingale density process $\mathcal{E}_t(L) = dQ/dP|_{\mathcal{F}_t}$. We may then apply the Girsanov-Cameron-Martin theorem. Therefore, it is important to get conditions for $\mathcal{E}(L)$ to be a martingale.

**Proposition 1.2.13** *(Novikov's condition)*
*Let $L$ be a continuous local martingale, $L_0 = 0$, such that*

$$E\left[ \exp\left( \frac{1}{2} < L, L >_{\bar{T}} \right) \right] < \infty.$$

*Then $L$ and $\mathcal{E}(L)$ are uniformly integrable martingales with $E[\exp(L_{\bar{T}}/2)] < \infty$.*

**Case of Brownian motion**

Consider a $d$-dimensional Brownian motion $W = (W^1, \ldots, W^d)$ with respect to $(\mathcal{F}_t)_{t \in \mathbb{T}}$. A common choice of local martingale $L$ is

$$L_t = \int_0^t \lambda_s . dW_s = \sum_{i=1}^d \int_0^t \lambda_s^i dW_s^i,$$

where $\lambda = (\lambda^1, \ldots, \lambda^d)$ lies in $L^2_{loc}(W)$. The associated process $\mathcal{E}(L)$ is

$$\mathcal{E}_t(L) = \exp\left( \int_0^t \lambda_u . dW_u - \frac{1}{2} \int_0^t |\lambda_u|^2 du \right), \quad t \in \mathbb{T}. \tag{1.10}$$

When $\mathcal{E}(L)$ is a martingale so that one can define probability measure $Q \sim P$ with martingale density process $\mathcal{E}(L)$, the Girsanov-Cameron-Martin theorem asserts that $W^Q = (W^{Q,1}, \ldots, W^{Q,d})$ with

$$W^{Q,i} = W^i - \int \lambda^i dt, \quad i = 1, \ldots, d \tag{1.11}$$

is a $Q$-Brownian motion. Notice that in the case where $\mathbb{F}$ is the Brownian filtration and from the Itô martingale representation theorem 1.2.9, any probability $Q \sim P$ has a martingale density process in the form (1.10). Finally, the Novikov condition ensuring that $\mathcal{E}(L)$ is a martingale is written as

$$E\left[ \exp\left( \frac{1}{2} \int_0^{\bar{T}} |\lambda_t|^2 dt \right) \right] < \infty.$$

We end this section by stating a representation theorem for Brownian martingales under a change of probability measure.

**Theorem 1.2.14** *(Itô's representation under a change of probability)*
*Assume that $\mathbb{F}$ is the natural (augmented) filtration of a standard $d$-dimensional Brownian motion $W = (W^1, \ldots, W^d)$. Let $Q \sim P$ with martingale density process $Z$ given by (1.10) and $W^Q$ the Brownian motion under $Q$ given by (1.11). Let $M = (M_t)_{t \in \mathbb{T}}$ a càd-làg $Q$-local martingale. Then there exists $\alpha = (\alpha^1, \ldots, \alpha^d) \in L^2_{loc}(W^Q)$ such that:*

$$M_t = M_0 + \int_0^t \alpha_u . dW_u^Q = M_0 + \sum_{i=1}^d \int_0^t \alpha_t^i dW_t^{Q,i}, \quad t \in \mathbb{T}, \quad a.s.$$

This result is the basis for the replication problem in complete markets in finance. It will also be used in the sequel of this book. Notice that it cannot be deduced as a direct consequence of Itô's representation theorem 1.2.9 applied to the $Q$-martingale $M$ with respect to the filtration $\mathbb{F}$. Indeed, the filtration $\mathbb{F}$ is the natural filtration of $W$ and not of $W^Q$. In order to show Theorem 1.2.14, one first boils down to the probability measure $P$ from the Bayes formula by writing that $N = ZM$ is a local martingale. We may then apply Itô's representation theorem 1.2.9 (under $P$) to $N$, and then go back to $M = N/Z$ from Itô's formula (see the details in the proof of Proposition 5.8.6 in Karatzas and Shreve [KaSh88]).

## 1.3 Stochastic differential equations

We recall in this section some results about stochastic differential equations (SDE) with random coefficients, with respect to a Brownian motion.

### 1.3.1 Strong solutions of SDE

We only introduce the concept of strong solutions of SDE. We fix a filtered probability space $(\Omega, \mathcal{F}, \mathbb{F} = (\mathcal{F}_t)_{t \in \mathbb{T}}, P)$ satisfying the usual conditions and a $d$-dimensional Brownian motion $W = (W^1, \ldots, W^d)$ with respect to $\mathbb{F}$. We are given functions $b(t, x, \omega) = (b_i(t, x, \omega))_{1 \le i \le d}$, $\sigma(t, x, \omega) = (\sigma_{ij}(t, x, \omega))_{1 \le i \le n, 1 \le j \le d}$ defined on $\mathbb{T} \times \mathbb{R}^n \times \Omega$, and valued respectively in $\mathbb{R}^n$ and $\mathbb{R}^{n \times d}$. We assume that for all $\omega$, the functions $b(., ., \omega)$ and $\sigma(., ., \omega)$ are Borelian on $\mathbb{T} \times \mathbb{R}^n$ and for all $x \in \mathbb{R}^n$, the processes $b(., x, .)$ and $\sigma(., x, .)$, written $b(., x)$ and $\sigma(., x)$ for simplification, are progressively measurable. We then consider the SDE valued in $\mathbb{R}^n$:

$$dX_t = b(t, X_t)dt + \sigma(t, X_t)dW_t \tag{1.12}$$

which is also written componentwise:

$$dX_t^i = b_i(t, X_t)dt + \sum_{j=1}^{d} \sigma_{ij}(t, X_t)dW_t^j, \quad 1 \le i \le d. \tag{1.13}$$

In the applications for this book, we shall mainly consider two types of situations:

(1) $b$ and $\sigma$ are deterministic Borelian functions $b(t, x)$, $\sigma(t, x)$ of $t$ and $x$, and we speak about diffusion for the SDE (1.12).

(2) The random coefficients $b(t, x, \omega)$ and $\sigma(t, x, \omega)$ are in the form $\tilde{b}(t, x, \alpha_t(\omega))$, $\tilde{\sigma}(t, x, \alpha_t(\omega))$ where $\tilde{b}$, $\tilde{\sigma}$ are deterministic Borelian functions on $\mathbb{T} \times \mathbb{R}^n \times A$, $A$ set of $\mathbb{R}^m$, and $\alpha = (\alpha_t)_{t \in \mathbb{T}}$ is a progressively measurable process valued in $A$. This case arises in stochastic control problems studied in Chapters 3 and 4, and we say that the SDE (1.12) is a controlled diffusion by $\alpha$.

**Definition 1.3.12** *(Strong solution of SDE)*
*A strong solution of the SDE (1.12) starting at time $t$ is a vectorial progressively measurable process $X = (X^1, \ldots, X^n)$ such that*

$$\int_t^s |b(u, X_u)|du + \int_t^s |\sigma(u, X_u)|^2 du < \infty, \quad a.s., \ \forall t \le s \in \mathbb{T},$$

*and the following relations:*

$$X_s = X_t + \int_t^s b(u, X_u)du + \int_t^s \sigma(u, X_u)dW_u, \quad t \le s \in \mathbb{T},$$

*i.e.*

$$X_s^i = X_t^i + \int_t^s b_i(u, X_u)du + \sum_{j=1}^{d} \int_t^s \sigma_{ij}(u, X_u)dW_u^j, \quad t \le s \in \mathbb{T}, \ 1 \le i \le d,$$

*hold true a.s.*

Notice that a strong solution of SDE (1.12) is a continuous process (up to an indistinguishable process).

We mention that there exists another concept of solution to the SDE (1.12), called weak, where the filtered probability space $(\Omega, \mathcal{F}, \mathbb{F} = (\mathcal{F}_t)_{t \in \mathbb{T}}, P)$ and the Brownian motion $W$ are part of the unknown of the SDE, in addition to $X$.

Existence and uniqueness of a strong solution to the SDE (1.12) is ensured by the following Lipschitz and linear growth conditions: there exists a (deterministic) constant $K$ and a real-valued process $\kappa$ such that for all $t \in \mathbb{T}, \omega \in \Omega, x, y \in \mathbb{R}^n$

$$|b(t, x, \omega) - b(t, y, \omega)| + |\sigma(t, x, \omega) - \sigma(t, y, \omega)| \leq K|x - y|, \tag{1.14}$$

$$|b(t, x, \omega)| + |\sigma(t, x, \omega)| \leq \kappa_t(\omega) + K|x|, \tag{1.15}$$

with

$$E\left[\int_0^t |\kappa_u|^2 du\right] < \infty, \quad \forall t \in \mathbb{T}. \tag{1.16}$$

Under (1.14), a natural choice for $\kappa$ is $\kappa_t = |b(t, 0)| + |\sigma(t, 0)|$ once it satisfies the condition (1.16).

**Theorem 1.3.15** *Under conditions* (1.14), (1.15) *and* (1.16), *there exists for all $t \in \mathbb{T}$, a strong solution to the SDE* (1.12) *starting at time $t$. Moreover, for any $\xi$ $\mathcal{F}_t$-measurable random variable valued in $\mathbb{R}^n$, such that $E[|\xi|^p] < \infty$, for some $p > 1$, there is uniqueness of a strong solution $X$ starting from $\xi$ at time $t$, i.e. $X_t = \xi$. The uniqueness is pathwise and means that if $X$ and $Y$ are two such strong solutions, we have $P[X_s = Y_s, \forall t \leq s \in \mathbb{T}] = 1$. This solution is square integrable: for all $T > t$, there exists a constant $C_T$ such that*

$$E\left[\sup_{t \leq s \leq T} |X_s|^p\right] \leq C_T(1 + E[|\xi|^p]).$$

This result is standard and one can find a proof in the books of Gihman and Skorohod [GS72], Ikeda and Watanabe [IW81], Krylov [Kry80] or Protter [Pro90]. We usually denote by $X^{t,\xi} = \{X_s^{t,\xi}, t \leq s \in \mathbb{T}\}$ the strong solution to the SDE (1.12) starting from $\xi$ at time $t$. When $t = 0$, we simply write $X^\xi = X^{0,\xi}$. By pathwise uniqueness, we notice that for all $t \leq \theta$ in $\mathbb{T}$, and $x \in \mathbb{R}^n$, we have a.s.

$$X_s^{t,x} = X_s^{\theta, X_\theta^{t,x}}, \quad \forall \theta \leq s \in \mathbb{T}.$$

When $b$ and $\sigma$ are deterministic functions of $t$ and $x$, we also know that the strong solution to the SDE (1.12) is adapted with respect to the natural filtration of $W$. We also have the Markov property of any strong solution to the SDE (1.12): for any Borelian bounded function $g$ on $\mathbb{R}^d$, for all $t \leq \theta$ in $\mathbb{T}$, we have

$$E\left[g(X_\theta)| \mathcal{F}_t\right] = \varphi_\theta(t, X_t)$$

where $\varphi_\theta$ is the function defined on $\mathbb{T} \times \mathbb{R}^d$ by

$$\varphi_\theta(t, x) = E\left[g(X_\theta^{t,x})\right].$$

Given a strong solution to the SDE (1.12) starting at time $t$, and a function $f$ of class $C^{1,2}$ on $\mathbb{T} \times \mathbb{R}^n$, Itô's formula to $f(s, X_s)$, $t \leq s$ in $\mathbb{T}$, is written as

$$f(s, X_s) = f(t, X_t) + \int_t^s \frac{\partial f}{\partial t}(u, X_u) + \mathcal{L}_u f(u, X_u) du$$

$$+ \int_t^s D_x f(u, X_u)' \sigma(u, X_u) dW_u$$

where

$$\mathcal{L}_t(\omega) f(t, x) = b(t, x, \omega).D_x f(t, x) + \frac{1}{2} \text{tr}(\sigma(t, x, \omega)) \sigma'(t, x, \omega) D_x^2 f(t, x))$$

$$= \sum_{i=1}^n b_i(t, x, \omega)) \frac{\partial f}{\partial x_i}(t, x) + \sum_{i,k=1}^n \gamma_{ik}(t, x, \omega) \frac{\partial^2 f}{\partial x_i \partial x_k}(t, x),$$

and $\gamma(t, x) = \sigma(t, x)\sigma'(t, x)$ is $n \times n$ matrix-valued with components

$$\gamma_{ik}(t, x) = \sum_{j=1}^d \sigma_{ij}(t, x)\sigma_{kj}(t, x).$$

### 1.3.2 Estimates on the moments of solutions to SDE

In this section, we give some estimates on the moments of solutions to the SDE (1.12). Notice that under condition (1.14), there exists a finite positive constant $\beta_0$ such that for all $t \in \mathbb{T}$, $\omega \in \Omega$, $x, y \in \mathbb{R}^n$

$$(b(t, x, \omega) - b(t, y, \omega)).(x - y) + \frac{1}{2}|\sigma(t, x, \omega) - \sigma(t, y, \omega)|^2 \leq \beta_0 |x - y|^2.$$

The following estimates are essentially based on the Burkholder-Davis-Gundy and Doob inequalities, Itô's formula and Gronwall's lemma that we recall here.

**Lemma 1.3.1** *(Gronwall)*
*Let $g$ be a continuous positive function on $\mathbb{R}_+$ such that*

$$g(t) \leq h(t) + C \int_0^t g(s)ds, \quad 0 \leq t \leq T,$$

*where $C$ is a positive constant, and $h$ is an integrable function on $[0, T]$, $T > 0$. Then*

$$g(t) \leq h(t) + C \int_0^t h(s)e^{C(t-s)}ds, \quad 0 \leq t \leq T.$$

**Theorem 1.3.16** *Assume that conditions (1.14), (1.15) and (1.16) hold.*
*(1) There exists a constant $C$ (depending on $K$) such that for all $t \leq \theta$ in $\mathbb{T}$ and $x \in \mathbb{R}^n$*

$$E\left[\sup_{t \leq s \leq \theta} |X_s^{t,x}|^2\right] \leq C|x|^2 + Ce^{C(\theta-t)} E\left[\int_t^\theta |x|^2 + |\kappa_u|^2 du\right] \qquad (1.17)$$

$$E\left[\sup_{t \leq s \leq \theta} |X_s^{t,x} - x|^2\right] \leq Ce^{C(\theta-t)} E\left[\int_t^\theta |x|^2 + |\kappa_u|^2 du\right]. \qquad (1.18)$$

*(2) For all $0 \leq t \leq s$ in $\mathbb{T}$ and $x, y \in \mathbb{R}^n$*

$$E\left[\sup_{t \leq u \leq s} \left|X_u^{t,x} - X_u^{t,y}\right|^2\right] \leq e^{2\beta_0(s-t)}|x - y|^2. \tag{1.19}$$

We may find a proof of these estimates in Krylov [Kry80], ch. 2. Notice that the inequality (1.18) implies in particular that

$$\lim_{h \downarrow 0} E\left[\sup_{s \in [t, t+h]} \left|X_s^{t,x} - x\right|^2\right] = 0.$$

### 1.3.3 Feynman-Kac formula

In this section, we consider the SDE (1.12) with deterministic coefficients $b(t,x)$ and $\sigma(t,x)$. For all $t \in \mathbb{T}$, we introduce the differential operator of second order:

$$(\mathcal{L}_t\varphi)(x) = b(t,x).D_x\varphi(x) + \frac{1}{2}\text{tr}(\sigma(t,x)\sigma'(t,x)D_x^2\varphi(x)), \quad \varphi \in C^2(\mathbb{R}^n).$$

$\mathcal{L}_t$ is called the infinitesimal generator of the diffusion (1.12). If $X$ is a solution to the SDE (1.12), $v(t,x)$ a (real-valued) function of class $C^{1,2}$ on $\mathbb{T} \times \mathbb{R}^n$ and $r(t,x)$ a continuous function on $\mathbb{T} \times \mathbb{R}^d$, we obtain by Itô's formula

$$M_t := e^{-\int_0^t r(s,X_s)ds}v(t,X_t) - \int_0^t e^{-\int_0^s r(u,X_u)du}\left(\frac{\partial v}{\partial t} + \mathcal{L}_s v - rv\right)(s,X_s)ds$$

$$= v(0,X_0) + \int_0^t e^{-\int_0^s r(u,X_u)du}D_x v(s,X_s)'\sigma(s,X_s)dW_s. \tag{1.20}$$

The process $M$ is thus a continuous local martingale.

On a finite horizon interval $\mathbb{T} = [0,T]$, we consider the Cauchy linear parabolic partial differential equation (PDE):

$$rv - \frac{\partial v}{\partial t} - \mathcal{L}_t v = f, \qquad \text{on } [0,T) \times \mathbb{R}^n \tag{1.21}$$

$$v(T,.) = g, \qquad \text{on } \mathbb{R}^n, \tag{1.22}$$

where $f$ (resp. $g$) is a continuous function from $[0,T] \times \mathbb{R}^n$ (resp. $\mathbb{R}^n$) into $\mathbb{R}$. We also assume that the function $r$ is nonnegative. We give here a simple version of the Feynman-Kac representation theorem. Recall that $X^{t,x}$ denotes the soluton to the diffusion (1.12) starting from $x$ at time $t$.

**Theorem 1.3.17** *(Feynman-Kac representation)*
*Let $v$ be a function $C^{1,2}([0,T[\times\mathbb{R}^d) \cap C^0([0,T] \times \mathbb{R}^d)$ with dervative in $x$ bounded and solution to the Cauchy problem (1.21)-(1.22). Then $v$ admits the representation*

$$v(t,x) = E\left[\int_t^T e^{-\int_t^s r(u,X_u^{t,x})du}f(s,X_s^{t,x})ds + e^{-\int_t^T r(u,X_u^{t,x})du}g(X_T^{t,x})\right], \tag{1.23}$$

*for all $(t,x) \in [0,T] \times \mathbb{R}^d$.*

The proof of this result follows from the fact that when $v$ has a bounded derivative in $x$, the integrand of the stochastic integral in (1.20) lies in $L^2(W)$ from the linear growth condition in $x$ of $\sigma$ and the estimate (1.17). Hence, $M$ is a (square integrable) martingale and the representation (1.23) is simply derived by writing that $E[M_T] = E[M_t]$. We may also obtain this Feynman-Kac representation under other conditions on $v$, for example with $v$ satisfying a quadratic growth condition. We shall show such a result in Chapter 3 in the more general case of controlled diffusion (see Theorem 3.5.2 and Remark 3.5.5). We shall also see a version of the Feynman-Kac theorem on an infinite horizon for elliptic PDE problems (see Theorem 3.5.3 and Remark 3.5.6).

The application of the previous theorem requires the existence of a smooth solution $v$ to the Cauchy problem (1.21)-(1.22). This type of result is typically obtained under an assumption of uniform ellipticity on the operator $\mathcal{L}_t$:

$$\exists \varepsilon > 0, \ \forall \, (t,x) \in [0,T] \times \mathbb{R}^n, \ \forall y \in \mathbb{R}^n, \ y'\sigma\sigma'(t,x)y \ \geq \ \varepsilon |y|^2, \qquad (1.24)$$

boundedness conditions on $b$, $\sigma$, and polynomial growth condition on $f$ and $g$ (see Friedman [Fr75] p. 147). There also exist other sufficient conditions relaxing the uniform ellipticity condition (1.24) but requiring stronger regularity conditions on the coefficients (see Krylov [Kry80] p. 118).

In the general case where there is no smooth solution to the Cauchy problem (1.21)-(1.22), this PDE may be formulated by means of the concept of weak solution, called viscosity solution. This notion will be developed in Chapter 4 in the more general framework of controlled diffusion, leading to a nonlinear PDE called Hamilton-Jacobi-Bellman.

# 2

# Stochastic optimization problems. Examples in finance

## 2.1 Introduction

In this chapter, we outline the basic structure of a stochastic optimization problem in continuous time, and we illustrate it through several examples from mathematical finance. The solution to these problems will be detailed later.

In general terms, a stochastic optimization problem is formulated with the following features:

- **State of the system**: We consider a dynamic system caracterized by its state at any time, and evolving in an uncertain environment formalized by a probability space $(\Omega, \mathcal{F}, P)$. The state of the system represents the set of quantitative variables needed to describe the problem. We denote by $X_t(\omega)$ the state of the system at time $t$ in a world scenario $\omega \in \Omega$. We then have to describe the (continuous-time) dynamics of the state system, i.e. the mapping $t \mapsto X_t(\omega)$ for all $\omega$, through a stochastic differential equation or process.

- **Control**: The dynamics $t \to X_t$ of the system is typically influenced by a control modeled as a process $\alpha = (\alpha_t)_t$ whose value is decided at any time $t$ in function of the available information. The control $\alpha$ should satisfy some constraints, and is called admissible control. We denote by $\mathcal{A}$ the set of admissible controls.

- **Performance/cost criterion**: The objective is to maximize (or minimize) over all admissible controls a functional $J(X, \alpha)$. We shall consider typically objective functionals in the form

$$E\left[ \int_0^T f(X_t, \omega, \alpha_t)dt + g(X_T, \omega)\right], \qquad \text{on a finite horizon } T < \infty,$$

and

$$E\left[ \int_0^\infty e^{-\beta t} f(X_t, \omega, \alpha_t)dt\right], \qquad \text{on an infinite horizon.}$$

The function $f$ is a running profit function, $g$ is a terminal reward function, and $\beta > 0$ is a discount factor. In other situations, the controller may also decide directly the horizon or ending time of his objective. The corresponding optimization problem is called optimal

H. Pham, *Continuous-time Stochastic Control and Optimization with Financial Applications*, Stochastic Modelling and Applied Probability 61, DOI 10.1007/978-3-540-89500-8_2, © Springer-Verlag Berlin Heidelberg 2009

stopping time. Extensions of optimal stopping problems with multiple decision times will be considered in Chapter 5. In a general formulation, the control can be mixed, composed of a pair control/stopping time $(\alpha, \tau)$, and the objective functional is in the form

$$J(X, \alpha, \tau) = E\left[ \int_0^\tau f(X_t, \alpha_t) dt + g(X_\tau) \right].$$

The maximum value, called he value function, is defined by

$$v = \sup_{\alpha, \tau} J(X, \alpha, \tau).$$

The main goal in a stochastic optimization problem is to find the maximizing control process and/or stopping time attaining the value function to be determined.

## 2.2 Examples

We present several examples of stochastic optimization problems arising in economics and finance. Other examples and solutions will be provided in the later chapters by different approaches.

### 2.2.1 Portfolio allocation

We consider a financial market consisting of a riskless asset with strictly positive price process $S^0$ representing the savings account, and $n$ risky assets of price process $S$ representing stocks. An agent may invest in this market at any time $t$, with a number of shares $\alpha_t$ in the $n$ risky assets. By denoting by $X_t$ its wealth at time $t$, the number of shares invested in the savings account at time $t$ is $(X_t - \alpha_t.S_t)/S_t^0$. The self-financed wealth process evolves according to

$$dX_t = (X_t - \alpha_t.S_t)\frac{dS_t^0}{S_t^0} + \alpha_t dS_t.$$

The control is the process $\alpha$ valued in $A$, subset of $\mathbb{R}^n$. The portfolio allocation problem is to choose the best investment in these assets. Classical modeling for describing the behavior and preferences of agents and investors are: expected utility criterion and mean-variance criterion.

In the first criterion relying on the theory of choice in uncertainty, the agent compares random incomes for which he knows the probability distributions. Under some conditions on the preferences, Von Neumann and Morgenstern show that they can be represented through the expectation of some function, called utility. By denoting by $U$ the utility function of the agent, the random income $X$ is preferred to a random income $X'$ if $E[U(X)] \geq E[U(X')]$. The utility function $U$ is nondecreasing and concave, this last feature formulating the risk aversion of the agent. We refer to Föllmer and Schied [FoS02] for a longer discussion on the preferences representation of agents. In this portfolio allocation context, the criterion consists of maximizing the expected utility from terminal wealth on a finite horizon $T < \infty$:

$$\sup_{\alpha} E\big[U(X_T)\big]. \tag{2.1}$$

The problem originally studied by Merton [Mer69] is a particular case of the model described above with a Black-Scholes model for the risky asset:

$$dS_t^0 = rS_t^0 dt,$$
$$dS_t = \mu S_t dt + \sigma S_t dW_t.$$

where $\mu$, $r$, $\sigma > 0$ are constants, $W$ is a Brownian motion on a filtered probability space $(\Omega, \mathcal{F}, \mathbb{F} = (\mathcal{F}_t)_{0 \le t \le T}, P)$, and with a utility function in the form

$$U(x) = \begin{cases} \frac{x^p - 1}{p}, & x \ge 0 \\ -\infty & x < 0, \end{cases}$$

with $p < 1$, $p \ne 0$, the limiting case $p = 0$ corresponding to a logarithmic utility function: $U(x) = \ln x$, $x > 0$. These popular utility functions are called CRRA (Constant Relative Risk Aversion) since the relative risk aversion defined by $\eta = -xU''(x)/U'(x)$, is constant in this case, equal to $1 - p$.

The mean-variance criterion, introduced by Markowitz [Ma52], relies on the assumption that the preferences of the agent depend only on the expectation and variance of his random incomes. To formulate the feature that the agent likes wealth and is risk-averse, the mean-variance criterion focuses on MV-efficient portfolios, i.e. minimizing the variance given an expectation. In our context, the optimization problem is written as

$$\inf_{\alpha} \{ \mathrm{Var}(X_T) \; : \; E(X_T) = m \}.$$

We shall see that this problem may be reduced to the resolution of a problem in the form (2.1) for some quadratic utility function:

$$U(x) = \lambda - x^2, \quad \lambda \in \mathbb{R}.$$

### 2.2.2 Production-consumption model

We consider the following model for a production unit. The capital value $K_t$ at time $t$ evolves according to the investment rate $I_t$ in capital and the price $S_t$ per unit of capital by

$$dK_t = K_t \frac{dS_t}{S_t} + I_t dt.$$

The debt $L_t$ of this production unit evolves in terms of the interest rate $r$, the consumption rate $C_t$ and the productivity rate $P_t$ of capital:

$$dL_t = rL_t dt - \frac{K_t}{S_t} dP_t + (I_t + C_t) dt.$$

We choose a dynamics model for $(S_t, P_t)$:

$$dS_t = \mu S_t dt + \sigma_1 S_t dW_t^1,$$
$$dP_t = bdt + \sigma_2 dW_t^2,$$

where $(W^1, W^2)$ is a two-dimensional Brownian motion on a filtered probability space $(\Omega, \mathcal{F}, \mathbb{F} = (\mathcal{F}_t)_t, P)$ and $\mu$, $b$, $\sigma_1$, $\sigma_2$ are constants, $\sigma_1, \sigma_2 > 0$. The net vaue of this production unit is

$$X_t = K_t - L_t.$$

We impose the constraints

$$K_t \geq 0, \ C_t \geq 0, \ X_t > 0, \ t \geq 0.$$

We denote by $k_t = K_t/X_t$ and $c_t = C_t/X_t$ the control variables for investment and consumption. The dynamics of the controlled system is then governed by:

$$dX_t = X_t \left[ k_t \left( \mu - r + \frac{b}{S_t} \right) + (r - c_t) \right] dt$$
$$+ k_t X_t \sigma_1 dW_t^1 + k_t \frac{X_t}{S_t} \sigma_2 dW_t^2,$$
$$dS_t = \mu S_t dt + \sigma_1 S_t dW_t^1.$$

Given a discount factor $\beta > 0$ and a utility function $U$, the objective is to determine the optimal investment and consumption for the production unit:

$$\sup_{(k,c)} E\left[ \int_0^\infty e^{-\beta t} U(c_t X_t) dt \right].$$

### 2.2.3 Irreversible investment model

We consider a firm with production goods (electricity, oil, etc.) The firm may increase its production capacity by transferring capital from an activity sector to another one. The controlled dynamics of its production capacity then evolves according to

$$dX_t = X_t(-\delta dt + \sigma dW_t) + \alpha_t dt.$$

$\delta \geq 0$ is the depreciation rate of the production, $\sigma > 0$ is the volatility, $\alpha_t dt$ is the capital-unit number obtained by the firm for a cost $\lambda \alpha_t dt$. $\lambda > 0$ is interpreted as a conversion factor from an activity sector to another one. The control $\alpha$ is valued in $\mathbb{R}_+$. This is an irreversible model for the capital expansion of the firm. The profit function of the company is an increasing, concave function $\Pi$ from $\mathbb{R}_+$ into $\mathbb{R}$, and the optimization problem for the firm is

$$\sup_\alpha E\left[ \int_0^\infty e^{-\beta t} \big( \Pi(X_t) dt \ - \ \lambda \alpha_t dt \big) \right].$$

### 2.2.4 Quadratic hedging of options

We consider risky assets with price process $S$ and a riskless asset with strictly positive price process $S^0$. A derivative asset with underlying $S$ and maturity $T$ is a financial contract whose value at expiration $T$ (also called payoff) is determined explicitly in terms of the underlying asset price $S$ at (or until) $T$. Options are the most traded ones among the derivative products in financial markets: those are financial instruments giving the right to make a transaction on the underlying at a date and price fixed in advance. Standard options on the underlying $S$ are European call options and put options, which can be exercised only at the maturity $T$, at an exercise price $K$ (called strike), and whose payoffs are respectively given by $g(S_T) = (S_T - K)_+$ and $(K - S_T)_+$. More generally, we define a contingent claim as a financial contract characterized by its payoff $H$ at maturity $T$, where $H$ is a $\mathcal{F}_T$-measurable random variable. The principle of hedging and valuation by arbitrage of an option or contingent claim consists of finding a self-financing portfolio strategy based on the basic assets $S^0$ and $S$ of the market, and such that wealth at maturity $T$ coincides with the option payoff (we speak about perfect replication). This is always possible in the framework of complete market, typically the Black-Scholes model, but in general we are in the context of incomplete markets where perfect replication is not attainable. In this case, various criteria are used for hedging and pricing options. We consider the popular quadratic hedging criterion, and in the next section the superreplication criterion, which both lead to interesting stochastic optimization problems.

An agent invests a number of shares $\alpha_t$ in the risky assets at time $t$. Its self-financed wealth process is then governed by

$$dX_t = \alpha_t dS_t + (X_t - \alpha_t S_t)\frac{dS_t^0}{S_t^0}.$$

We are given a contingent claim of maturity $T$, represented by a random variable $H$ $\mathcal{F}_T$-measurable. The quadratic hedging criterion consists of determining the portfolio strategy $\alpha^*$ minimizing the residual hedging error for the quadratic norm:

$$\inf_\alpha E\left[H - X_T\right]^2.$$

### 2.2.5 Superreplication cost in uncertain volatility

We consider the price of an asset whose volatility $\alpha_t$ is unknown but we know a priori that it is valued in an interval $A = [\underline{\sigma}, \bar{\sigma}]$. The dynamics of the risky asset price under a martingale probability measure is

$$dX_t = \alpha_t X_t dW_t.$$

Given a payoff option $g(X_T)$ at maturity $T$, its superreplication cost is given by

$$\sup_\alpha E[g(X_T)].$$

One can show by duality methods that this cost is the minimal initial capital, which allows us to find a self-financed portfolio strategy with a terminal wealth dominating the option payoff $g(S_T)$ for any possible realizations of the volatility, see for example Denis and Martini [DeMa06].

### 2.2.6 Optimal selling of an asset

We consider an agent of a firm owning an asset or good with price process $X_t$ that she wants to sell. Suppose that at the selling of this asset, there is a fixed transaction fee $K > 0$. The agent is looking for the optimal time to sell the asset and her objective is to solve the optimal stopping problem:

$$\sup_{\tau} E[e^{-\beta \tau}(X_\tau - K)],$$

where the supremum runs over all stopping times on $[0, \infty]$, and $\beta > 0$ is a positive discount factor.

### 2.2.7 Valuation of natural resources

We consider a firm producing some natural resources (oil, gas, etc.) with price process $X$. The running profit of the production is given by a nondecreasing function $f$ depending on the price, and the firm may decide at any time to stop the production at a fixed constant cost $-K$. The real options value of the firm is then given by the optimal stopping problem:

$$\sup_{\tau} E\Big[ \int_0^\tau e^{-\beta t} f(X_t)dt + e^{-\beta \tau} K \Big].$$

## 2.3 Other optimization problems in finance

### 2.3.1 Ergodic and risk-sensitive control problems

Some stochastic systems may exhibit in the long term a stationary behavior characterized by an invariant measure. This measure, when it exists, is obtained by calculating the average of the state of the system over a long time. An ergodic control problem then consists in optimizing over the long term some criterion taking into account this invariant measure.

A standard formulation resulting from the criteria described above consists in optimizing over controls $\alpha$ a functional in the form

$$\limsup_{T \to \infty} \frac{1}{T} E\Big[ \int_0^T f(X_t, \alpha_t)dt \Big],$$

or

$$\limsup_{T \to \infty} \frac{1}{T} \ln E\Big[ \exp \Big( \int_0^T f(X_t, \alpha_t)dt \Big) \Big].$$

This last formulation is called risk-sensitive control in the literature, and was recently largely used in mathematical finance as a criterion for portfolio management over the long term.

Another criterion based on the large-deviation behavior type of the system: $P[X_T/T \geq c] \simeq e^{-I(c)T}$, when $T$ goes to infnity, consists in maximizing over controls $\alpha$ a functional in the form

$$\limsup_{T\to\infty} \frac{1}{T}\ln P\Big[\frac{X_T}{T}\geq c\Big].$$

This large-deviation control problem is interpreted in finance as the asymptotic version of the quantile or value-at-risk (VaR) criterion consisting in maximizing the probability that the terminal value $X_T$ of a portfolio overperforms some benchmark.

### 2.3.2 Superreplication under gamma constraints

We consider a Black-Scholes model for a financial market with one stock of price process $S$ and one riskless bond supposed for simplicity equal to a constant:

$$dS_t = \mu S_t dt + \sigma S_t dW_t.$$

The wealth process $X$ of an agent starting from an initial capital $x$ and investing a number of shares $\alpha_t$ at time $t$ in the stock is given by

$$dX_t = \alpha_t dS_t, \qquad X_0 = x.$$

The control process $\alpha$ is assumed to be a semimartingale in the form

$$d\alpha_t = dB_t + \gamma_t dS_t$$

where $B$ is a finite variation process and $\gamma$ is an adapted process constrained to take values in an interval $[\underline{\gamma},\bar{\gamma}]$.

Given an option payoff $g(S_T)$, the superreplication problem is

$$v = \inf\{x \ : \exists\,(\alpha,B,\gamma), \text{ such that } X_T \geq g(S_T) \ a.s.\}.$$

The new feature is that the constraint carries not directly on $\alpha$ but on the "derivative" of $\alpha$ with respect to the variations $dS$ of the price.

### 2.3.3 Robust utility maximization problem and risk measures

In the examples described in Section 2.2.1 and 2.2.4, the utility of the portfolio and/or option payoff to be optimized, is measured according to the Von Neumann-Morgenstern criterion. The risk of the uncertain payoff $-X$ is valued by the risk measure $\rho(-X)$ written as an expectation $-EU(-X)$ under some fixed probability measure and for a given utility function $U$. Other risk measures were recently proposed such as

$$\rho(-X) = \sup_{Q\in\mathcal{Q}} E^Q[X] \tag{2.2}$$

or

$$\rho(-X) = -\inf_{Q\in\mathcal{Q}} E^Q[U(-X)], \tag{2.3}$$

where $\mathcal{Q}$ is a set of probability measures. The representation (2.2) is inspired from the theory of coherent risk measures initiated by Artzner et al. [ADEH99]. For specific choices of $\mathcal{Q}$, this embeds in particular the well-known measures of VaR type. Robust utility functionals of the form (2.3) were suggested by Gilboa and Schmeidler [GS89]. Risk measures in the form

$$-\rho(-X) = \int_0^\infty \psi(P[U(-X) \geq x])dx,$$

where $\psi$ is a probability distortion, i.e. a function from $[0, 1]$ into $[0, 1]$, and $U$ is a utility function from $\mathbb{R}_+$ into $\mathbb{R}_+$, were recently considered as a portfolio choice model, and include Yaari's criterion, Choquet expected utility, and the behavorial portfolio selection model.

### 2.3.4 Forward performance criterion

In the traditional portfolio choice criterion, the utility function is given exogenously and the performance of the portfolio problem is measured in backward time by the value function expressed as the supremum over portfolio strategies of the expected utility from terminal wealth (in the finite horizon case). Recently, Musiela and Zariphopoulou [MuZa07] introduced a new class of performance criterion. It is characterized by a deterministic datum $u_0(x)$ at the beginning of the period, and a family of adapted processes $U_t(x)$, $x \in \mathbb{R}$, such that: $U_0(x) = u_0(x)$, $U_t(X_t)$ is a supermartingale for any wealth process, and there exists a wealth process $X_t^*$ s.t. $U_t(X_t^*)$ is a martingale. Thus, this criterion is defined for all bounded trading horizons, and is not exogenously given.

## 2.4 Bibliographical remarks

The first papers on stochastic control appeared in the 1960s with linear state processes and quadratic costs. Such problems, called stochastic linear regulators, arise in engineering applications, see e.g. Davis [Da77] for an introduction to this topic.

Applications of stochastic optimal control to management and finance problems were developed from the 1970s, especially with the seminal paper by Merton [Mer69] on portfolio selection. The model and results of Merton were then extended by many authors, among them Zariphopoulou [Zar88], Davis and Norman [DN90], Øksendal and Sulem [OS02]. These problems are also studied in the monograph by Karatzas and Shreve [KaSh98]. The model of consumption-production in Section 2.2.2 is formulated in Fleming and Pang [FP05]. Investment models for firms are considered in the book of Dixit and Pindick [DP94]. The example of irreversible investment problem described in Section 2.2.3 is a simplified version of a model studied in Guo and Pham [GP05]. The mean-variance criterion for portfolio selection was historically formulated by Markowitz [Ma52] in a static framework over one period. The quadratic hedging criterion presented in Section 2.2.4 was introduced by Föllmer and Sondermann [FoS86]. The superreplication problem under uncertain volatility was initially studied by Avellaneda, Levy and Paras [ALP95].

Optimal stopping problems were largely studied in the literature, initially in the 1960s, and then with a renewed interest due to the various applications in finance, in particular American options and real options. We mention the works by Dynkin [Dyn63],

Van Morbecke [Van76], Shiryaev [Sh78], El Karoui [Elk81], or more recently the book by Peskir and Shiryaev [PeSh06]. There are many papers about the pricing of American options, and we refer to the article by Lamberton [Lam98] for an overview on this subject. The example of optimal selling of an asset in paragraph 2.2.6 is outlined in the book [Oks00]. The application of optimal stopping problem to the valuation of natural resources in Section 2.2.7 is studied in [KMZ98]. Optimal stopping models arising in firm production are studied for instance in Duckworth and Zervos [DZ00].

Ergodic control problems were studied by Lasry [Las74], Karatzas [Kar80], Bensoussan and Nagai [BN91], Fleming and McEneaney [FM95]. Applications of risk-sensitive control problems in finance are developed in the papers by Bielecki and Pliska [BP99], Fleming and Sheu [FS00], or Nagai [Na03].

The large -deviation control problem in finance was introduced and developed in Pham [Pha03a], [Pha03b]. Some variation was recently studied in Hata, Nagai and Sheu [HaNaSh08].

The superreplication problem under gamma constraints was developed by Soner and Touzi [ST00], and Cheridito, Soner and Touzi [CST05].

Coherent risk measures initiated by Artzner et al [ADEH99] were then developed by many other authors. We mention in particular convex risk measures introduced by Föllmer and Schied [FoS02], and Frittelli and Rosazza Gianin [FG04]. Optimization problems in finance with risk measures are recently studied in the papers by Schied [Schi05], Gundel [Gu05], Barrieu and El Karoui [BElk04]. The Choquet expected utility criterion is studied in Carlier and Dana [CaDa06], and behavorial portfolio selection is developed in Jin and Zhou [JiZh08]. The forward performance criterion for investment was introduced by Musiela and Zariphopoulou [MuZa07].

The list of control models outlined in this book is of course not exhaustive. There are other interesting optimization problems in finance: we may cite partial observation contro problems (see the book by Bensoussan [Ben92]), or impulse control problems (see Jeanblanc-Picqué and Shiryaev [JS95] or Øksendal and Sulem [OS04]). For others examples of control models in finance, we may refer to the books of Kamien and Schwartz [KS81], Seierstad and Sydsaeter [SS87], Sethi and Zhang [SZ94] or Korn [Kor97]. We also mention a recent book by Schmidli [Schm08] on stochastic control in insurance.

# 3

## The classical PDE approach to dynamic programming

### 3.1 Introduction

In this chapter, we use the dynamic programming method for solving stochastic control problems. We consider in Section 3.2 the framework of controlled diffusion and the problem is formulated on finite or infinite horizon. The basic idea of the approach is to consider a family of control problems by varying the initial state values, and to derive some relations between the associated value functions. This principle, called the dynamic programming principle and initiated in the 1950s by Bellman, is stated precisely in Section 3.3. This approach yields a certain partial differential equation (PDE), of second order and nonlinear, called Hamilton-Jacobi-Bellman (HJB), and formally derived in Section 3.4. When this PDE can be solved by the explicit or theoretical achievement of a smooth solution, the verification theorem proved in Section 3.5, validates the optimality of the candidate solution to the HJB equation. This classical approach to the dynamic programming is called the verification step. We illustrate this method in Section 3.6 by solving three examples in finance. The main drawback of this approach is to suppose the existence of a regular solution to the HJB equation. This is not the case in general, and we give in Section 3.7 a simple example inspired by finance pointing out this feature.

### 3.2 Controlled diffusion processes

We consider a control model where the state of the system is governed by a stochastic differential equation (SDE) valued in $\mathbb{R}^n$:

$$dX_s = b(X_s, \alpha_s)ds + \sigma(X_s, \alpha_s)dW_s, \tag{3.1}$$

where $W$ is a $d$-dimensional Brownian motion on a filtered probability space $(\Omega, \mathcal{F}, \mathbb{F} = (\mathcal{F}_t)_{t \geq 0}, P)$ satisfying the usual conditions. We can consider coefficients $b(t, x, a)$ and $\sigma(t, x, a)$ depending on time $t$, but in the case of infinite horizon problems described below, it is important that coefficients do not depend on time in order to get the stationarity of the problem, and so a value function independent of time.

The control $\alpha = (\alpha_s)$ is a progressively measurable (with respect to $\mathbb{F}$) process, valued in $A$, subset of $\mathbb{R}^m$.

H. Pham, *Continuous-time Stochastic Control and Optimization with Financial Applications*, Stochastic Modelling and Applied Probability 61, DOI 10.1007/978-3-540-89500-8_3, © Springer-Verlag Berlin Heidelberg 2009

The measurable functions $b : \mathbb{R}^n \times A \to \mathbb{R}^n$ and $\sigma : \mathbb{R}^n \times A \to \mathbb{R}^{n \times d}$ satisfy a uniform Lipschitz condition in $A$: $\exists\, K \geq 0$, $\forall\, x, y \in \mathbb{R}^n$, $\forall\, a \in A$,

$$|b(x, a) - b(y, a)| + |\sigma(x, a) - \sigma(y, a)| \leq K|x - y|. \tag{3.2}$$

In the sequel, for $0 \leq t \leq T \leq \infty$, we denote by $\mathcal{T}_{t,T}$ the set of stopping times valued in $[t, T]$. When $t = 0$ and $T = \infty$, we simply write $\mathcal{T} = \mathcal{T}_{0,\infty}$.

**Finite horizon problem.**
We fix a finite horizon $0 < T < \infty$. We denote by $\mathcal{A}$ the set of control processes $\alpha$ such that

$$E\left[ \int_0^T |b(0, \alpha_t)|^2 + |\sigma(0, \alpha_t)|^2 dt \right] < \infty. \tag{3.3}$$

In the above condition (3.3), the element $x = 0$ is an arbitrary value of the diffusion and if this element does not lie in the support of the diffusion, we may take any arbitrary value in this support. From Section 1.3 in Chapter 1, the conditions (3.2) and (3.3) ensure for all $\alpha \in \mathcal{A}$ and for any initial condition $(t, x) \in [0, T] \times \mathbb{R}^n$, the existence and uniqueness of a strong solution to the SDE (with random coefficients) (3.1) starting from $x$ at $s = t$. We then denote by $\{X_s^{t,x}, t \leq s \leq T\}$ this solution with a.s. continuous paths. We also recall that under these conditions on $b$, $\sigma$ and $\alpha$, we have

$$E\left[ \sup_{t \leq s \leq T} |X_s^{t,x}|^2 \right] < \infty. \tag{3.4}$$

$$\lim_{h \downarrow 0^+} E\left[ \sup_{s \in [t, t+h]} \left| X_s^{t,x} - x \right|^2 \right] = 0. \tag{3.5}$$

*Functional objective.*
Let $f : [0, T] \times \mathbb{R}^n \times A \to \mathbb{R}$ and $g : \mathbb{R}^n \to \mathbb{R}$ two measurable functions. We suppose that:

**(Hg)** (*i*) $g$ is lower-bounded
  or (*ii*) $g$ satisfies a quadratic growth condition: $|g(x)| \leq C(1 + |x|^2)$, $\forall x \in \mathbb{R}^n$,
    for some constant $C$ independent of $x$.

For $(t, x) \in [0, T] \times \mathbb{R}^n$, we denote by $\mathcal{A}(t, x)$ the subset of controls $\alpha$ in $\mathcal{A}$ such that

$$E\left[ \int_t^T |f(s, X_s^{t,x}, \alpha_s)|\, ds \right] < \infty, \tag{3.6}$$

and we assume that $\mathcal{A}(t, x)$ is not empty for all $(t, x) \in [0, T] \times \mathbb{R}^n$. We can then define under **(Hg)** the gain function:

$$J(t, x, \alpha) = E\left[ \int_t^T f(s, X_s^{t,x}, \alpha_s) ds + g(X_T^{t,x}) \right],$$

for all $(t, x) \in [0, T] \times \mathbb{R}^n$ and $\alpha \in \mathcal{A}(t, x)$. The objective is to maximize over control processes the gain function $J$, and we introduce the associated value function:

$$v(t, x) = \sup_{\alpha \in \mathcal{A}(t,x)} J(t, x, \alpha). \tag{3.7}$$

• Given an initial condition $(t, x) \in [0, T) \times \mathbb{R}^n$, we say that $\hat{\alpha} \in \mathcal{A}(t, x)$ is an optimal control if $v(t, x) = J(t, x, \hat{\alpha})$.

• A control process $\alpha$ in the form $\alpha_s = a(s, X_s^{t,x})$ for some measurable function $a$ from $[0, T] \times \mathbb{R}^n$ into $A$, is called Markovian control.

In the sequel, we shall implicitly assume that the value function $v$ is measurable in its arguments. This point is not trivial a priori, and we refer to measurable section theorems (see Appendix in Chapter III in Dellacherie and Meyer [DM75]) for sufficient conditions.

**Remark 3.2.1** When $f$ satisfies a quadratic growth condition in $x$, i.e. there exist a positive constant $C$ and a positive function $\kappa : A \to \mathbb{R}_+$ such that

$$|f(t, x, a)| \le C(1 + |x|^2) + \kappa(a), \quad \forall (t, x, a) \in [0, T] \times \mathbb{R}^n \times A, \tag{3.8}$$

then the estimate (3.4) shows that for all $(t, x) \in [0, T] \times \mathbb{R}^n$, for any constant control $\alpha = a$ in $A$

$$E\left[ \int_t^T |f(s, X_s^{t,x}, a)| \, ds \right] < \infty.$$

Hence, the constant controls in $A$ lie in $\mathcal{A}(t, x)$. Moreover, if there exists a positive constant $C$ such that $\kappa(a) \le C(1 + |b(0, a)|^2 + |\sigma(0, a)|^2)$, for all $a$ in $A$, then the conditions (3.3) and (3.4) show that for all $(t, x) \in [0, T] \times \mathbb{R}^n$, for any control $\alpha \in \mathcal{A}$

$$E\left[ \int_t^T |f(s, X_s^{t,x}, \alpha_s)| \, ds \right] < \infty.$$

In other words, in this case, $\mathcal{A}(t, x) = \mathcal{A}$.

**Infinite horizon problem.**

We denote by $\mathcal{A}_0$ the set of control processes $\alpha$ such that

$$E\left[ \int_0^T |b(0, \alpha_t)|^2 + |\sigma(0, \alpha_t)|^2 dt \right] < \infty, \quad \forall T > 0. \tag{3.9}$$

Given an initial condition $t = 0$, $x \in \mathbb{R}^n$, and a control $\alpha \in \mathcal{A}_0$, there exists a unique strong solution, denoted by $\{X_s^x, s \ge 0\}$, to (3.1) starting from $x$ at $t = 0$. We recall the following estimate (see Theorem 1.3.16):

$$E\left[ |X_s^x|^2 \right] \le C|x|^2 + C e^{Cs} E\left[ \int_0^s |x|^2 + |b(0, \alpha_u)|^2 + |\sigma(0, \alpha_u)|^2 du \right], \tag{3.10}$$

for some constant $C$ independent of $s$, $x$ and $\alpha$.

*Functional objective.*

Let $\beta > 0$ and $f : \mathbb{R}^n \times A \to \mathbb{R}$ a measurable function. For $x \in \mathbb{R}^n$, we denote by $\mathcal{A}(x)$ the subset of controls $\alpha$ in $\mathcal{A}_0$ such that

$$E\left[ \int_0^\infty e^{-\beta s} |f(X_s^x, \alpha_s)| \, ds \right] < \infty, \tag{3.11}$$

and we assume that $\mathcal{A}(x)$ is not empty for all $x \in \mathbb{R}^n$. We then define the gain function:

$$J(x,\alpha) = E\Big[\int_0^\infty e^{-\beta s} f(X_s^x, \alpha_s) ds\Big],$$

for all $x \in \mathbb{R}^n$ and $\alpha \in \mathcal{A}(x)$, and the associated value function

$$v(x) = \sup_{\alpha \in \mathcal{A}(x)} J(x,\alpha). \tag{3.12}$$

• Given an initial condition $x \in \mathbb{R}^n$, we say that $\hat{\alpha} \in \mathcal{A}(x)$ is an optimal control if $v(x) = J(x, \hat{\alpha})$.

• A control process $\alpha$ in the form $\alpha_s = a(s, X_s^x)$ for some measurable function $a$ from $\mathbb{R}_+ \times \mathbb{R}^n$ into $A$, is called Markovian control.

Notice that it is important to suppose here that the function $f(x,a)$ does not depend on time in order to get the stationarity of the problem, i.e. the value function does not depend on the initial date at which the optimization problem is considered.

**Remark 3.2.2** When $f$ satisfies a quadratic growth condition in $x$, i.e. there exist a positive constant $C$ and a positive function $\kappa : A \to \mathbb{R}_+$ such that

$$|f(x,a)| \le C(1 + |x|^2) + \kappa(a), \quad \forall (x,a) \in \mathbb{R}^n \times A, \tag{3.13}$$

then the estimate (3.10) shows that for $\beta > 0$ large enough, for all $x \in \mathbb{R}^n$, $a \in A$

$$E\Big[\int_0^\infty e^{-\beta s} |f(X_s^x, a)| ds\Big] < \infty.$$

Hence, the constant controls in $A$ belong to $\mathcal{A}(x)$.

## 3.3 Dynamic programming principle

The dynamic programming principle (DPP) is a fundamental principle in the theory of stochastic control. In the context of controlled diffusion processes described in the previous section, and in fact more generally for controlled Markov processes, it is formulated as follows:

**Theorem 3.3.1** *(Dynamic programming principle)*
*(1) Finite horizon: let* $(t,x) \in [0,T] \times \mathbb{R}^n$. *Then we have*

$$v(t,x) = \sup_{\alpha \in \mathcal{A}(t,x)} \sup_{\theta \in \mathcal{T}_{t,T}} E\Big[\int_t^\theta f(s, X_s^{t,x}, \alpha_s) ds + v(\theta, X_\theta^{t,x})\Big] \tag{3.14}$$

$$= \sup_{\alpha \in \mathcal{A}(t,x)} \inf_{\theta \in \mathcal{T}_{t,T}} E\Big[\int_t^\theta f(s, X_s^{t,x}, \alpha_s) ds + v(\theta, X_\theta^{t,x})\Big]. \tag{3.15}$$

*(2) Infinite horizon: let* $x \in \mathbb{R}^n$. *Then we have*

$$v(x) = \sup_{\alpha \in \mathcal{A}(x)} \sup_{\theta \in \mathcal{T}} E\Big[\int_0^\theta e^{-\beta s} f(X_s^x, \alpha_s) ds + e^{-\beta\theta} v(X_\theta^x)\Big] \tag{3.16}$$

$$= \sup_{\alpha \in \mathcal{A}(x)} \inf_{\theta \in \mathcal{T}} E\Big[\int_0^\theta e^{-\beta s} f(X_s^x, \alpha_s) ds + e^{-\beta\theta} v(X_\theta^x)\Big], \tag{3.17}$$

*with the convention that* $e^{-\beta\theta(\omega)} = 0$ *when* $\theta(\omega) = \infty$.

**Remark 3.3.3** In the sequel, we shall often use the following equivalent formulation (in the finite horizon case) of the dynamic programming principle:
(i) For all $\alpha \in \mathcal{A}(t,x)$ and $\theta \in \mathcal{T}_{t,T}$:

$$v(t,x) \geq E\Big[\int_t^\theta f(s,X_s^{t,x},\alpha_s)ds + v(\theta,X_\theta^{t,x})\Big]. \tag{3.18}$$

(ii) For all $\varepsilon > 0$, there exists $\alpha \in \mathcal{A}(t,x)$ such that for all $\theta \in \mathcal{T}_{t,T}$

$$v(t,x) - \varepsilon \leq E\Big[\int_t^\theta f(s,X_s^{t,x},\alpha_s)ds + v(\theta,X_\theta^{t,x})\Big]. \tag{3.19}$$

This is a stronger version than the usual version of the DPP, which is written in the finite horizon case as

$$v(t,x) = \sup_{\alpha\in\mathcal{A}(t,x)} E\Big[\int_t^\theta f(s,X_s^{t,x},\alpha_s)ds + v(\theta,X_\theta^{t,x})\Big], \tag{3.20}$$

for any stopping time $\theta \in \mathcal{T}_{t,T}$. We have a similar remark in the infinite horizon case.

The interpretation of the DPP is that the optimization problem can be split in two parts: an optimal control on the whole time interval $[t,T]$ may be obtained by first searching for an optimal control from time $\theta$ given the state value $X_\theta^{t,x}$, i.e. compute $v(\theta,X_\theta^{t,x})$, and then maximizing over controls on $[t,\theta]$ the quantity

$$E\Big[\int_t^\theta f(s,X_s^{t,x},\alpha_s)ds + v(\theta,X_\theta^{t,x})\Big].$$

**Proof of the DPP.**
We consider the finite horizon case.
**1.** Given an admissible control $\alpha \in \mathcal{A}(t,x)$, we have by pathwise uniqueness of the flow of the SDE for $X$, the Markovian structure

$$X_s^{t,x} = X_s^{\theta,X_\theta^{t,x}}, \quad s \geq \theta,$$

for any stopping time $\theta$ valued in $[t,T]$. By the law of iterated conditional expectation, we then get

$$J(t,x,\alpha) = E\Big[\int_t^\theta f(s,X_s^{t,x},\alpha_s)ds + J(\theta,X_\theta^{t,x},\alpha)\Big],$$

and since $J(.,.,\alpha) \leq v$, $\theta$ is arbitrary in $\mathcal{T}_{t,T}$

$$J(t,x,\alpha) \leq \inf_{\theta\in\mathcal{T}_{t,T}} E\Big[\int_t^\theta f(s,X_s^{t,x},\alpha_s)ds + v(\theta,X_\theta^{t,x})\Big]$$
$$\leq \sup_{\alpha\in\mathcal{A}(t,x)} \inf_{\theta\in\mathcal{T}_{t,T}} E\Big[\int_t^\theta f(s,X_s^{t,x},\alpha_s)ds + v(\theta,X_\theta^{t,x})\Big].$$

By taking the supremum over $\alpha$ in the left hand side term, we obtain the inequality:

$$v(t,x) \leq \sup_{\alpha \in \mathcal{A}(t,x)} \inf_{\theta \in \mathcal{T}_{t,T}} E\left[ \int_t^\theta f(s, X_s^{t,x}, \alpha_s)ds + v(\theta, X_\theta^{t,x}) \right]. \qquad (3.21)$$

**2.** Fix some arbitrary control $\alpha \in \mathcal{A}(t,x)$ and $\theta \in \mathcal{T}_{t,T}$. By definition of the value functions, for any $\varepsilon > 0$ and $\omega \in \Omega$, there exists $\alpha^{\varepsilon,\omega} \in \mathcal{A}(\theta(\omega), X_{\theta(\omega)}^{t,x}(\omega))$, which is an $\varepsilon$-optimal control for $v(\theta(\omega), X_{\theta(\omega)}^{t,x}(\omega))$, i.e.

$$v(\theta(\omega), X_{\theta(\omega)}^{t,x}(\omega)) - \varepsilon \leq J(\theta(\omega), X_{\theta(\omega)}^{t,x}(\omega), \alpha^{\varepsilon,\omega}). \qquad (3.22)$$

Let us now define the process

$$\hat{\alpha}_s(\omega) = \begin{cases} \alpha_s(\omega), & s \in [0, \theta(\omega)] \\ \alpha_s^{\varepsilon,\omega}(\omega), & s \in [\theta(\omega), T]. \end{cases}$$

There are delicate measurability questions here, but it can be shown by the measurable selection theorem (see e.g. Chapter 7 in [BeSh78]) that the process $\hat{\alpha}$ is progressively measurable, and so lies in $\mathcal{A}(t,x)$. By using again the law of iterated conditional expectation, and from (3.22), we then get

$$v(t,x) \geq J(t,x,\hat{\alpha}) = E\left[ \int_t^\theta f(s, X_s^{t,x}, \alpha_s)ds + J(\theta, X_\theta^{t,x}, \alpha^\varepsilon) \right]$$

$$\geq E\left[ \int_t^\theta f(s, X_s^{t,x}, \alpha_s)ds + v(\theta, X_\theta^{t,x}) \right] - \varepsilon.$$

From the arbitrariness of $\alpha \in \mathcal{A}(t,x)$, $\theta \in \mathcal{T}_{t,T}$ and $\varepsilon > 0$, we obtain the inequality

$$v(t,x) \geq \sup_{\alpha \in \mathcal{A}(t,x)} \sup_{\theta \in \mathcal{T}_{t,T}} E\left[ \int_t^\theta f(s, X_s^{t,x}, \alpha_s)ds + v(\theta, X_\theta^{t,x}) \right]. \qquad (3.23)$$

By combining the two relations (3.21) and (3.23), we get the required result.

## 3.4 Hamilton-Jacobi-Bellman equation

The Hamilton-Jacobi-Bellman equation (HJB) is the infinitesimal version of the dynamic programming principle: it describes the local behavior of the value function when we send the stopping time $\theta$ in (3.20) to $t$. The HJB equation is also called dynamic programming equation. In this chapter, we shall use the HJB equation in a classical way as follows:

- Derive the HJB equation formally
- Obtain or try to show the existence of a smooth solution by PDE techniques
- Verification step: show that the smooth solution is the value function by Itô's formula
- As a byproduct, we obtain an optimal feedback control.

### 3.4.1 Formal derivation of HJB

**Finite horizon problem.**

Let us consider the time $\theta = t + h$ and a constant control $\alpha_s = a$, for some arbitrary $a$ in $A$, in the relation (3.18) of the dynamic programming principle:

$$v(t,x) \geq E\left[ \int_t^{t+h} f(s, X_s^{t,x}, a) ds + v(t+h, X_{t+h}^{t,x}) \right]. \tag{3.24}$$

By assuming that $v$ is smooth enough, we may apply Itô's formula between $t$ and $t+h$:

$$v(t+h, X_{t+h}^{t,x}) = v(t,x) + \int_t^{t+h} \left( \frac{\partial v}{\partial t} + \mathcal{L}^a v \right)(s, X_s^{t,x}) ds \quad + \quad \text{(local) martingale},$$

where $\mathcal{L}^a$ is the operator associated to the diffusion (3.1) for the constant control $a$, and defined by (see Section 1.3.1)

$$\mathcal{L}^a v = b(x,a).D_x v + \frac{1}{2} \text{tr} \left( \sigma(x,a)\sigma'(x,a) D_x^2 v \right).$$

By substituting into (3.24), we then get

$$0 \geq E\left[ \int_t^{t+h} (\frac{\partial v}{\partial t} + \mathcal{L}^a v)(s, X_s^{t,x}) + f(s, X_s^{t,x}, a) ds \right].$$

Dividing by $h$ and sending $h$ to 0, this yields by the mean-value theorem

$$0 \geq \frac{\partial v}{\partial t}(t,x) + \mathcal{L}^a v(t,x) + f(t,x,a).$$

Since this holds true for any $a \in A$, we obtain the inequality

$$-\frac{\partial v}{\partial t}(t,x) - \sup_{a \in A} \left[ \mathcal{L}^a v(t,x) + f(t,x,a) \right] \geq 0. \tag{3.25}$$

On the other hand, suppose that $\alpha^*$ is an optimal control. Then in (3.20), we have

$$v(t,x) = E\left[ \int_t^{t+h} f(s, X_s^*, \alpha_s^*) ds + v(t+h, X_{t+h}^*) \right],$$

where $X^*$ is the state system solution to (3.1) starting from $x$ at $t$, with the control $\alpha^*$. By similar arguments as above, we then get

$$-\frac{\partial v}{\partial t}(t,x) - \mathcal{L}^{\alpha_t^*} v(t,x) - f(t,x,\alpha_t^*) = 0, \tag{3.26}$$

which combined with (3.25), suggests that $v$ should satisfy

$$-\frac{\partial v}{\partial t}(t,x) - \sup_{a \in A} \left[ \mathcal{L}^a v(t,x) + f(t,x,a) \right] = 0, \quad \forall (t,x) \in [0,T) \times \mathbb{R}^n, \tag{3.27}$$

if the above supremum in $a$ is finite. We shall see later how to deal with the case when this supremum is infinite, which may arise typically when the control space $A$ is unbounded. We often rewrite this PDE in the form

$$-\frac{\partial v}{\partial t}(t,x) - H(t,x,D_x v(t,x), D_x^2 v(t,x)) = 0, \quad \forall (t,x) \in [0,T) \times \mathbb{R}^n, \tag{3.28}$$

where for $(t, x, p, M) \in [0, T] \times \mathbb{R}^n \times \mathbb{R}^n \times \mathcal{S}_n$

$$H(t, x, p, M) = \sup_{a \in A} \left[ b(x, a).p + \frac{1}{2} \mathrm{tr} \left( \sigma \sigma'(x, a) M \right) + f(t, x, a) \right].$$

This function $H$ is called the Hamiltonian of the associated control problem. The equation (3.28) is called the dynamic programing equation or Hamilton-Jacobi-Bellman (HJB) equation. The regular terminal condition associated to this PDE is

$$v(T, x) = g(x), \quad \forall x \in \mathbb{R}^n, \tag{3.29}$$

which results immediately from the very definition (3.7) of the value function $v$ considered at the horizon date $T$.

**Remark 3.4.4** (1) When the control space $A$ is reduced to a singleton $\{a_0\}$, i.e. there is no control on the state process, the HJB equation is reduced to the linear Cauchy problem:

$$-\frac{\partial v}{\partial t}(t, x) - \mathcal{L}^{a_0} v(t, x) = f(t, x, a_0), \quad \forall (t, x) \in [0, T) \times \mathbb{R}^n \tag{3.30}$$

$$v(T, x) = g(x), \quad \forall x \in \mathbb{R}^n. \tag{3.31}$$

2) The optimality argument of the dynamic programming principle suggests that if one can find a measurable function $\alpha^*(t, x)$ such that

$$\sup_{a \in A} \left[ \mathcal{L}^a v(t, x) f(t, x, a) \right] = \mathcal{L}^{\alpha^*(t,x)} v(t, x) + f(t, x, \alpha^*(t, x)),$$

i.e.

$$\alpha^*(t, x) \in \arg\max_{a \in A} \left[ \mathcal{L}^a v(t, x) + f(t, x, a) \right],$$

then we would get

$$-\frac{\partial v}{\partial t} - \mathcal{L}^{\alpha^*(t,x)} v(t, x) - f(t, x, \alpha^*(t, x)) = 0, \quad v(T, .) = g,$$

and so by Feynman-Kac formula

$$v(t, x) = E \left[ \int_t^T f(X_s^*, \alpha^*(s, X_s^*)) ds + g(X_T^*) \right],$$

where $X^*$ is the solution to the SDE

$$dX_s^* = b(X_s^*, \alpha^*(s, X_s^*)) + \sigma(X_s^*, \alpha^*(s, X_s^*)) dW_s, \quad t \le s \le T$$

$$X_t^* = x.$$

As a byproduct, this shows that the process $\alpha^*(s, X_s^*)$ is an optimal Markovian control.

**Infinite horizon problem.**

By using similar arguments as in the finite horizon case, we derive formally the HJB equation for the value function in (3.12):

$$\beta v(x) - \sup_{a\in A}\left[\mathcal{L}^a v(x) + f(x,a)\right] = 0, \quad \forall x \in \mathbb{R}^n,$$

which may be written also as

$$\beta v(x) - H(x, Dv(x), D^2 v(x)) = 0, \quad \forall x \in \mathbb{R}^n,$$

where for $(x,p,M) \in \mathbb{R}^n \times \mathbb{R}^n \times \mathcal{S}_n$

$$H(x,p,M) = \sup_{a\in A}\left[b(x,a).p + \frac{1}{2}\mathrm{tr}\left(\sigma\sigma'(x,a)M\right) + f(x,a)\right].$$

### 3.4.2 Remarks and extensions

**1.** The results in the finite horizon case are easily extended when the gain function $J$ to be maximized has the general form

$$J(t,x,\alpha) = E\left[\int_t^T \Gamma(t,s)f(s,X_s^{t,x},\alpha_s)ds + \Gamma(t,T)g(X_T^{t,x})\right],$$

where

$$\Gamma(t,s) - \exp\left(-\int_t^s \beta(u,X_u^{t,x},\alpha_u)du\right), \quad t \le s \le T,$$

and $\beta$ is a measurable function on $[0,T]\times\mathbb{R}^n\times A$. In this case, the Hamiltonian associated to the stochastic control problem is

$$H(t,x,v,p,M) = \sup_{a\in A}\left[-\beta(t,x,a)v + b(x,a).p + \frac{1}{2}\mathrm{tr}\left(\sigma(x,a)\sigma'(x,a)M\right) + f(t,x,a)\right].$$

**2.** When the control space $A$ is unbounded, the Hamiltonian

$$H(t,x,p,M) = \sup_{a\in A}\left[b(x,a).p + \frac{1}{2}\mathrm{tr}\left(\sigma\sigma'(x,a)M\right) + f(t,x,a)\right].$$

may take the value $\infty$ in some domain of $(t,x,p,M)$. More precisely, assume that there exists a continuous function $G(t,x,p,M)$ on $[0,T]\times\mathbb{R}^n\times\mathbb{R}^n\times\mathcal{S}_n$ such that

$$H(t,x,p,M) < \infty \Longleftrightarrow G(t,x,p,M) \ge 0.$$

Then from the arguments leading to the HJB equation (3.28), we must have

$$G(t,x,D_xv(t,x),D_x^2v(t,x)) \ge 0, \tag{3.32}$$

$$\text{and} \quad -\frac{\partial v}{\partial t}(t,x) - H(t,x,D_xv(t,x),D_x^2v(t,x)) \ge 0 \tag{3.33}$$

Moreover, if inequality (3.32) is strict at some $(t,x) \in [0,T)\times\mathbb{R}^n$, then there exists a neighborhood of $(t,x,D_xv(t,x),D_x^2v(t,x))$ for which $H$ is finite. Then, formally the optimal control should be attained in a neighborhood of $(t,x)$, and by similar argument as in (3.28), we should have equality in (3.33). We then obtain a variational inequality for the dynamic programming equation:

$$\min \Big[ -\frac{\partial v}{\partial t}(t,x) - H(t,x,D_x v(t,x), D_x^2 v(t,x)),$$
$$G(t,x,D_x v(t,x), D_x^2 v(t,x))\Big] = 0. \qquad (3.34)$$

In this case, we say that the control problem is singular, in contrast with the regular case of HJB equation (3.28). A typical case of a singular problem arises when the control influences linearly the dynamics of the system and the gain function. For example, in the one-dimensional case $n = 1$, $A = \mathbb{R}_+$, and

$$b(x,a) \; = \; \hat{b}(x) + a, \quad \sigma(x,a) \; = \; \hat{\sigma}(x), \quad f(t,x,a) \; = \; \hat{f}(t,x) - a,$$

we have

$$H(t,x,p,M) = \begin{cases} \hat{b}(x)p + \frac{1}{2}\hat{\sigma}(x)^2 M + \hat{f}(t,x) & \text{if } -p+1 \geq 0 \\ \infty & \text{if } -p+1 < 0. \end{cases}$$

The variational inequality is then written as

$$\min \Big[ -\frac{\partial v}{\partial t}(t,x) - \hat{b}(x)\frac{\partial v}{\partial x}(t,x) - \frac{1}{2}\hat{\sigma}(x)^2 \frac{\partial^2 v}{\partial x^2}(t,x) \, , \, -\frac{\partial v}{\partial x}(t,x) + 1 \Big] = 0.$$

We shall give in Section 3.7. another example of singular control arising in finance and where the control is in the diffusion term. In singular control problem, the value function is in general discontinuous in $T$ so that (3.29) is not the relevant terminal condition. We shall study in the next chapter how to derive rigorously the HJB equation (or variational inequality) with the concept of viscosity solutions to handle the lack a priori of regularity of the value function, and also to determine the correct terminal condition.

**3.** When looking at the minimization problem

$$v(t,x) = \sup_{\alpha \in \mathcal{A}(t,x)} E\Big[ \int_t^T f(s, X_s^{t,x}, \alpha_s)ds + g(X_T^{t,x})\Big],$$

we can go back to a maximization problem by considering the value function $-v$. This is equivalent to considering a Hamiltonian function:

$$H(t,x,p,M) = \inf_{a \in A} \Big[ b(x,a).p + \frac{1}{2}\mathrm{tr}\,(\sigma(x,a)\sigma'(x,a)M) + f(t,x,a)\Big],$$

with an HJB equation:

$$-\frac{\partial v}{\partial t}(t,x) - H(t,x,D_x v(t,x), D_x^2 v(t,x)) = 0.$$

When $H$ may take the value $-\infty$, and assuming that there exists a continuous function $G(t,x,p,M)$ on $[0,T] \times \mathbb{R}^n \times \mathbb{R}^n \times \mathcal{S}_n$ such that

$$H(t,x,p,M) > -\infty \iff G(t,x,p,M) \leq 0,$$

the HJB variational inequality is written as

$$\max \Big[ -\frac{\partial v}{\partial t}(t,x) - H(t,x,D_x v(t,x), D_x^2 v(t,x)) \, , \, G(t,x,D_x v(t,x), D_x^2 v(t,x))\Big] = 0.$$

We have an analogous remark in the case of the infinite horizon minimization problem.

## 3.5 Verification theorem

The crucial step in the classical approach to dynamic programming consists in proving that, given a smooth solution to the HJB equation, this candidate coincides with the value function. This result is called the verification theorem, and allows us to exhibit as byproduct an optimal Markovian control. The proof relies essentially on Itô's formula. The assertions on the sufficient conditions may slightly vary from a problem to another one, and should be adapted in the context of the considered problem. We formulate a general version of the verification theorem given the assumptions defined in the previous paragraphs.

**Theorem 3.5.2** *(Finite horizon)*
*Let $w$ be a function in $C^{1,2}([0,T) \times \mathbb{R}^n) \cap C^0([0,T] \times \mathbb{R}^n)$, and satisfying a quadratic growth condition, i.e. there exists a constant $C$ such that*

$$|w(t,x)| \leq C(1+|x|^2), \quad \forall (t,x) \in [0,T] \times \mathbb{R}^n.$$

*(i) Suppose that*

$$-\frac{\partial w}{\partial t}(t,x) - \sup_{a \in A} \left[ \mathcal{L}^a w(t,x) + f(t,x,a) \right] \geq 0, \quad (t,x) \in [0,T) \times \mathbb{R}^n, \qquad (3.35)$$

$$w(T,x) \geq g(x), \quad x \in \mathbb{R}^n. \qquad (3.36)$$

*Then $w \geq v$ on $[0,T] \times \mathbb{R}^n$.*
*(ii) Suppose further that $w(T.) = g$, and there exists a measurable function $\hat{\alpha}(t,x)$, $(t,x) \in [0,T) \times \mathbb{R}^n$, valued in $A$ such that*

$$-\frac{\partial w}{\partial t}(t,x) - \sup_{a \in A} \left[ \mathcal{L}^a w(t,x) + f(t,x,a) \right] = -\frac{\partial w}{\partial t}(t,x) - \mathcal{L}^{\hat{\alpha}(t,x)} w(t,x) - f(t,x,\hat{\alpha}(t,x))$$

$$= 0,$$

*the SDE*

$$dX_s = b(X_s, \hat{\alpha}(s, X_s))ds + \sigma(X_s, \hat{\alpha}(s, X_s))dW_s$$

*admits a unique solution, denoted by $\hat{X}_s^{t,x}$, given an initial condition $X_t = x$, and the process $\{\hat{\alpha}(s, \hat{X}_s^{t,x}) \; t \leq s \leq T\}$ lies in $\mathcal{A}(t,x)$. Then*

$$w = v \quad on \;\; [0,T] \times \mathbb{R}^n,$$

*and $\hat{\alpha}$ is an optimal Markovian control.*

**Proof.** (i) Since $w \in C^{1,2}([0,T) \times \mathbb{R}^n)$, we have for all $(t,x) \in [0,T) \times \mathbb{R}^n$, $\alpha \in \mathcal{A}(t,x)$, $s \in [t,T)$, and any stopping time $\tau$ valued in $[t,\infty)$, by Itô's formula

$$w(s \wedge \tau, X_{s \wedge \tau}^{t,x}) = w(t,x) + \int_t^{s \wedge \tau} \frac{\partial w}{\partial t}(u, X_u^{t,x}) + \mathcal{L}^{\alpha_u} w(u, X_u^{t,x}) du$$

$$+ \int_t^{s \wedge \tau} D_x w(u, X_u^{t,x})' \sigma(X_u^{t,x}, \alpha_u) dW_u.$$

We choose $\tau = \tau_n = \inf\{s \geq t : \int_t^s |D_x w(u, X_u^{t,x})' \sigma(X_u^{t,x}, \alpha_u)|^2 du \geq n\}$, and we notice that $\tau_n \nearrow \infty$ when $n$ goes to infinity. The stopped process $\{\int_t^{s \wedge \tau_n} D_x w(u, X_u^{t,x})' \sigma(X_u^{t,x}, \alpha_u) dW_u, t \leq s \leq T\}$ is then a martingale, and by taking the expectation, we get

$$E\big[w(s \wedge \tau_n, X_{s \wedge \tau_n}^{t,x})\big] = w(t,x) + E\Big[\int_t^{s \wedge \tau_n} \frac{\partial w}{\partial t}(u, X_u^{t,x}) + \mathcal{L}^{\alpha_u} w(u, X_u^{t,x}) du\Big].$$

Since $w$ satisfies (3.35), we have

$$\frac{\partial w}{\partial t}(u, X_u^{t,x}) + \mathcal{L}^{\alpha_u} w(u, X_u^{t,x}) + f(X_u^{t,x}, \alpha_u) \leq 0, \quad \forall \alpha \in \mathcal{A}(t,x),$$

and so

$$E\big[w(s \wedge \tau_n, X_{s \wedge \tau_n}^{t,x})\big] \leq w(t,x) - E\Big[\int_t^{s \wedge \tau_n} f(X_u^{t,x}, \alpha_u) du\Big], \ \forall \alpha \in \mathcal{A}(t,x). \quad (3.37)$$

We have

$$\Big|\int_t^{s \wedge \tau_n} f(X_u^{t,x}, \alpha_u) du\Big| \leq \int_t^T \big|f(X_u^{t,x}, \alpha_u)\big| du,$$

and the right-hand-side term is integrable by the integrability condition on $\mathcal{A}(t,x)$. Since $w$ satisfies a quadratic growth condition, we have

$$\big|w(s \wedge \tau_n, X_{s \wedge \tau_n}^{t,x})\big| \leq C(1 + \sup_{s \in [t,T]} |X_s^{t,x}|^2),$$

and the right-hand-side term is integrable from (3.4). We can then apply the dominated convergence theorem, and send $n$ to infinity into (3.37):

$$E\big[w(s, X_s^{t,x})\big] \leq w(t,x) - E\Big[\int_t^s f(X_u^{t,x}, \alpha_u) du\Big], \quad \forall \alpha \in \mathcal{A}(t,x).$$

Since $w$ is continuous on $[0,T] \times \mathbb{R}^n$, by sending $s$ to $T$, we obtain by the dominated convergence theorem and by (3.36)

$$E\big[g(X_T^{t,x})\big] \leq w(t,x) - E\Big[\int_t^T f(X_u^{t,x}, \alpha_u) du\Big], \quad \forall \alpha \in \mathcal{A}(t,x).$$

From the arbitrariness of $\alpha \in \mathcal{A}(t,x)$, we deduce that $w(t,x) \leq v(t,x)$, for all $(t,x) \in [0,T] \times \mathbb{R}^n$.

(ii) We apply Itô's formula to $w(u, \hat{X}_s^{t,x})$ between $t \in [0,T)$ and $s \in [t,T)$ (after an eventual localization for removing the stochastic integral term in the expectation ):

$$E\Big[w(s, \hat{X}_s^{t,x})\Big] = w(t,x) + E\Big[\int_t^s \frac{\partial w}{\partial t}(u, \hat{X}_u^{t,x}) + \mathcal{L}^{\hat{\alpha}(u, \hat{X}_u^{t,x})} w(u, \hat{X}_u^{t,x}) du\Big].$$

Now, by definition of $\hat{\alpha}(t,x)$, we have

$$-\frac{\partial w}{\partial t} - \mathcal{L}^{\hat{\alpha}(t,x)} w(t,x) - f(t,x,\hat{\alpha}(t,x)) = 0,$$

and so

$$E\left[w(s, \hat{X}_s^{t,x})\right] = w(t, x) - E\left[\int_t^s f(\hat{X}_u^{t,x}, \hat{\alpha}(u, \hat{X}_u^{t,x}))du\right].$$

By sending $s$ to $T$, we then obtain

$$w(t, x) = E\left[\int_t^T f(\hat{X}_u^{t,x}, \hat{\alpha}(u, \hat{X}_u^{t,x}))du + g(\hat{X}_T^{t,x})\right] = J(t, x, \hat{\alpha}).$$

This shows that $w(t, x) = J(t, x, \hat{\alpha}) \leq v(t, x)$, and finally that $w = v$ with $\hat{\alpha}$ as an optimal Markovian control. $\qquad\qquad\qquad\qquad\qquad\qquad\qquad\qquad\qquad\qquad\qquad\qquad\qquad\square$

**Remark 3.5.5** In the particular case where the control space $A$ is reduced to a singleton $\{a_0\}$, this verification theorem is a version of the Feynman-Kac formula: it states that if $w$ is a function in $C^{1,2}([0, T) \times \mathbb{R}^n) \cap C^0([0, T) \times \mathbb{R}^n)$ with a quadratic growth condition, the solution to the Cauchy problem (3.30)-(3.31), then $w$ admits the representation

$$w(t, x) = E\left[\int_t^T f(X_s^{t,x}, a_0)ds + g(X_T^{t,x})\right].$$

**Theorem 3.5.3** *(Infinite horizon)*
*Let $w \in C^2(\mathbb{R}^n)$, and satisfies a quadratic growth condition.*
*(i) Suppose that*

$$\beta w(x) - \sup_{a \in A} [\mathcal{L}^a w(x) + f(x, a)] \geq 0, \quad x \in \mathbb{R}^n, \tag{3.38}$$

$$\limsup_{T \to \infty} e^{-\beta T} E[w(X_T^x)] \geq 0, \quad \forall x \in \mathbb{R}^n, \forall \alpha \in \mathcal{A}(x), \tag{3.39}$$

*Then $w \geq v$ on $\mathbb{R}^n$.*
*(ii) Suppose further that for all $x \in \mathbb{R}^n$, there exists a measurable function $\hat{\alpha}(x)$, $x \in \mathbb{R}^n$, valued in $A$ such that*

$$\beta w(x) - \sup_{a \in A} [\mathcal{L}^a w(x) + f(x, a)] = \beta w(x) - \mathcal{L}^{\hat{\alpha}(x)} w(x) - f(x, \hat{\alpha}(x))$$
$$= 0,$$

*the SDE*

$$dX_s = b(X_s, \hat{\alpha}(X_s))ds + \sigma(X_s, \hat{\alpha}(X_s))dW_s$$

*admits a unique solution, denoted by $\hat{X}_s^x$, given an initial condition $X_0 = x$, satisfying*

$$\liminf_{T \to \infty} e^{-\beta T} E[w(\hat{X}_T^x)] \leq 0, \tag{3.40}$$

*and the process $\{\hat{\alpha}(\hat{X}_s^x), s \geq 0\}$, lies in $\mathcal{A}(x)$. Then*

$$w(x) = v(x), \quad \forall x \in \mathbb{R}^n$$

*and $\hat{\alpha}$ is an optimal Markovian control.*

**Proof.** (i) Let $w \in C^2(\mathbb{R}^n)$ and $\alpha \in \mathcal{A}(x)$. By Itô's formula applied to $e^{-\beta t} w(X_t^x)$ between 0 and $T \wedge \tau_n$, we have

$$e^{-\beta(T \wedge \tau_n)} w(X_{T \wedge \tau_n}^x) = w(x) + \int_0^{T \wedge \tau_n} e^{-\beta u} \left[ \mathcal{L}^{\alpha_u} w(X_u^x) - \beta w(X_u^x) \right] du$$

$$+ \int_0^{T \wedge \tau_n} e^{-\beta u} Dw(X_u^x)' \sigma(X_u^x, \alpha_u) dW_u.$$

Here, $\tau_n$ is the stopping time: $\tau_n = \inf\{t \geq 0 : \int_0^t |D_x w(u, X_u^x)' \sigma(X_u^x, \alpha_u)|^2 du \geq n\}$, so that the stopped stochastic integral is a martingale, and by taking the expectation

$$E\left[ e^{-\beta(T \wedge \tau_n)} w(X_{T \wedge \tau_n}^x) \right] = w(x) + E\left[ \int_0^{T \wedge \tau_n} e^{-\beta u} \left( -\beta w + \mathcal{L}^{\alpha_u} w \right) (X_u^x) du \right]$$

$$\leq w(x) - E\left[ \int_0^{T \wedge \tau_n} e^{-\beta u} f(X_u^x, \alpha_u) du \right], \qquad (3.41)$$

from (3.38). By the quadratic growth condition on $w$ and the integrability condition (3.11), we may apply the dominated convergence theorem and send $n$ to infinity:

$$E\left[ e^{-\beta T} w(X_T^x) \right] \leq w(x) - E\left[ \int_0^T e^{-\beta u} f(X_u^x, \alpha_u) du \right], \quad \forall \alpha \in \mathcal{A}(x). \qquad (3.42)$$

By sending $T$ to infinity, we have by (3.39) and the dominated convergence theorem

$$w(x) \geq E\left[ \int_0^\infty e^{-\beta t} f(X_t^x, \alpha_t) dt \right], \quad \forall \alpha \in \mathcal{A}(x),$$

and so $w(x) \geq v(x)$, $\forall x \in \mathbb{R}^n$.

(ii) By repeating the above arguments and observing that the control $\hat{\alpha}$ achieves equality in (3.42), we have

$$E\left[ e^{-\beta T} w(\hat{X}_T^x) \right] = w(x) - E\left[ \int_0^T e^{-\beta u} f(\hat{X}_u^x, \hat{\alpha}(\hat{X}_u^x)) du \right].$$

By sending $T$ to infinity and from (3.40), we then deduce

$$w(x) \leq J(x, \hat{\alpha}) = E\left[ \int_0^\infty e^{-\beta u} f(\hat{X}_u^x, \hat{\alpha}(\hat{X}_u^x)) du \right],$$

and therefore $w(x) = v(x) = J(x, \hat{\alpha})$.     □

**Remark 3.5.6** In the particular case where the control space $A$ is reduced to a singleton $\{a_0\}$, this verification theorem is a version of the Feynman-Kac formula on an infinite horizon: it states that if $w$ is a function $C^2(\mathbb{R}^n)$ with a quadratic growth condition, the solution to the linear elliptic PDE

$$\beta w(x) - \mathcal{L}^{a_0} w(x) - f(x, a_0) = 0, \quad x \in \mathbb{R}^n,$$
$$\lim_{T \to \infty} e^{-\beta T} E[w(X_T^x)] = 0, \quad x \in \mathbb{R}^n,$$

then $w$ admits the representation

$$w(t, x) = E\left[ \int_0^\infty e^{-\beta t} f(X_t^x, a_0) dt \right].$$

The two previous theorems suggest the following strategy for solving stochastic control problems. In the finite horizon case, we first try to solve the nonlinear HJB equation:

$$-\frac{\partial w}{\partial t} - \sup_{a \in A} \left[ \mathcal{L}^a w(t, x) + f(t, x, a) \right] = 0, \quad (t, x) \in [0, T) \times \mathbb{R}^n, \qquad (3.43)$$

with the terminal condition $w(T, x) = g(x)$. Then, fix $(t, x) \in [0, T) \times \mathbb{R}^n$, and solve $\sup_{a \in A}$ $\left[ \mathcal{L}^a w(t, x) + f(t, x, a) \right]$ as a maximum problem in $a \in A$. Denote by $a^*(t, x)$ the value of $a$ that realizes this maximum. If this nonlinear PDE with a terminal condition admits a smooth solution $w$, then $w$ is the value function to the stochastic control problem, and $a^*$ is an optimal Markovian control. This approach is valid once the HJB equation (3.43) has the solution $C^{1,2}$ satisfying the conditions for applying the verification theorem. Existence results for smooth solutions to parabolic PDEs of HJB type are provided in Fleming and Rishel [FR75], Gilbarg and Trudinger [GT85] or Krylov [Kry87]. The main required condition is a uniform ellipticity condition:

there exists a constant $c > 0$ such that
$$y' \sigma(x, a) \sigma'(x, a) y \geq c|y|^2, \qquad \forall x, y \in \mathbb{R}^n, \forall a \in A.$$

We also mention that in the verification of conditions (ii) in Theorems 3.5.2 and 3.5.3, it is not always easy to obtain the existence of a solution to the SDE associated to the candidate $\hat{a}$ for the optimal control.

## 3.6 Applications

### 3.6.1 Merton portfolio allocation problem in finite horizon

We consider again the example described in Section 2.2.1 in the framework of the Black-Scholes-Merton model over a finite horizon $T$. An agent invests at any time $t$ a proportion $\alpha_t$ of his wealth in a stock of price $S$ (governed by geometric Brownian motion) and $1 - \alpha_t$ in a bond of price $S^0$ with interest rate $r$. The investor faces the portfolio constraint that at any time $t$, $\alpha_t$ is valued in $A$ closed convex subset of $\mathbb{R}$. His wealth process evolves according to

$$dX_t = \frac{X_t \alpha_t}{S_t} dS_t + \frac{X_t (1 - \alpha_t)}{S_t^0} dS_t^0$$
$$= X_t \left( \alpha_t \mu + (1 - \alpha_t) r \right) dt + X_t \alpha_t \sigma dW_t.$$

We denote by $\mathcal{A}$ the set of progressively measurable processes $\alpha$ valued in $A$, and such that $\int_0^T |\alpha_s|^2 ds < \infty$ a.s. This integrability condition ensures the existence and uniqueness of a stong solution to the SDE governing the wealth process controlled by $\alpha \in \mathcal{A}$ (to see this consider the logarithm of the positive wealth process). Given a portfolio strategy $\alpha \in \mathcal{A}$, we denote by $X^{t,x}$ the corresponding wealth process starting from an initial capital $X_t = x > 0$ at time $t$. The agent wants to maximize the expected utility from terminal wealth at horizon $T$. The value function of the utility maximization problem is then defined by

$$v(t, x) = \sup_{\alpha \in \mathcal{A}} E\left[ U(X_T^{t,x}) \right], \quad (t, x) \in [0, T] \times \mathbb{R}_+. \qquad (3.44)$$

The utility function $U$ is increasing and concave on $\mathbb{R}_+$. Let us check that for all $t \in [0, T]$, $v(t, .)$ is also increasing and concave in $x$. Fix some arbitrary $0 < x \leq y$ and $\alpha$ a control process in $\mathcal{A}$. We write $Z_s = X_s^{t,x} - X_s^{t,y}$. Then the process $Z$ satisfies the following SDE: $dZ_s = Z_s\big[(\alpha_s\mu + (1 - \alpha_s)r)ds + \alpha_s\sigma dW_s\big]$, $Z_t = y - x \geq 0$, and so $Z_s \geq 0$, i.e. $X_s^{t,y} \geq X_s^{t,x}$ for all $s \geq t$. Since $U$ is increasing, we have $U(X_T^{t,x}) \leq U(X_T^{t,y})$, and thus

$$E\big[U(X_T^{t,x})\big] \leq E\big[U(X_T^{t,y})\big] \leq v(t, y), \quad \forall \alpha \in \mathcal{A}.$$

This shows $v(t, x) \leq v(t, y)$. Now, let $0 < x_1, x_2$, $\alpha^1, \alpha^2$ two control processes in $\mathcal{A}$, and $\lambda \in [0, 1]$. We write $x_\lambda = \lambda x_1 + (1 - \lambda)x_2$, $X^{t,x_i}$ the wealth process starting from $x_i$ at time $t$, and controlled by $\alpha^i$, $i = 1, 2$. We write

$$\alpha_s^\lambda = \frac{\lambda X_s^{t,x_1}\alpha_s^1 + (1 - \lambda)X_s^{t,x_2}\alpha_s^2}{\lambda X_s^{t,x_1} + (1 - \lambda)X_s^{t,x_2}}.$$

Observe that by convexity of $A$, the process $\alpha^\lambda$ lies in $\mathcal{A}$. Moreover, from the linear dynamics of the wealth process, we see that $X^\lambda := \lambda X^{t,x_1} + (1 - \lambda)X^{t,x_2}$ is governed by

$$dX_s^\lambda = X_s^\lambda\left(\alpha_s^\lambda\mu + (1 - \alpha_s^\lambda)r\right)ds + X_s^\lambda\alpha_s^\lambda\sigma dW_s, \quad s \geq t,$$
$$X_t^\lambda = x_\lambda.$$

This shows that $\lambda X^{t,x_1} + (1 - \lambda)X^{t,x_2}$ is a wealth process starting from $x_\lambda$ at $t$, and controlled by $\alpha^\lambda$. By the concavity of the utility function $U$, we have

$$U\left(\lambda X_T^{t,x_1} + (1 - \lambda)X_T^{t,x_2}\right) \geq \lambda U(X_T^{t,x_1}) + (1 - \lambda)U(X_T^{t,x_2}),$$

and so

$$v(\lambda x_1 + (1 - \lambda)x_2) \geq \lambda E\left[U(X_T^{t,x_1})\right] + (1 - \lambda)E\left[U(X_T^{t,x_2})\right].$$

Since this holds true for any $\alpha^1, \alpha^2$ in $\mathcal{A}$, we deduce that

$$v(\lambda x_1 + (1 - \lambda)x_2) \geq \lambda v(x_1) + (1 - \lambda)v(x_2).$$

Actually, if $U$ is strictly concave and if there exists an optimal control, the above arguments show that the value function $v$ is also strictly concave in $x$.

The HJB equation for the stochastic control problem (3.44) is

$$-\frac{\partial w}{\partial t} - \sup_{a \in A}\left[\mathcal{L}^a w(t, x)\right] = 0, \tag{3.45}$$

together with the terminal condition

$$w(T, x) = U(x), \quad x \in \mathbb{R}_+. \tag{3.46}$$

Here, $\mathcal{L}^a w(t, x) = x(a\mu + (1 - a)r)\dfrac{\partial w}{\partial x} + \frac{1}{2}x^2a^2\sigma^2\dfrac{\partial^2 w}{\partial x^2}$. It turns out that for the particular case of power utility functions of CRRA type, as considered originally in Merton

$$U(x) = \frac{x^p}{p}, \quad x \geq 0, \; p < 1, \; p \neq 0,$$

one can find explicitly a smooth solution to (3.45)-(3.46). We are looking for a candidate solution in the form

$$w(t, x) = \phi(t)U(x),$$

for some positive function $\phi$. By substituting into (3.45)-(3.46), we derive that $\phi$ should satisfy the ordinary differential equation

$$\phi'(t) + \rho\phi(t) = 0, \quad \phi(T) = 1,$$

where

$$\rho = p \sup_{a \in A} \left[ a(\mu - r) + r - \frac{1}{2}a^2(1-p)\sigma^2 \right]. \tag{3.47}$$

We then obtain $\phi(t) = \exp(\rho(T - t))$. Hence, the function given by

$$w(t, x) = \exp(\rho(T - t))U(x), \quad (t, x) \in [0, T] \times \mathbb{R}_+, \tag{3.48}$$

is strictly increasing and concave in $x$, and is a smooth solution to (3.45)-(3.46). Furthermore, the function $a \in A \mapsto a(\mu - r) + r - \frac{1}{2}a^2(1-p)\sigma^2$ is strictly concave on the closed convex set $A$, and thus attains its maximum at some constant $\hat{a}$. By construction, $\hat{a}$ attains the supremum of $\sup_{a \in A}[\mathcal{L}^a w(t, x)]$. Moreover, the wealth process associated to the constant control $\hat{a}$

$$dX_t = X_t(\hat{a}\mu + (1 - \hat{a})r)\, dt + X_t \hat{a}\sigma dW_t,$$

admits a unique solution given an initial condition. From the verification Theorem 3.5.2, this proves that the value function to the utility maximization problem (3.44) is equal to (3.48), and the optimal proportion of wealth to invest in stock is constant given by $\hat{a}$. Finally, notice that when $A = \mathbb{R}$, the values of $\hat{a}$ and $\rho$ are explicitly given by

$$\hat{a} = \frac{\mu - r}{\sigma^2(1 - p)}, \tag{3.49}$$

and

$$\rho = \frac{(\mu - r)^2}{2\sigma^2} \frac{p}{1 - p} + rp.$$

### 3.6.2 Investment-consumption problem with random time horizon

We consider a setup with intertemporal utility from lifetime consumption. We use a similar framework as in the previous section for the model on asset prices. A control is a pair of progressively measurable process $(\alpha, c)$ valued in $A \times \mathbb{R}_+$ for some closed convex subset $A$ of $\mathbb{R}$ such that $\int_0^\infty |\alpha_t|^2 dt + \int_0^\infty c_t dt < \infty$ a.s. We denote by $\mathcal{A} \times \mathcal{C}$ the set of control processes. The quantity $\alpha_t$ represents the proportion of wealth invested in stock, and $c_t$ is the consumption per unit of wealth. Given $(\alpha, c) \in \mathcal{A} \times \mathcal{C}$, there exists a unique solution, denoted by $X^x$, to the SDE governing the wealth process

$$dX_t = X_t(\alpha_t\mu + (1 - \alpha_t)r - c_t)\, dt + X_t\alpha_t\sigma dW_t,$$

given an initial condition $X_0 = x \geq 0$. The agent's investment-consumption problem is to maximize over strategies $(\alpha, c)$ the expected utility from intertemporal consumption up to a random horizon $\tau$. The random factors, which may affect the horizon $\tau$ of the investor, are for example death, changes in the opportunity set, or time of an exogenous shock to the investment-consumption process (e.g. purchasing or selling a house). Given a utility function $u$ for consumption, we then consider the corresponding value function:

$$v(x) = \sup_{(\alpha,c)\in\mathcal{A}\times\mathcal{C}} E\Big[\int_0^\tau e^{-\beta t}u(c_t X_t^x)dt\Big].$$

We assume that the random time $\tau$ is independent of the securities market model, i.e. $\tau$ is independent of $\mathcal{F}_\infty$, and we denote by $F(t) = P[\tau \leq t] = P[\tau \leq t|\mathcal{F}_\infty]$ the distribution function of $\tau$. The value function can then be written as

$$\begin{aligned}v(x) &= \sup_{(\alpha,c)\in\mathcal{A}\times\mathcal{C}} E\Big[\int_0^\infty e^{-\beta t}u(c_t X_t^x)1_{t<\tau}dt\Big]\\ &= \sup_{(\alpha,c)\in\mathcal{A}\times\mathcal{C}} E\Big[\int_0^\infty e^{-\beta t}u(c_t X_t^x)(1 - F(t))dt\Big].\end{aligned}$$

We also specialize our setup by assuming an exponential distribution for the random time horizon: $1 - F(t) = e^{-\lambda t}$ for some positive constant $\lambda$, also called intensity. In this case, the intertemporal consumption utility problem with random horizon is turned into an infinite horizon problem:

$$v(x) = \sup_{(\alpha,c)\in\mathcal{A}\times\mathcal{C}} E\Big[\int_0^\tau e^{-(\beta+\lambda)t}u(c_t X_t^x)dt\Big],$$

with adjusted discount factor $\hat{\beta} = \beta + \lambda$. The HJB equation associated to this control problem is

$$\hat{\beta}w(x) - \sup_{a\in A, c\geq 0}\big[\mathcal{L}^{a,c}w(x) + u(cx)\big] = 0, \quad x \geq 0,$$

where $\mathcal{L}^{a,c}w(x) = x(a\mu + (1-a)r - c)w' + \frac{1}{2}x^2a^2\sigma^2 w''$. By defining $\tilde{u}(z) = \sup_{C\geq 0}[u(C) - Cz]$, the Legendre transform of $u$, this HJB equation may be written as

$$\hat{\beta}w(x) - \sup_{a\in A}\big[\mathcal{L}^a w(x)\big] - \tilde{u}(w') = 0, \quad x \geq 0, \tag{3.50}$$

where $\mathcal{L}^a w(x) = x(a\mu + (1-a)r)w' + \frac{1}{2}x^2a^2\sigma^2 w''$. As in the previous section, we consider a power utility function $u(C) = C^p/p$ for $p < 1$, $p \neq 0$, so that $\tilde{u}(z) = z^{-q}/q$ with $q = p/(1-p)$, and a maximum attained in $C^*(z) = z^{\frac{1}{p-1}}$. We are then looking for a candidate solution to the HJB equation in the form

$$w(x) = Ku(x),$$

for some positive constant $K$. By substituting into the HJB equation (3.50), we derive an equation satisfied by the unknown $K$:

$$(\hat{\beta} - \rho)\frac{K}{p} - \frac{1}{q}K^{-q} = 0,$$

where $\rho$ is as in (3.47). Such an equation admits a positive solution for $K$ if and only if $\hat{\beta} > \rho$, i.e. the discount factor $\beta$ satisfies

$$\beta > \rho - \lambda,$$

and in this case, $K$ is given by

$$K = \left(\frac{1-p}{\beta+\lambda-\rho}\right)^{1-p}.$$

Hence, with this value of $K$, the function $w(x) = Ku(x)$ solves the HJB equation (3.50), with an argument max in the HJB equation given by

$$\hat{a} = \arg\max_{a \in A}[a(\mu - r) + r - \frac{1}{2}a^2(1-p)\sigma^2]$$

$$\hat{c} = \frac{1}{x}(w'(x))^{\frac{1}{p-1}} = K^{\frac{1}{p-1}}.$$

Moreover, the wealth process associated to these constant controls $(\hat{a}, \hat{c})$

$$dX_t = X_t\big(\hat{a}\mu + (1-\hat{a})r - \hat{c}\big)dt + X_t\hat{a}\sigma dW_t,$$

is a Brownian geometric motion, and so admits a unique solution, denoted by $\hat{X}^x$, given an initial condition $\hat{X}_0^x = x$. Finally, a straightforward calculation shows that

$$e^{-\hat{\beta}T}E[w(\hat{X}_T^x)] = Ku(x)\exp\big(-(\hat{\beta}-\rho)T - \hat{c}pT\big)$$
$$\to 0, \quad \text{as } T \text{ goes to infinity,}$$

since $\hat{\beta} > \rho$. From the verification theorem 3.5.3, this proves that $v(x) = w(x) = Ku(x)$, and the optimal controls are constant given by $(\hat{a}, \hat{c})$. We observe that the optimal proportion of wealth is not affected by the uncertainty on the time horizon.

### 3.6.3 A model of production-consumption on infinite horizon

We develop the example of Section 2.2.2. We consider the following model for a production firm. Its capital value $K_t$ evolves according to the investment rate $I_t$ and the price $S_t$ per unit of capital:

$$dK_t = K_t\frac{dS_t}{S_t} + I_tdt.$$

The debt $L_t$ of this firm is determined by the interest rate $r$, the consumption $C_t$, and the productivity rate $P_t$ of capital:

$$dL_t = rL_tdt - \frac{K_t}{S_t}dP_t + (I_t + C_t)dt.$$

We assume a model for the dynamics of $(Y_t = \ln S_t, P_t)$ as

$$dY_t = \left(\mu - \frac{\sigma_1^2}{2}\right)dt + \sigma_1 dW_t^1$$
$$dP_t = bdt + \sigma_2 dW_t^2,$$

where $(W^1, W^2)$ is a two-dimensional Brownian motion on a filtered probability space $(\Omega, \mathcal{F}, \mathbb{F} = (\mathcal{F}_t)_{t \geq 0}, P)$, and $\mu$, $b$, $\sigma_1$, $\sigma_2$ are constants, $\sigma_1, \sigma_2 > 0$. The net value of the firm is

$$X_t = K_t - L_t,$$

and the following constraints are required:

$$K_t \geq 0, \ C_t \geq 0, \ X_t > 0, \ t \geq 0.$$

We denote by $k_t = K_t/X_t$ and $c_t = C_t/X_t$ the control variables of investment and consumption. The dynamics of the controlled system is then governed by

$$dX_t = X_t \left[ k_t(\mu - r + be^{-Y_t}) + (r - c_t) \right] dt$$
$$+ k_t X_t \sigma_1 dW_t^1 + k_t X_t e^{-Y_t} \sigma_2 dW_t^2 \qquad (3.51)$$

$$dY_t = \left( \mu - \frac{\sigma_1^2}{2} \right) dt + \sigma_1 dW_t^1. \qquad (3.52)$$

Given a discount factor $\beta > 0$, and a power utility function:

$$U(C) = \frac{C^\gamma}{\gamma}, \quad C \geq 0, \quad \text{with } 0 < \gamma < 1,$$

we denote by $\mathcal{A}(x, y)$ the set of progressively measurable processes $(k, c)$ valued in $\mathbb{R}_+ \times \mathbb{R}_+$ such that

$$\int_0^T k_t^2 dt + \int_0^T c_t^2 dt < \infty, \quad a.s., \quad \forall T > 0,$$

$$E \left[ \int_0^\infty e^{-\beta t} U(c_t X_t^{x,y}) dt \right] < \infty,$$

where $(X^{x,y}, Y^y)$ is the solution to the SDE (3.51)-(3.52) starting from $(x, y)$ at $t = 0$. The objective is to find the optimal investment and production for this production firm, and we study the stochastic optimal control on an infinite horizon:

$$v(x, y) = \sup_{(k,c) \in \mathcal{A}(x,y)} E \left[ \int_0^\infty e^{-\beta t} U(c_t X_t^{x,y}) dt \right]. \qquad (3.53)$$

The associated HJB equation is

$$\beta v - \left( \mu - \frac{\sigma_1^2}{2} \right) \frac{\partial v}{\partial y} - rx \frac{\partial v}{\partial x} - \frac{\sigma_1^2}{2} \frac{\partial^2 v}{\partial y^2} - \sup_{c \geq 0} \left[ U(cx) - cx \frac{\partial v}{\partial x} \right] \qquad (3.54)$$

$$- \sup_{k \geq 0} \left[ k(\mu - r + be^{-y})x \frac{\partial v}{\partial x} + \frac{1}{2} k^2 x^2 (\sigma_1^2 + e^{-2y}\sigma_2^2) \frac{\partial^2 v}{\partial x^2} + kx\sigma_1^2 \frac{\partial^2 v}{\partial x \partial y} \right] = 0.$$

By observing that $X^{x,y}$ is expressed in the form $X^{x,y} = x \exp(Z(y))$ where $Z(y)$ is written as an exponential of process in terms of $(k, c)$ and $Y^y$, we deduce that

$$v(x, y) = x^\gamma v(1, y).$$

We are then looking for a candidate solution to HJB (3.54) in the form

$$v(x,y) = \frac{x^{\gamma}}{\gamma} \exp\left(\varphi(y)\right),$$

for some function $\varphi(y)$. By substituting into (3.54), we obtain the ordinary differential equation (ODE) satisfied by $\varphi$:

$$\beta - \gamma r - \frac{\sigma_1^2}{2}\left(\varphi_{yy} + \varphi_y^2\right) - \left(\mu - \frac{\sigma_1^2}{2}\right)\varphi_y$$
$$- \sup_{c \geq 0}\left[c^{\gamma}e^{-\varphi} - c\gamma\right] - \gamma \sup_{k \geq 0} G(y, \varphi_y, k) = 0, \qquad (3.55)$$

where

$$G(y, p, k) = -\frac{k^2}{2}(1 - \gamma)(\sigma_1^2 + \sigma_2^2 e^{-2y}) + k(\mu - r + be^{-y} + \sigma_1^2 p).$$

One can show that for $\beta$ large enough, there exists a unique smooth $C^2$ bounded solution $\varphi$ to the ODE (3.55). We refer to [FP05] for the details. Since $G(y, p, 0) = 0$, we derive that at any extremum point $y$ of $\varphi_y$, i.e. $\varphi_{yy}(y) = 0$:

$$0 \leq \beta - \gamma r - \left(\mu - \frac{\sigma_1^2}{2}\right)\varphi_y(y) - \frac{\sigma_1^2}{2}\varphi_y^2(y) - (1 - \gamma)e^{\frac{\gamma\varphi(y)}{\gamma-1}}.$$

Since $\varphi$ is bounded, this shows that $\varphi_y$ is also bounded on $\mathbb{R}$.

By construction, the positive function $w(x,y) = (x^{\gamma}/\gamma)e^{\varphi(y)}$ is a solution to the HJB equation (3.54). From the first part of the verification theorem 3.5.3, we deduce that $w \geq v$. On the other hand, let us consider the functions

$$\hat{k}(y) = \left(\frac{be^{-y} + \mu - r + \sigma_1^2 \varphi_y}{(1 - \gamma)(\sigma_1^2 + \sigma_2^2 e^{-2y})}\right)_+ \in \arg\min_{k \geq 0} G(y, \varphi_y, k)$$

$$\hat{c}(y) = \exp\left(\frac{\varphi(y)}{\gamma - 1}\right) \in \arg\min_{c \geq 0}\left[c\gamma - c^{\gamma}e^{-\varphi}\right].$$

Since $\varphi$ and $\varphi_y$ are bounded, this implies that the functions $\hat{c}$, $\hat{k}$, $e^{-y}\hat{k}$ are also bounded in $y$. We deduce that there exists a constant $M > 0$ such that

$$E\left[|\hat{X}_t^{x,y}|^2\right] \leq x^2 \exp(Mt), \quad \forall t > 0,$$

where we denote by $\hat{X}^{x,y}$ the solution to (3.51) controlled by $(\hat{k}(Y_t^y), \hat{c}(Y_t^y))_{t \geq 0}$. This shows that the control $(\hat{k}(Y_t^y), \hat{c}(Y_t^y))_{t \geq 0}$ lies in $\mathcal{A}(x,y)$:

$$E\left[\int_0^{\infty} e^{-\beta t}U(\hat{c}(Y_t^y)\hat{X}_t^{x,y})dt\right] < \infty, \qquad (3.56)$$

Moreover, since $\varphi$ is bounded, the function $\hat{c}(y)$ is lower-bounded by a strictly positive constant. We then get the existence of a constant $B > 0$ such that:

$$0 \leq e^{-\beta T}w(\hat{X}_T^{x,y}, Y_T^y) \leq Be^{-\beta T}U(\hat{c}(Y_T^y)\hat{X}_T^{x,y}),$$

which, combined with (3.56), yields

$$\lim_{T \to \infty} E\left[e^{-\beta T}w(\hat{X}_T^{x,y}, Y_T^y)\right] = 0.$$

We conclude with the second part of the verification theorem 3.5.3 that $w = v$ and $(\hat{k}(Y_t^y), \hat{c}(Y_t^y))_{t \geq 0}$ is an optimal Markovian control.

## 3.7 Example of singular stochastic control problem

We consider a controlled process governed by

$$dX_s = \alpha_s dW_s,$$

where $\alpha$ is a progressively measurable process valued in $A = \mathbb{R}$ and such that $E[\int_0^T |\alpha_s|^2 ds]$ $< \infty$. We denote by $\mathcal{A}$ this set of control processes. Let $g$ be a nonnegative measurable function with linear growth condition on $\mathbb{R}$, and consider the stochastic control problem

$$v(t,x) = \sup_{\alpha \in \mathcal{A}} E[g(X_T^{t,x})], \quad (t,x) \in [0,T] \times \mathbb{R}. \tag{3.57}$$

We now see that for a large choice of functions $g$, the value function $v$ is not smooth.

From the dynamic programming principle, we have for any stopping time $\theta \in \mathcal{T}_{t,T}$, and any constant control $\alpha_t = a \in \mathbb{R}$

$$v(t,x) \geq E\left[v(\theta, X_\theta^{t,x})\right]. \tag{3.58}$$

Suppose that $v$ is smooth $C^{1,2}$, and apply Itô's formula to $v(s, X_s^{t,x})$ between $s = t \in [0,T)$ and $s = \theta = (t+h) \wedge \tau$ where $\tau = \inf\{s \geq t : |X_s^{t,x} - x| \geq 1\}$. By observing that the stochastic integral appearing in Itô's formula is a stopped martingale, we obtain by substituting into (3.58)

$$0 \geq E\left[\frac{1}{h} \int_t^{(t+h)\wedge\tau} \left(\frac{\partial v}{\partial t} + a^2 \frac{\partial^2 v}{\partial x^2}\right)(s, X_s^{t,x})ds\right]. \tag{3.59}$$

Notice that by continuity a.s. of the path $X_s^{t,x}$, we have a.s. $\theta = t + h$ for $h \leq \bar{h}(\omega)$ small enough. We deduce by the mean-value theorem that the random variable inside the expectation in (3.59) converges a.s. to $\left(\frac{\partial v}{\partial t} + a^2 \frac{\partial^2 v}{\partial x^2}\right)(t,x)$ when $h$ goes to zero. Moreover, since this random variable is bounded by a constant independent of $h$, we can apply the dominated convergence theorem, and obtain that when $h$ goes to zero

$$0 \geq \frac{\partial v}{\partial t}(t,x) + a^2 \frac{\partial^2 v}{\partial x^2}(t,x), \quad \forall(t,x) \in [0,T) \times \mathbb{R}. \tag{3.60}$$

Since this inequality holds true for any $a$ in $\mathbb{R}$, this implies in particular that: $\frac{\partial^2 v}{\partial x^2} \leq 0$ on $[0,T) \times \mathbb{R}$, and so

$$v(t,.) \text{ is concave on } \mathbb{R} \text{ for all } t \in [0,T). \tag{3.61}$$

On the other hand, since the zero constant control lies in $\mathcal{A}$, it is immediate by definition of the value function that

$$v(t,x) \geq g(x), \quad \forall(t,x) \in [0,T] \times \mathbb{R}.$$

Denoting by $g^{con}$ the concave envelope of $g$, i.e. the smallest concave function above $g$, we deduce with (3.61) that

$$v(t,x) \geq g^{con}(x), \quad \forall(t,x) \in [0,T] \times \mathbb{R}. \tag{3.62}$$

Now, using the fact that $g^{con} \geq g$, Jensen's inequality and the martingale property of $X_s^{t,x}$ for $\alpha \in \mathcal{A}$, we get:

$$v(t,x) \leq \sup_{\alpha \in \mathcal{A}} E[g^{con}(X_T^{t,x})]$$

$$\leq \sup_{\alpha \in \mathcal{A}} g^{con}\left(E[X_T^{t,x}]\right) = g^{con}(x).$$

By combining with (3.62), we deduce that

$$v(t,x) = g^{con}(x), \quad \forall (t,x) \in [0,T) \times \mathbb{R}. \tag{3.63}$$

We then get a contradiction whenever the function $g^{con}$ is not $C^2$, for example if $g(x) = \max(x - \kappa, 0) = g^{con}(x)$.

**Remark 3.7.7** We shall see in the next chapter by means of the theory of viscosity solutions that although the inequality (3.60) cannot be interpreted in the classical sense, the relation (3.63) remains valid. In particular, we see that the value function $v$ is discontinuous in $T$ once $g \neq g^{con}$. Actually, the Hamiltonian for this singular control problem (3.57) is $H(M) = \sup_{a \in \mathbb{R}}[\frac{1}{2}a^2 M]$. Thus, $H(M) < \infty$ if and only if $-M \geq 0$, and in this case $H(M) = 0$. We shall prove in the next chapter that $v$ is a viscosity solution to the HJB variational inequality

$$\min\left[-\frac{\partial v}{\partial t}, -\frac{\partial^2 v}{\partial x^2}\right] = 0.$$

## 3.8 Bibliographical remarks

The principle of optimality of the dynamic programming principle was initiated by Bellman in the 1950s [Be57]. Simple and intuitive in its formulation, the rigorous proof is very technical, and was studied by several authors following various methods: Bensoussan and J.L. Lions [BL78], Krylov [Kry80], Nisio [Nis81], P.L. Lions [Lio83], Borkar [Bor89], Fleming and Soner [FSo93], or Yong and Zhou [YZ00].

The classical approach to optimal control for diffusion processes via a verification theorem on the HJB equation is dealt with in many books: Fleming and Rishel [FR75], Krylov [Kry80], Fleming and Soner [FSo93], Yong and Zhou [YZ00]. The case of controlled jump-diffusion processes is considered in the recent book by Oksendal and Sulem [OS04]. Other aspects of dynamic programming are studied in a general framework in the lecture notes of St-Flour by El Karoui [Elk81].

The example of the investment-consumption problem on a random time horizon was developed in Blanchet-Scalliet et al. [BElJM08]. The example of singular stochastic control problem is inspired by the superreplication problem in the uncertain volatility model, and it will be developed in the next chapter.

We formulated stochastic control problems in a standard form, where the goal is to optimize a functional in an expectation form. Recently, motivated by the superreplication problem under gamma constraints, Soner and Touzi [ST00], [ST02] developed other formulations of control problems, called stochastic target problems, and proved the corresponding dynamic programming principle.

The dynamic programming approach provides a characterization of the value function and the optimal control when this value function is smooth enough. The general existence problem of an optimal control is not considered here. We refer to the works by Kushner [Ku75] and El Karoui, Huu Nguyen, and Jeanblanc-Picqué [ElkNJ87] for results in this direction.

# 4

## The viscosity solutions approach to stochastic control problems

### 4.1 Introduction

As outlined in the previous chapter, the dynamic programming method is a powerful tool to study stochastic control problems by means of the Hamilton-Jacobi-Bellman equation. However, in the classical approach, the method is used only when it is assumed a priori that the value function is smooth enough. This is not necessarily true even in very simple cases.

To circumvent this difficulty, Crandall and Lions introduced in the 1980s the notion of viscosity solutions for first-order equations. This theory was then generalized to second-order equations. The viscosity solutions approach provides very powerful means to study in great generality stochastic control problems and gives a rigorous formulation of the HJB equation for functions that are only assumed to be locally bounded. By combining these results with comparison principles for viscosity solutions, we characterize the value function as the unique viscosity solution of the associated dynamic programming equation, and this can then be used to obtain further results.

This chapter is an introduction to the notion of viscosity solutions and to the essential tools to study stochastic control problems. There is a large literature on the theory of viscosity solutions and we will refer for instance to Crandall, Ishii and Lions [CIL92] for a seminal reference on this topic. In Section 4.2, we define the notion of viscosity solutions and give some basic properties. We show, in Section 4.3, how to derive rigorously and in great generality, from the dynamic programming principle, the Hamilton-Jacobi-Bellman equation for the value function in the viscosity sense. We state comparison principles and uniqueness results for viscosity solutions in Section 4.4. Finally, Sections 4.5 and 4.6 show how to use the viscosity solutions approach to solve two stochastic control problems arising in finance.

### 4.2 Definition of viscosity solutions

We consider the following nonlinear second-order partial differential equations:

$$F(x, w(x), Dw(x), D^2w(x)) = 0, \quad x \in \mathcal{O}, \tag{4.1}$$

H. Pham, *Continuous-time Stochastic Control and Optimization with Financial Applications*, Stochastic Modelling and Applied Probability 61, DOI 10.1007/978-3-540-89500-8_4, © Springer-Verlag Berlin Heidelberg 2009

in which $\mathcal{O}$ is an open subset of $\mathbb{R}^N$ and $F$ is a continuous function of $\mathcal{O} \times \mathbb{R} \times \mathbb{R}^N \times \mathcal{S}_N$ taking values in $\mathbb{R}$. The function $F$ is assumed to satisfy the ellipticity condition: for all $x \in \mathcal{O}$, $r \in \mathbb{R}$, $p \in \mathbb{R}^N$, $M, \widehat{M} \in \mathcal{S}_N$,

$$M \leq \widehat{M} \implies F(x, r, p, M) \geq F(x, r, p, \widehat{M}). \tag{4.2}$$

For time-dependent problems, a point in $\mathbb{R}^N$ must be understood as a time variable $t$ and a space variable $x$ in $\mathbb{R}^n$ ($N = n+1$). Furthermore, $\mathcal{O}$ must be an open subset of the form $[0, T) \times \mathcal{O}_n$ in which $\mathcal{O}_n$ is an open subset of $\mathbb{R}^n$ and $F(t, x, r, q, p, M)$ must satisfy the following parabolicity condition: for all $t \in [0, T)$, $x \in \mathcal{O}_n$, $r \in \mathbb{R}$, $q, \hat{q} \in \mathbb{R}$, $p \in \mathbb{R}^n$, $M \in \mathcal{S}_n$,

$$q \leq \hat{q} \implies F(t, x, r, q, p, M) \geq F(t, x, r, \hat{q}, p, M). \tag{4.3}$$

This last condition means that we are dealing with forward PDE, i.e. (4.1) holds for time $t < T$, and the terminal condition is for $t = T$. This is in accordance with control problem in finite horizon and HJB equation formulated in the previous chapter, and corresponding to

$$F(t, x, q, p, M) = -q - \sup_{a \in A} \left[ b(x, a).p + \frac{1}{2}\mathrm{tr}\left(\sigma\sigma'(x, a)M\right) + f(t, x, a) \right], \tag{4.4}$$

while the case of HJB equations for infinite horizon problem corresponds to

$$F(x, r, p, M) = \beta r - \sup_{a \in A} \left[ b(x, a).p + \frac{1}{2}\mathrm{tr}\left(\sigma\sigma'(x, a)M\right) + f(x, a) \right]. \tag{4.5}$$

The ellipticity condition (4.2) is obviously satisfied since the matrix $\sigma\sigma'$ is positive definite.

The conditions (4.2)-(4.3) and the notion of viscosity solutions are motivated by the following arguments: let us assume that $v$ is smooth, and is a classical supersolution to (4.1), i.e. relation (4.1) holds with $\geq$ in the whole domain $\mathcal{O}$. Let $\varphi$ a smooth function on $\mathcal{O} = [0, T) \times \mathcal{O}_n$, and $(\bar{t}, \bar{x}) \in [0, T) \times \mathcal{O}_n$ be a minimum point of $v - \varphi$. In this case, the first- and second-order optimality conditions imply

$$\frac{\partial(w - \varphi)}{\partial t}(\bar{t}, \bar{x}) \geq 0 \quad (= 0 \text{ if } \bar{t} > 0)$$
$$D_x w(\bar{t}, \bar{x}) = D_x \varphi(\bar{t}, \bar{x}) \text{ and } D_x^2 w(\bar{t}, \bar{x}) \geq D_x^2 \varphi(\bar{t}, \bar{x}).$$

From the conditions (4.2) and (4.3), we deduce that

$$F(\bar{t}, \bar{x}, w(\bar{t}, \bar{x}), \frac{\partial \varphi}{\partial t}(\bar{t}, \bar{x}), D_x \varphi(\bar{t}, \bar{x}), D_x^2 \varphi(\bar{t}, \bar{x}))$$

$$\geq F(\bar{t}, \bar{x}, w(\bar{t}, \bar{x}), \frac{\partial w}{\partial t}(\bar{t}, \bar{x}), D_x w(\bar{t}, \bar{x}), D_x^2 w(\bar{t}, \bar{x})) \geq 0,$$

Similarly, if $v$ is a classical subsolution to (4.1), i.e. relation (4.1) holds with $\leq$ in the whole domain $\mathcal{O}$, then for all smooth functions $\varphi$ on $\mathcal{O}$, and $(\bar{t}, \bar{x}) \in \mathcal{O} = [0, T) \times \mathcal{O}_n$ such that $(\bar{t}, \bar{x})$ is a maximum point of $v - \varphi$, we have

$$F(\bar{t}, \bar{x}, w(\bar{t}, \bar{x}), \frac{\partial \varphi}{\partial t}(\bar{t}, \bar{x}), D_x \varphi(\bar{t}, \bar{x}), D_x^2 \varphi(\bar{t}, \bar{x})) \le 0.$$

The above arguments lead to the following notion of viscosity solutions. We first introduce some additional notations. Given a locally bounded function $w$ from $\mathcal{O}$ to $\mathbb{R}$ (i.e. for all $x$ in $\mathcal{O}$, there exists a compact neighborhood $V_x$ of $x$ such that $w$ is bounded on $V_x$), we define its upper-semicontinuous envelope $w^*$ and lower-semicontinuous envelope $w_*$ on $\bar{\mathcal{O}}$ by

$$w^*(x) = \limsup_{x' \to x} w(x'), \quad w_*(x) = \liminf_{x' \to x} w(x').$$

Recall that $w^*$ (resp. $w_*$) is the smallest (resp. largest) upper-semicontinuous function (u.s.c.) above (resp. lower-semicontinuous function (l.s.c.) below) $w$ on $\mathcal{O}$. Note that a locally bounded function $w$ on $\mathcal{O}$ is lower-semicontinuous (resp. upper-semicontinuous) if and only if $w = w_*$ (resp. $w^*$) on $\mathcal{O}$, and it is continuous if (and only if) $w = w_* = w^*$ on $\mathcal{O}$.

**Definition 4.2.1** *Let $w : \mathcal{O} \to \mathbb{R}$ be locally bounded.*
*(i) $w$ is a (discontinuous) viscosity subsolution of (4.1) on $\mathcal{O}$ if*

$$F(\bar{x}, w^*(\bar{x}), D\varphi(\bar{x}), D^2\varphi(\bar{x})) \le 0,$$

*for all $\bar{x} \in \mathcal{O}$ and for all $\varphi \in C^2(\mathcal{O})$ such that $\bar{x}$ is a maximum point of $w^* - \varphi$.*
*(ii) $w$ is a (discontinuous) viscosity supersolution of (4.1) on $\mathcal{O}$ if*

$$F(\bar{x}, w_*(\bar{x}), D\varphi(\bar{x}), D^2\varphi(\bar{x})) \ge 0,$$

*for all $\bar{x} \in \mathcal{O}$ and for all $\varphi \in C^2(\mathcal{O})$ such that $\bar{x}$ is a minimum point of $w_* - \varphi$.*
*(iii) We say that $w$ is a (discontinuous) viscosity solution of (4.1) on $\mathcal{O}$ if it is both a subsolution and supersolution of (4.1).*

**Remark 4.2.1** **1.** The above definition is unchanged if the maximum or minimum point $\bar{x}$ is local and/or strict.
**2.** Without loss of generality we can also assume in the above definition that $\varphi(\bar{x}) = w^*(\bar{x})$ (resp. $\varphi(\bar{x}) = w_*(\bar{x})$). Indeed, it suffices otherwise to consider the smooth function $\psi(x) = \varphi(x) + w^*(\bar{x}) - \varphi(\bar{x})$ (resp. $\psi(x) = \varphi(x) + w_*(\bar{x}) - \varphi(\bar{x})$). Then $\bar{x}$ is a local maximum (resp. minimum) of $w^* - \psi$ (resp. $w_* - \psi$) and $\psi(\bar{x}) = w^*(\bar{x})$ (resp. $w_*(\bar{x})$).

**Remark 4.2.2** $v$ is a viscosity subsolution (resp. supersolution) of (4.1) if and only if $v^*$ (resp $v_*$) is a u.s.c. viscosity subsolution (resp. l.s.c. viscosity supersolution) of (4.1).

**Remark 4.2.3** In the general discontinuous viscosity solutions approach, there is no need to prove a priori the continuity of the value function $v$, since we work with the l.s.c. and u.s.c. envelopes of the value function. The continuity will actually follow from a strong comparison principle stated in Section 4.4, which, under suitable conditions, implies that $v_* \ge v^*$, and so $v_* = v^* = v$ is continuous inside its domain, see also Remark 4.4.8.

## 4.3 From dynamic programming to viscosity solutions of HJB equations

We return to the framework of stochastic control problems for diffusions formulated in Section 3.2 of Chapter 3. The chief goal of this section is to characterize the value function as a viscosity solution to the associated HJB equation. We also determine the relevant terminal condition. As in the smooth case, the proofs rely crucially on the dynamic programming principle, and there are some modifications for handling with viscosity solutions and eventual singularity of the HJB equation, see Remark **2** in Section 3.4.2. Recall the Hamiltonian for our stochastic control problems:

$$H(t, x, p, M) = \sup_{a \in A} \left[ b(x, a).p + \frac{1}{2} \mathrm{tr} \left( \sigma(x, a) \sigma'(x, a) M \right) + f(t, x, a) \right]. \quad (4.6)$$

In the infinite horizon case, $f$ (hence also $H$) does not depend on $t$. We present a unifying result for taking into account the possible singularity of the Hamiltonian $H$ when the control space $A$ is unbounded. We then introduce

$$\mathrm{dom}(H) = \{(t, x, p, M) \in [0, T) \times \mathbb{R}^n \times \mathbb{R}^n \times \mathcal{S}_n : H(t, x, p, M) < \infty \},$$

and make the following hypothesis:

$$H \text{ is continuous on } \mathrm{int}(\mathrm{dom}(H))$$
$$\text{and there exists } G : [0, T) \times \mathbb{R}^n \times \mathbb{R}^n \times \mathcal{S}_n \text{ continuous such that}$$
$$(t, x, p, M) \in \mathrm{dom}(H) \iff G(t, x, p, M) \geq 0. \quad (4.7)$$

Evidently, in the case of infinite horizon problems, the function $H$ (hence also $G$) does not depend on $t$.

### 4.3.1 Viscosity properties inside the domain

The smoothness condition of the value function can be relaxed in the theory of viscosity solutions and we shall prove that the value function satisfies the HJB variational inequality (3.34) in the viscosity sense. We separate the proof of viscosity supersolution and subsolution properties, which are different. The supersolution property follows from the first part (3.18) of the dynamic programming principle. The subsolution property is more delicate and should take into account the possible singular part of the Hamiltonian. The derivation is obtained from the second part (3.19) of the DPP and a contraposition argument.

**Viscosity supersolution property**

**Proposition 4.3.1** *(1) Finite horizon: Suppose the value function $v$ is locally bounded on $[0, T) \times \mathbb{R}^n$, that the function $f$ has quadratic growth in the sense of (3.8), and that $f(., ., a)$ is continuous in $(t, x)$ for all $a \in A$. Then $v$ is a viscosity supersolution of the HJB equation:*

$$-\frac{\partial v}{\partial t}(t, x) - H(t, x, D_x v(t, x), D_x^2 v(t, x)) = 0, \quad (t, x) \in [0, T) \times \mathbb{R}^n. \quad (4.8)$$

(2) Infinite horizon: Suppose the value function $v$ is locally bounded, that the function $f$ has quadratic growth in the sense of (3.13), and that $f(.,a)$ is continuous in $x$ for all $a \in A$. Then for all $\beta > 0$ large enough, $v$ is a viscosity supersolution of the HJB equation:

$$\beta v(x) - H(x, Dv(x), D^2 v(x)) = 0, \quad x \in \mathbb{R}^n. \tag{4.9}$$

**Proof.** We show the result in the finite horizon case. Let $(\bar{t}, \bar{x}) \in [0,T) \times \mathbb{R}^n$ and let $\varphi \in C^2([0,T) \times \mathbb{R}^n)$ be a test function such that

$$0 = (v_* - \varphi)(\bar{t}, \bar{x}) = \min_{(t,x) \in [0,T) \times \mathbb{R}^n} (v_* - \varphi)(t, x). \tag{4.10}$$

By definition of $v_*(\bar{t}, \bar{x})$, there exists a sequence $(t_m, x_m)$ in $[0,T) \times \mathbb{R}^n$ such that

$$(t_m, x_m) \to (\bar{t}, \bar{x}) \text{ and } v(t_m, x_m) \to v_*(\bar{t}, \bar{x}),$$

when $m$ goes to infinity. By the continuity of $\varphi$ and by (4.10) we also have that

$$\gamma_m := v(t_m, x_m) - \varphi(t_m, x_m) \to 0,$$

when $m$ goes to infinity.

Let $a \in A$ and $\alpha$ the control identically equal to $a$. Then $\alpha$ is in $\mathcal{A}(t_m, x_m) = \mathcal{A}$ according to Remark 3.2.1. We denote by $X_s^{t_m, x_m}$ the associated controlled process. Let $\tau_m$ be the stopping time given by $\tau_m = \inf\{s \geq t_m : |X_s^{t_m, x_m} - x_m| \geq \eta\}$ in which $\eta > 0$ is a fixed constant. Let $(h_m)$ be a strictly positive sequence such that

$$h_m \to 0 \text{ and } \frac{\gamma_m}{h_m} \to 0,$$

when $m$ converges to infinity. We apply the first part of the dynamic programming principle (3.18) for $v(t_m, x_m)$ to $\theta_m := \tau_m \wedge (t_m + h_m)$ and get

$$v(t_m, x_m) \geq E\left[ \int_{t_m}^{\theta_m} f(s, X_s^{t_m, x_m}, a)ds + v(\theta_m, X_{\theta_m}^{t_m, x_m}) \right].$$

Equation (4.10) implies that $v \geq v_* \geq \varphi$, thus

$$\varphi(t_m, x_m) + \gamma_m \geq E\left[ \int_{t_m}^{\theta_m} f(s, X_s^{t_m, x_m}, a)ds + \varphi(\theta_m, X_{\theta_m}^{t_m, x_m}) \right].$$

Applying Itô's formula to $\varphi(s, X_s^{t_m, x_m})$ between $t_m$ and $\theta_m$, we obtain

$$\frac{\gamma_m}{h_m} + E\left[ \frac{1}{h_m} \int_{t_m}^{\theta_m} \left( -\frac{\partial \varphi}{\partial t} - \mathcal{L}^a \varphi - f \right)(s, X_s^{t_m, x_m}, a)ds \right] \geq 0 \tag{4.11}$$

after noting that the stochastic integral term cancels out by taking expectations since the integrand is bounded. By a.s. continuity of the trajectory $X_s^{t_m, x_m}$, it follows that for $m$ sufficiently large $(m \geq N(\omega))$, $\theta_m(\omega) = t_m + h_m$ a.s. Thus, by the mean value theorem, the random variable inside the expectation in (4.11) converges a.s. to $-\dfrac{\partial \varphi}{\partial t}(\bar{t}, \bar{x})$

$-\mathcal{L}^a \varphi(\bar{t}, \bar{x}) - f(\bar{t}, \bar{x}, a)$ when $m$ converges to infinity. Moreover, this random variable is bounded by a constant independent of $m$. We then obtain

$$-\frac{\partial \varphi}{\partial t}(\bar{t}, \bar{x}) - \mathcal{L}^a \varphi(\bar{t}, \bar{x}) - f(\bar{t}, \bar{x}, a) \geq 0,$$

when $m$ goes to infinity by the dominated convergence theorem. We conclude from the arbitrariness of $a \in A$.

In the infinite horizon case, we assume that $\beta > 0$ is large enough so that constant controls are in $\mathcal{A}(x)$, for all $x$ in $\mathbb{R}^n$ (see Remark 3.2.2). We then use the same argument as above.                                                                          □

**Viscosity subsolution property**

**Proposition 4.3.2** *(1) Finite horizon: Assume that (4.7) is satisfied and that the value function $v$ is locally bounded on $[0, T[ \times \mathbb{R}^n$. Then $v$ is a viscosity subsolution of the Hamilton-Jacobi-Bellman variational inequality:*

$$\min \Big\{ -\frac{\partial v}{\partial t}(t, x) - H(t, x, D_x v(t, x), D_x^2 v(t, x)),$$
$$G(t, x, D_x v(t, x), D_x^2 v(t, x)) \Big\} = 0, \quad (t, x) \in [0, T) \times \mathbb{R}^n. \ (4.12)$$

*(2) Infinite horizon: Assume that (4.7) is satisfied and that the value function $v$ is locally bounded on $\mathbb{R}^n$. Then $v$ is a viscosity subsolution of*

$$\min \Big\{ \beta v(x) - H(x, Dv(x), D^2 v(x)),$$
$$G(x, Dv(x), D^2 v(x)) \Big\} = 0, \ x \in \mathbb{R}^n. \quad (4.13)$$

**Proof.** We show the result in the infinite horizon case. Let $\bar{x} \in \mathbb{R}^n$ and let $\varphi \in C^2(\mathbb{R}^n)$ be a test function such that

$$0 = (v^* - \varphi)(\bar{x}) = \max_{x \in \mathbb{R}^n} (v^* - \varphi)(x). \quad (4.14)$$

We will show the result by contradiction. Assume on the contrary that

$$\beta \varphi(\bar{x}) - H(\bar{x}, D\varphi(\bar{x}), D^2 \varphi(\bar{x})) > 0,$$
$$\text{and } G(\bar{x}, D\varphi(\bar{x}), D^2 \varphi(\bar{x})) > 0.$$

Then by the continuity of the function $G$, and the continuity of $H$ on the interior of its domain, there exist $\eta > 0$ and $\varepsilon > 0$ such that

$$\beta \varphi(y) - H(y, D\varphi(y), D^2 \varphi(y)) \geq \varepsilon,$$

for all $y \in B(\bar{x}, \eta) = \{y \in \mathbb{R}^n : |\bar{x} - y| < \eta\}$. By definition of $v^*(\bar{x})$, there exists a sequence $(x_m)$ taking values in $B(\bar{x}, \eta)$ such that

$$x_m \to \bar{x} \text{ and } v(x_m) \to v^*(\bar{x}),$$

when $m$ goes to infinity. By continuity of $\varphi$ and using (4.14), we also find that

$$\gamma_m := v(x_m) - \varphi(x_m) \to 0,$$

when $m$ goes to infinity. Let $(h_m)$ be a strictly positive sequence such that

$$h_m \to 0 \text{ and } \frac{\gamma_m}{h_m} \to 0.$$

Then, according to the second part of the dynamic programming principle (3.19) and using (4.14), there is an $\hat{\alpha}^m \in \mathcal{A}(x_m)$ such that

$$\varphi(x_m) + \gamma_m - \frac{\varepsilon h_m}{2} \leq E\left[\int_0^{\theta_m} e^{-\beta s} f(X_s^{x_m}, \hat{\alpha}_s^m)ds + e^{-\beta \theta_m}\varphi(X_{\theta_m}^{x_m})\right],$$

in which we took $\theta_m := \tau_m \wedge h_m$, $\tau_m = \inf\{s \geq 0 : |X_s^{x_m} - x_m| \geq \eta'\}$ and $0 < \eta' < \eta$. Since $(x_m)$ converges to $\bar{x}$, we can always assume that $B(x_m, \eta') \subset B(\bar{x}, \eta)$, in such a way that for $0 \leq s < \theta_m$, $X_s^{x_m} \in B(\bar{x}, \eta)$. Here $X_s^{x_m}$ corresponds to the diffusion controlled by $\hat{\alpha}^m$. By Itô's formula applied to $e^{-\beta s}\varphi(X_s^{x_m})$ between $s = 0$ and $s = \theta_m$, we get

$$0 \geq \frac{\gamma_m}{h_m} - \frac{\varepsilon}{2} + E\left[\frac{1}{h_m}\int_0^{\theta_m} L(X_s^{x_m}, \hat{\alpha}_s^m)ds\right]$$

$$- E\left[\int_0^{\theta_m} D_x\varphi(X_s^{x_m})'\sigma(X_s^{x_m}, \hat{\alpha}_s^m)dW_s\right] \qquad (4.15)$$

with

$$L(x, a) = \beta\varphi(x) - \mathcal{L}^a\varphi(x) - f(x, a).$$

We note that by condition (3.2) on $\sigma$, the integrand in the above stochastic integral is bounded on $[0, \theta_m]$ by

$$|D_x\varphi(X_s^{x_m})'\sigma(X_s^{x_m}, \hat{\alpha}_s^m)| \leq C_\eta(1 + |\sigma(0, \hat{\alpha}_s^m)|).$$

Using (3.3) on $\hat{\alpha}^m \in \mathcal{A}(x_m)$, we find that the expectation of the stochastic integral in (4.15) is equal to zero.

Moreover, noting that for $0 \leq s < \theta_m$

$$L(X_s^{x_m}, \hat{\alpha}_s^m) \geq \beta\varphi(X_s^{x_m}) - H(X_s^{x_m}, D\varphi(X_s^{x_m}), D^2\varphi(X_s^{x_m}))$$

$$\geq \varepsilon,$$

we find using (4.15) that

$$0 \geq \frac{\gamma_m}{h_m} - \varepsilon\left(\frac{1}{2} - \frac{1}{h_m}E[\theta_m]\right). \qquad (4.16)$$

By Tchebyshev's inequality and (3.5), we deduce that

$$P[\tau_m \leq h_m] \leq P\left[\sup_{s \in [0, h_m]} |X_s^{x_m} - x_m| \geq \eta\right]$$

$$\leq \frac{E\left|\sup_{s \in [0, h_m]} |X_s^{x_m} - x_m|^2\right|}{\eta^2} \to 0,$$

when $h_m$ goes to zero, i.e. when $m$ goes to infinity. Moreover, since

$$P[\tau_m > h_m] \leq \tfrac{1}{h_m} E[\theta_m] \leq 1,$$

this implies that $\tfrac{1}{h_m} E[\theta_m]$ converges to 1 when $h_m$ goes to zero. We thus get the desired contradiction by letting $m$ go to infinity in (4.16).                      □

**The viscosity solution property**

By combining the two previous propositions, we obtain the main result of this section.

**Theorem 4.3.1** *Under the assumptions of Propositions 4.3.1 and 4.3.2, the value function $v$ is a viscosity solution of the Hamilton-Jacobi-Bellman variational inequality:*
*(1) Finite horizon:*

$$\min \Big\{ -\frac{\partial v}{\partial t}(t,x) - H(t,x,D_x v(t,x), D_x^2 v(t,x)) \,,$$
$$G(t,x,D_x v(t,x), D_x^2 v(t,x)) \Big\} = 0, \quad (t,x) \in [0,T) \times \mathbb{R}^n. \ (4.17)$$

*(2) Infinite horizon:*

$$\min \Big\{ \beta v(x) - H(x, Dv(x), D^2 v(x)) \,,$$
$$G(x, Dv(x), D^2 v(x)) \Big\} = 0, \quad x \in \mathbb{R}^n. \tag{4.18}$$

**Proof.** The viscosity supersolution property of the value function in Proposition 4.3.1 of $v$ means that, in the finite horizon case,

$$-\frac{\partial \varphi}{\partial t}(\bar{t}, \bar{x}) - H(\bar{t}, \bar{x}, D_x \varphi(\bar{t}, \bar{x}), D_x^2 \varphi(\bar{t}, \bar{x})) \geq 0,$$

for all $(\bar{t}, \bar{x}) \in [0,T) \times \mathbb{R}^n$ and $\varphi \in C^2([0,T) \times \mathbb{R}^n)$ such that $(\bar{t}, \bar{x})$ is a minimum of $v_* - \varphi$. With condition (4.7), this implies that

$$G(\bar{t}, \bar{x}, D_x \varphi(\bar{t}, \bar{x}), D_x^2 \varphi(\bar{t}, \bar{x})) \geq 0,$$

and hence

$$\min \Big\{ -\frac{\partial \varphi}{\partial t}(\bar{t}, \bar{x}) + H(\bar{t}, \bar{x}, D_x \varphi(\bar{t}, \bar{x}), D_x^2 \varphi(\bar{t}, \bar{x})), G(\bar{t}, \bar{x}, D_x \varphi(\bar{t}, \bar{x}), D_x^2 \varphi(\bar{t}, \bar{x})) \Big\} \geq 0.$$

In other words, $v$ is a viscosity supersolution of the variational inequality (4.17). The final result follows from the viscosity subsolution property of Proposition 4.3.2.          □

**Remark 4.3.4** In the regular case, i.e. when the Hamiltonian $H$ is finite on the whole domain $[0,T] \times \mathbb{R}^n \times \mathbb{R}^n \times \mathcal{S}_n$ (this occurs typically when the control space is compact), the condition (4.7) is satisfied with any choice of strictly positive continuous function $G$. In this case, the HJB variational inequality is reduced to the regular HJB equation:

$$-\frac{\partial v}{\partial t}(t,x) - H(t,x,D_x v(t,x), D_x^2 v(t,x)) = 0, \quad (t,x) \in [0,T) \times \mathbb{R}^n,$$

which the value function satisfies in the viscosity sense. Hence, Theorem 4.3.1 states a general viscosity property including both the regular and singular case. We shall see later how to use this viscosity property for deriving further results, and to solve a stochastic control problem in some examples.

### 4.3.2 Terminal condition

In the finite horizon case, the parabolic PDE (4.17) should be completed with a terminal condition, in order to fully characterize the value function of the stochastic control problem. By the very definition of the value function, we have

$$v(T, x) = g(x), \quad x \in \mathbb{R}^d. \tag{4.19}$$

However, due to the possible singularity of the Hamiltonian, the value function may be discontinuous at $T$. In this case, (4.19) is not the relevant terminal condition associated to the HJB variational inequality. We need actually to determine $v(T^-, x) := \lim_{t \nearrow T} v(t, x)$ if it exists. This is achieved by the following result:

**Theorem 4.3.2** *Assume that $f$ and $g$ are lower-bounded or satisfy a linear growth condition, and (4.7) holds.*
*(i) Suppose that $g$ is lower-semicontinuous. Then $v_*(T, .)$ is a viscosity supersolution of*

$$\min \left[ v_*(T, x) - g(x) \,,\, G(T, x, D_x v_*(T, x), D_x^2 v_*(T, x)) \right] = 0, \quad on \ \mathbb{R}^n. \tag{4.20}$$

*(ii) Suppose that $g$ is upper-semicontinuous. Then $v^*(T, .)$ is a viscosity subsolution of*

$$\min \left[ v^*(T, x) - g(x) \,,\, G(T, x, D_x v^*(T, x), D_x^2 v^*(T, x)) \right] = 0, \quad on \ \mathbb{R}^n. \tag{4.21}$$

**Remark 4.3.5** In usual cases, there is a comparison principle for the PDE arising in the above theorem, meaning that a u.s.c. subsolution is not greater than a l.s.c. supersolution. Therefore, under the conditions of Theorem 4.3.2, we have $v^*(T, .) \le v_*(T, .)$ and so $v^*(T, .) = v_*(T, .)$. This means that $v(T^-, .)$ exists, equal to $v^*(T, .) = v_*(T, .)$ and is a viscosity solution to

$$\min \left[ v(T^-, x) - g(x) \,,\, G(T, x, D_x v(T^-, x), D_x^2 v(T^-, x)) \right] = 0, \quad on \ \mathbb{R}^n. \tag{4.22}$$

Denote by $\hat{g}$ the upper $G$-envelope of $g$, defined as the smallest function above $g$ and viscosity supersolution to

$$G(T, x, D\hat{g}(x), D^2 \hat{g}(x)) = 0, \quad on \ \mathbb{R}^n, \tag{4.23}$$

when it exists and is finite. Such a function may be calculated in a number of examples, see e.g. Section 4.6. Since $v(T^-, .)$ is a viscosity supersolution to (4.22), it is greater than $g$ and is a viscosity supersolution to the same PDE as $\hat{g}$. Hence, by definition of $\hat{g}$, we have $v(T^-, ) \ge \hat{g}$. On the other hand, $\hat{g}$ is a viscosity supersolution to the PDE (4.22), and so by a comparison principle, the subsolution $v(T^-, )$ of (4.22) is not greater than $\hat{g}$. We have then determined explicitly the terminal data:

$$v(T^-, x) = \hat{g}(x).$$

Recall that in the regular case, we may take for $G$ a positive constant function, so that obviously $\hat{g} = g$. Therefore, in this case, $v$ is continuous in $T$ and $v(T^-, x) = v(T, x) = g(x)$. In the singular case, $\hat{g}$ is in general different from $g$ and so $v$ is discontinuous in $T$. The effect of the singularity is to lift up, via the $G$ operator, the terminal function $g$ to $\hat{g}$.

The rest of this section is devoted to the (technical) proof of Theorem 4.3.2, which requires several lemmas. We start with the following result.

**Lemma 4.3.1** *Suppose that $f$ and $g$ are lower-bounded or satisfy a quadratic growth condition, and $g$ is lower-semicontinuous. Then,*

$$v_*(T, x) \geq g(x), \quad \forall x \in \mathbb{R}^n.$$

**Proof.** Take some arbitrary sequence $(t_m, x_m) \to (T, x)$ with $t_m < T$ and fix some control $\alpha \in \mathcal{A}(t_m, x_m)$. By definition of the value function, we have:

$$v(t_m, x_m) \geq E\left[\int_{t_m}^{T} f(s, X_s^{t_m, x_m}, \alpha)ds + g(X_T^{t_m, x_m})\right].$$

Under the quadratic growth or lower-boundedness condition on $f$ and $g$, we may apply the dominated convergence theorem or Fatou's lemma, and so

$$\liminf_{m \to \infty} v(t_m, x_m) \geq E\left[\liminf_{m \to \infty} g(X_T^{t_m, x_m})\right]$$
$$\geq g(x),$$

by the lower-semicontinuity of $g$ and the continuity of the flow $X_T^{t,x}$ in $(t, x)$.    $\square$

The supersolution property (4.20) for the terminal condition is then obtained with the following result.

**Lemma 4.3.2** *Under (4.7), $v_*(T, .)$ is a viscosity supersolution of*

$$G(T, x, D_x v_*(T, x)(x), D_x^2 v_*(T, x)) = 0, \quad on \ \mathbb{R}^n.$$

**Proof.** Let $\bar{x} \in \mathbb{R}^n$ and $\psi$ a smooth function on $\mathbb{R}^n$ s.t.

$$0 = (v_*(T, .) - \psi)(\bar{x}) = \min_{\mathbb{R}^n}(v_*(T, .) - \psi). \tag{4.24}$$

By definition of $v_*(T, .)$, there exists a sequence $(s_m, y_m)$ converging to $(T, \bar{x})$ with $s_m < T$ and

$$\lim_{m \to \infty} v_*(s_m, y_m) = v_*(T, \bar{x}). \tag{4.25}$$

Consider the auxiliary test function:

$$\varphi_m(t, x) = \psi(x) - |x - \bar{x}|^4 + \frac{T - t}{(T - s_m)^2},$$

and choose $(t_m, x_m) \in [s_m, T] \times \bar{B}(\bar{x}, 1)$ as a minimum of $(v_* - \varphi_m)$ on $[s_m, T] \times \bar{B}(\bar{x}, 1)$.

*Step 1.* We claim that, for sufficiently large $m$, $t_m < T$ and $x_m$ converges to $\bar{x}$, so that $(t_m, x_m)$ is a local minimizer of $(v_* - \varphi_m)$. Indeed, recalling $v_*(T, \bar{x}) = \psi(\bar{x})$ and (4.25), we have for sufficiently large $m$

$$(v_* - \varphi_m)(s_m, y_m) \leq -\frac{1}{2(T - s_m)} < 0. \tag{4.26}$$

On the other hand, for any $x \in \mathbb{R}^n$, we have

$$(v_* - \varphi_m)(T, x) = v_*(T, x) - \psi(x) + |x - \bar{x}|^4 \geq v_*(T, x) - \psi(x) \geq 0, \quad (4.27)$$

by (4.24). The two inequalities (4.26)-(4.27) show that $t_m < T$ for large $m$. We can suppose that $x_m$ converges, up to a subsequence, to some $x_0 \in \bar{B}(\bar{x}, 1)$. From (4.24), since $s_m \leq t_m$ and $(t_m, x_m)$ is a minimum of $(v_* - \psi_m)$, we have

$$0 \leq (v_*(T, .) - \psi)(x_0) - (v_*(T, .) - \psi)(\bar{x})$$
$$\leq \liminf_{m \to \infty} \left[ (v_* - \varphi_m)(t_m, x_m) - (v_* - \varphi_m)(s_m, y_m) - |x_m - \bar{x}|^4 \right]$$
$$\leq -|x_0 - \bar{x}|^4,$$

which proves that $x_0 = \bar{x}$.

*Step 2.* Since $(t_m, x_m)$ is a local minimizer of $(v_* - \varphi_m)$, the viscosity supersolution property of $v_*$ holds at $(t_m, x_m)$ with the test function $\varphi_m$, and so for every $m$

$$G(t_m, x_m, D_x\varphi_m(t_m, x_m), D_x^2\varphi_m(t_m, x_m)) \geq 0. \quad (4.28)$$

Now, since $D_x\varphi_m(t_m, x_m) = D\psi(x_m) - 4(x_m - \bar{x})|x_m - \bar{x}|^2$, $D_x^2\varphi_m(t_m, x_m) = D^2\psi(x_m) - 4|x_m - \bar{x}|^2 I_n - 4(x_m - x)(x_m - \bar{x})'$, recalling that $G$ is continuous, and $(t_m, x_m)$ converges to $(T, \bar{x})$, we get from (4.28)

$$G(T, \bar{x}, D\psi(\bar{x}), D^2\psi(\bar{x})) > 0.$$

This is the required supersolution inequality.                                    □

We finally turn to the subsolution property for the terminal condition. As for the viscosity subsolution property inside the domain, the proof is based on a contraposition argument and the second part (3.19) of the dynamic programming principle. We then introduce for a given smooth function $\varphi$, the set in $[0, T] \times \mathbb{R}^n$:

$$\mathcal{M}(\varphi) = \Big\{ (t, x) \in [0, T] \times \mathbb{R}^n : G(t, x, D_x\varphi(t, x), D_x^2\varphi(t, x)) > 0$$

$$\text{and} \quad -\frac{\partial \varphi}{\partial t}(t, x) - H(t, x, D_x\varphi(t, x), D_x^2\varphi(t, x)) > 0 \Big\}.$$

The following lemma is a consequence of the second part (3.19) of the DPP.

**Lemma 4.3.3** *Let $\varphi$ be a smooth function on $[0, T] \times \mathbb{R}^n$, and suppose there exist $t_1 < t_2 \leq T$, $\bar{x} \in \mathbb{R}^n$ and $\eta > 0$ s.t.:*

$$[t_1, t_2] \times \bar{B}(\bar{x}, \eta) \in \mathcal{M}(\varphi).$$

*Then,*

$$\sup_{\partial_p([t_1, t_2] \times \bar{B}(\bar{x}, \eta))} (v - \varphi) = \max_{[t_1, t_2] \times \bar{B}(\bar{x}, \eta)} (v^* - \varphi),$$

*where $\partial_p([t_1, t_2] \times B(\bar{x}, \eta))$ is the forward parabolic boundary of $[t_1, t_2] \times \bar{B}(\bar{x}, \eta)$, i.e. $\partial_p([t_1, t_2] \times \bar{B}(\bar{x}, \eta)) = [t_1, t_2] \times \partial\bar{B}(\bar{x}, \eta) \cup \{t_2\} \times \bar{B}(\bar{x}, \eta)$.*

**Proof.** By definition of $\mathcal{M}(\varphi)$ and $H$, we have for all $a \in A$

$$-\frac{\partial\varphi}{\partial t}(t,x) - \mathcal{L}^a\varphi(t,x) - f(t,x,a) > 0, \quad \forall(t,x) \in [t_1,t_2] \times \bar{B}(\bar{x},\eta). \qquad (4.29)$$

We argue by contradiction and suppose on the contrary that

$$\max_{[t_1,t_2]\times\bar{B}(\bar{x},\eta)}(v^* - \varphi) - \sup_{\partial_p([t_1,t_2]\times\bar{B}(\bar{x},\eta))}(v - \varphi) := 2\delta.$$

We can choose $(t_0,x_0) \in (t_1,t_2) \times B(\bar{x},\eta)$ s.t. $(v-\varphi)(t_0,x_0) \geq -\delta + \max_{[t_1,t_2]\times\bar{B}(\bar{x},\eta)}(v^* - \varphi)$,

and so

$$(v - \varphi)(t_0,x_0) \geq \delta + \sup_{\partial_p([t_1,t_2]\times\bar{B}(\bar{x},\eta))}(v - \varphi). \qquad (4.30)$$

Fix now $\varepsilon = \delta/2$, and apply the second part (3.19) of DPP to $v(t_0,x_0)$: there exists $\hat{\alpha}^\varepsilon \in \mathcal{A}(t_0,x_0)$ s.t.

$$v(t_0,x_0) - \varepsilon \leq E\left[\int_{t_0}^\theta f(s,X_s^{t_0,x_0},\hat{\alpha}_s^\varepsilon)ds + v(\theta,X_\theta^{t_0,x_0})\right], \qquad (4.31)$$

where we choose

$$\theta = \inf\left\{s \geq t_0 : (s,X_s^{t_0,x_0}) \notin [t_1,t_2] \times \bar{B}(\bar{x},\eta)\right\} \wedge T.$$

First, notice that by continuity of $X^{t_0,x_0}$, we have $(\theta,X_\theta^{t_0,x_0}) \in \partial_p([t_1,t_2] \times B(\bar{x},\eta))$. Since from (4.30), we have $v \leq \varphi + (v - \varphi)(t_0,x_0) - \delta$ on $\partial_p([t_1,t_2] \times B(\bar{x},\eta))$, we get with (4.31)

$$-\varepsilon \leq E\left[\int_{t_0}^\theta f(s,X_s^{t_0,x_0},\hat{\alpha}_s^\varepsilon)ds + \varphi(\theta,X_\theta^{t_0,x_0}) - \varphi(t_0,x_0)\right] - \delta.$$

Applying Itô's formula to $\varphi(s,X_s^{t_0,x_0})$ between $s = t_0$ and $s = \theta$, we obtain

$$E\left[\int_{t_0}^\theta \left(-\frac{\partial\varphi}{\partial t}(s,X_s^{t_0,x_0}) - \mathcal{L}^{\hat{\alpha}_s^\varepsilon}\varphi(s,X_s^{t_0,x_0}) - f(s,X_s^{t_0,x_0},\hat{\alpha}_s^\varepsilon)\right)ds\right] \leq \varepsilon - \delta.$$

Since, by definition of $\theta$, $(s,X_s^{t_0,x_0})$ lies in $[t_1,t_2] \times \bar{B}(\bar{x},\eta)$ for all $t_0 \leq s \leq \theta$, we get with (4.29) the required contradiction: $0 \leq \varepsilon - \delta = -\delta/2$. $\qquad\square$

**Remark 4.3.6** The above lemma provides an immediate alternative proof to the viscosity subsolution property of the value function inside the domain stated in Proposition 4.3.2. Indeed, let $(\bar{t},\bar{x}) \in [0,T) \times \mathbb{R}^n$ and $\varphi$ a smooth test function s.t.

$$0 = (v^* - \varphi)(\bar{t},\bar{x}) = (\text{strict})\max_{[0,T)\times\mathbb{R}^n}(v^* - \varphi).$$

First, observe that by the continuity condition in (4.7), the set $\mathcal{M}(\varphi)$ is open. Since $(\bar{t},\bar{x})$ is a strict maximizer of $(v^* - \varphi)$, we then deduce by Lemma 4.3.3 that $(\bar{t},\bar{x}) \notin \mathcal{M}(\varphi)$. By definition of $\mathcal{M}(\varphi)$, this means:

$$\min \Big\{ -\frac{\partial \varphi}{\partial t}(t,x) - H(\bar{t}, \bar{x}, D_x \varphi(\bar{t}, \bar{x}), D_x^2 \varphi(\bar{t}, \bar{x})) \,,$$

$$G(\bar{t}, \bar{x}, D_x \varphi(\bar{t}, \bar{x}), D_x^2 \varphi(\bar{t}, \bar{x})) \Big\} \leq 0,$$

which is the required subsolution inequality.

We can finally prove the viscosity subsolution property for the terminal condition.

**Lemma 4.3.4** *Suppose that $g$ is upper-semicontinuous and (4.7) holds. Then, $v^*(T, .)$ is a viscosity subsolution of*

$$\min \big[ v^*(T,x) - g(x) \,,\; G(T, x, D_x v^*(T,x), D_x^2 v^*(T,x)) \big] = 0, \quad on \; \mathbb{R}^n.$$

**Proof.** Let $\bar{x} \in \mathbb{R}^n$ and $\psi$ a smooth function on $\mathbb{R}^n$ s.t.

$$0 = (v^*(T, .) - \psi)(\bar{x}) = \max_{\mathbb{R}^n} (v^*(T, .) - \psi). \qquad (4.32)$$

We have to show that whenever

$$v^*(T, \bar{x}) > g(\bar{x}), \qquad (4.33)$$

then

$$G(T, \bar{x}, D\psi(\bar{x}), D^2\psi(\bar{x})) \leq 0. \qquad (4.34)$$

So, suppose that (4.33) holds, and let us consider the auxiliary test function:

$$\varphi_m(t,x) = \psi(x) + |x - \bar{x}|^4 + m(T - t).$$

We argue by contradiction and suppose on the contrary that

$$G(T, \bar{x}, D\psi(\bar{x}), D^2\psi(\bar{x})) > 0.$$

Since $D_x \varphi_m(t,x) = D\psi(x) - 4(x-\bar{x})|x-\bar{x}|^2 \rightarrow D\psi(\bar{x})$, $D_x^2 \varphi_m(t,x) = D^2\psi(x) - 4I_n|x - \bar{x}|^2 - 4(x-\bar{x})(x-\bar{x})' \rightarrow D^2\psi(\bar{x})$ when $x$ tends to $\bar{x}$, there exists, by continuity of $G$, $s_0 < T$ and $\eta > 0$ s.t. for all $m$

$$G(t, x, D_x \varphi_m(t,x), D_x^2 \varphi_m(t,x)) > 0, \quad \forall (t,x) \in [s_0, T] \times \bar{B}(\bar{x}, \eta). \qquad (4.35)$$

Under condition (4.7), the function $H(t, x, D_x \varphi_m(t,x), D_x^2 \varphi_m(t,x))$ is then finite on the compact set $[s_0, T] \times \bar{B}(\bar{x}, \eta)$ and by continuity of $H$ on int(dom($H$)), there exists some constant $h_0$ (independent of $m$) s.t.

$$H(t, x, D_x \varphi_m(t,x), D_x^2 \varphi_m(t,x)) \leq h_0, \quad \forall (t,x) \in [s_0, T] \times \bar{B}(\bar{x}, \eta). \qquad (4.36)$$

*Step 1.* Since by definition $v^*(T, .) \geq v_*(T, .)$, we have from Lemma 4.3.1

$$v^*(T, .) \geq g. \qquad (4.37)$$

Hence, for all $x \in \mathbb{R}^n$,

$$(v - \varphi_m)(T, x) = (g - \psi)(x) - |x - \bar{x}|^4 \leq (v^*(T, .) - \psi)(x) - |x - \bar{x}|^4$$
$$\leq -|x - \bar{x}|^4 \leq 0 \qquad (4.38)$$

by (4.32). This implies $\sup_{B(\bar{x},\eta)}(v - \varphi_m)(T, .) \leq 0$. We claim that

$$\limsup_{m\to\infty} \sup_{B(\bar{x},\eta)} (v - \varphi_m)(T, .) < 0. \qquad (4.39)$$

On the contrary, there exists a subsequence of $(\varphi_m)$, still denoted by $(\varphi_m)$ s.t.

$$\lim_{m\to\infty} \sup_{B(\bar{x},\eta)} (v - \varphi_m)(T, .) = 0.$$

For each $m$, let $(x_m^k)_k$ be a maximizing sequence of $(v - \varphi_m)(T, .)$ on $B(\bar{x}, \eta)$, i.e.

$$\lim_{m\to\infty} \lim_{k\to\infty} (v - \varphi_m)(T, x_m^k) = 0.$$

Now, from (4.38), we have $(v - \varphi_m)(T, x_m^k) \leq -|x_m^k - \bar{x}|^4$, which combined with the above equality shows that

$$\lim_{m\to\infty} \lim_{k\to\infty} x_m^k = \bar{x}.$$

Hence,

$$0 = \lim_{m\to\infty} \lim_{k\to\infty} (v - \varphi_m)(T, x_m^k) = \lim_{m\to\infty} \lim_{k\to\infty} g(x_m^k) - \psi(\bar{x})$$
$$\leq g(\bar{x}) - \psi(\bar{x}) < (\bar{v} - \psi)(\bar{x}),$$

by the upper-semicontinuty of $g$ and (4.33). This contradicts $(v^*(T, .) - \psi)(\bar{x}) = 0$ in (4.32).

*Step 2.* Take a sequence $(s_m)$ converging to $T$ with $s_0 \leq s_m < T$. Let us consider a maximizing sequence $(t_m, x_m)$ of $v^* - \varphi_m$ on $[s_m, T] \times \partial \bar{B}(\bar{x}, \eta)$. Then

$$\limsup_{m\to\infty} \sup_{[s_m,T]\times\partial\bar{B}(\bar{x},\eta)} (v^* - \varphi_m) \leq \limsup_{m\to\infty}(v^*(t_m, x_m) - \psi(x_m)) - \eta^4.$$

Since $t_m$ converges to $T$ and $x_m$, up to a subsequence, converges to some $x_0 \in \partial\bar{B}(\bar{x}, \eta)$, we have by definition of $\bar{v}$

$$\limsup_{m\to\infty} \sup_{[s_m,T]\times\partial\bar{B}(\bar{x},\eta)} (v^* - \varphi_m) \leq (v^*(T, .) - \psi)(x_0) - \eta^4 \leq -\eta^4, \qquad (4.40)$$

by (4.32). Recall also from (4.32) that $(v^* - \varphi_m)(T, \bar{x}) = (v^*(T, .) - \psi)(\bar{x}) = 0$. Therefore, with (4.39) and (4.40), we deduce that for $m$ large enough

$$\sup_{[s_m,T]\times\partial\bar{B}(\bar{x},\eta)} (v - \psi_m) < 0 = (v^* - \varphi_m)(T, \bar{x}) \leq \max_{[s_m,T]\times\partial B(\bar{x},\eta)} (v^* - \varphi_m).$$

In view of Lemma 4.3.3, this proves that for $m$ large enough

$$[s_m, T] \times \bar{B}(\bar{x}, \eta) \text{ is not a subset of } \mathcal{M}(\varphi_m). \qquad (4.41)$$

*Step 3.* From (4.36), notice that for all $(t, x) \in [s_m, T] \times \bar{B}(\bar{x}, \eta)$, we have

$$-\frac{\partial \varphi_m}{\partial t}(t, x) - H(t, x, D_x \varphi_m(t, x), D_x^2 \varphi_m(t, x)) \geq m - h_0 > 0$$

for $m$ large enough. In view of (4.41) and by definition of $\mathcal{M}(\varphi_m)$, we then may find some element $(t, x) \in [s_m, T] \times \bar{B}(\bar{x}, \eta)$ s.t.

$$G(t, x, D_x \varphi_m(t, x), D_x^2 \varphi_m(t, x)) \leq 0.$$

This is in contradiction with (4.35). □

**Proof of Theorem 4.3.2.** Lemmas 4.3.1 and 4.3.2 prove the viscosity supersolution property (i), while Lemma 4.3.4 proves the viscosity subsolution property (ii).

## 4.4 Comparison principles and uniqueness results

In general terms, we say that a strong comparison principle (for discontinuous solutions) holds for the PDE (4.1) if the following statement is true:

If $v$ is a u.s.c. viscosity subsolution of (4.1) on $\mathcal{O}$ and $w$ is a l.s.c. viscosity supersolution of (4.1) on $\mathcal{O}$ such that $v \leq w$ on $\partial \mathcal{O}$, then $v \leq w$ on $\mathcal{O}$.

**Remark 4.4.7** In the case of an elliptic PDE (4.1) on the entire space $\mathcal{O} = \mathbb{R}^n$, the conditions for $v$, $w$ on the boundary $\partial \mathcal{O}$ are growth conditions at infinity on $x$, e.g. polynomial growth condition in $x$. In the case of a parabolic PDE on $\mathcal{O} = [0, T) \times \mathbb{R}^n$, the conditions for $v$, $w$ on the boundary $\partial \mathcal{O}$ are terminal conditions at $T$, in addition to the growth conditions at infinity in $x$.

**Remark 4.4.8** As for classical comparison principles (for continuous solutions), the strong comparison principle allows us to compare a subsolution and a supersolution on the entire domain from the comparison on the boundary of the domain. In particular, it proves the uniqueness of the viscosity solution of (4.1) from a condition on the boundary given by $v^* = v_* = g$ on $\partial \mathcal{O}$. Indeed, if $v$ and $w$ are both viscosity solutions of (4.1) with the same boundary condition, then we have by a strong comparison principle $v^* \leq w_*$ and $w^* \leq v_*$ on $\mathcal{O}$. By construction we already have $v_* \leq v^*$ and $w_* \leq w^*$, hence this implies the following equalities:

$$v_* = v^* = w_* = w^*.$$

This proves the uniqueness of a viscosity solution $v = w$ on $\mathcal{O}$. Furthermore, we obtain as a byproduct the continuity of $v$ on $\mathcal{O}$ since $v^* = v_*$.

Comparison principles for viscosity solutions of general nonlinear PDEs received a lot of interest in the PDE literature, and we refer to Crandall, Ishii and P.L. Lions [CIL92] or Barles [Ba95] for results in this direction. In this section, we mainly focus on the case of HJB equations arising from stochastic control problems, and explain the important techniques in the proofs of comparison principles. We first detail the classical arguments in the case of smooth solutions, and then outline the key tools used for dealing with (discontinuous) viscosity solutions.

### 4.4.1 Classical comparison principle

We consider HJB equations in the form

$$-\frac{\partial w}{\partial t} + \beta w - H(t, x, D_x w, D_x^2 w) = 0, \quad \text{on } [0, T) \times \mathbb{R}^n, \tag{4.42}$$

with a Hamiltonian

$$H(t, x, p, M) = \sup_{a \in A} \left[ b(x, a).p + \frac{1}{2} \text{tr}(\sigma \sigma'(x)M) + f(t, x, a) \right], \tag{4.43}$$

for $(t, x, p, M) \in [0, T] \times \mathbb{R}^n \times \mathbb{R}^n \times \mathcal{S}_n$, $A$ a subset of $\mathbb{R}^m$, and $\beta \in \mathbb{R}$. We assume that the coefficients $b$, $\sigma$ satisfy a linear growth condition in $x$, uniformly in $a \in A$.

**Theorem 4.4.3** *Let $U$ (resp. $V$) $\in C^{1,2}([0, T) \times \mathbb{R}^n) \cap C^0([0, T] \times \mathbb{R}^n)$ be a subsolution (resp. supersolution) with polynomial growth condition to (4.42). If $U(T, .) \le V(T, .)$ on $\mathbb{R}^n$, then $U \le V$ on $[0, T) \times \mathbb{R}^n$.*

**Proof.** *Step 1.* Let $\tilde{U}(t, x) = e^{\lambda t} U(x)$ and $\tilde{V}(t, x) = e^{\lambda t} V(x)$. Then a straightforward calculation shows that $\tilde{U}$ (resp. $\tilde{V}$) is a subsolution (resp. supersolution) to

$$-\frac{\partial w}{\partial t} + (\beta + \lambda)w - \tilde{H}(t, x, D_x w, D_x^2 w) = 0, \quad \text{on } [0, T) \times \mathbb{R}^n,$$

where $\tilde{H}$ has the same form as $H$ with $f$ replaced by $\tilde{f}(t, x) = e^{\lambda t} f(t, x)$. Therefore, by taking $\lambda$ so that $\beta + \lambda > 0$, and possibly replacing $(U, V)$ by $(\tilde{U}, \tilde{V})$, we can assume w.l.o.g. that $\beta > 0$.

*Step 2: penalization and perturbation of supersolution.* From the polynomial growth condition on $U$, $V$, we may choose an integer $p$ greater than 1 such that

$$\sup_{[0,T] \times \mathbb{R}^n} \frac{|U(t, x)| + |V(t, x)|}{1 + |x|^p} < \infty,$$

and we consider the function $\phi(t, x) = e^{-\lambda t}(1 + |x|^{2p}) =: e^{-\lambda t} \psi(x)$. From the linear growth condition on $b$, $\sigma$, a direct calculation shows that there exists some positive constant $c$ s.t.

$$-\frac{\partial \phi}{\partial t} + \beta \phi - \sup_{a \in A} \left[ b(x, a).D_x \phi + \frac{1}{2} \text{tr}(\sigma \sigma'(x) D_x^2 \phi) \right]$$

$$= e^{-\lambda t} \left\{ (\beta + \lambda)\psi - \sup_{a \in A} \left[ b(x, a).D_x \psi + \frac{1}{2} \text{tr}(\sigma \sigma'(x) D_x^2 \psi) \right] \right\}$$

$$\ge e^{-\lambda t} (\beta + \lambda - c)\psi \ge 0,$$

by taking $\lambda \ge c - \beta$. This implies that for all $\varepsilon > 0$, the function $V_\varepsilon = V + \varepsilon \phi$ is, as $V$, a supersolution to (4.42). Furthermore, from the growth conditions on $U, V, \phi$, we have for all $\varepsilon > 0$,

$$\lim_{|x| \to \infty} \sup_{[0,T]} (U - V_\varepsilon)(t, x) = -\infty. \tag{4.44}$$

*Step 3.* We finally argue by contradiction to show that $U - V_\varepsilon \leq 0$ on $[0, T] \times \mathbb{R}^n$ for all $\varepsilon$ $> 0$, which gives the required result by sending $\varepsilon$ to zero. On the contrary, by continuity of $U - V_\varepsilon$, and from (4.44), there would exist $\varepsilon > 0$, $(\bar{t}, \bar{x}) \in [0, T] \times \mathbb{R}^n$ such that

$$\sup_{[0,T] \times \mathbb{R}^n} (U - V_\varepsilon) = (U - V_\varepsilon)(\bar{t}, \bar{x}) > 0. \qquad (4.45)$$

Since $(U - V_\varepsilon)(T, .) \leq (U - V)(T, .) \leq 0$ on $\mathbb{R}^n$, we have $\bar{t} < T$. Thus, the first and second-order optimality conditions of (4.45) imply

$$\frac{\partial(U - V_\varepsilon)}{\partial t}(\bar{t}, \bar{x}) \leq 0 \quad (= 0 \text{ if } \bar{t} > 0) \qquad (4.46)$$

$$D_x(U - V_\varepsilon)(\bar{t}, \bar{x}) = 0 \text{ and } D_x^2(U - V_\varepsilon)(\bar{t}, \bar{x}) \leq 0. \qquad (4.47)$$

By writing that $U$ (resp. $V_\varepsilon$) is a subsolution (resp. supersolution) to (4.42), and recalling that $H$ is nondecreasing in its last argument, we then deduce that

$$\beta(U - V_\varepsilon)(\bar{t}, \bar{x})$$
$$\leq \frac{\partial(U - V_\varepsilon)}{\partial t}(\bar{t}, \bar{x}) + H(\bar{t}, \bar{x}, D_x U(\bar{t}, \bar{x}), D_x^2 U(\bar{t}, \bar{x})) - H(\bar{t}, \bar{x}, D_x V_\varepsilon(\bar{t}, \bar{x}), D_x^2 V_\varepsilon(\bar{t}, \bar{x}))$$
$$\leq 0,$$

and this contradicts (4.45).     □

### 4.4.2 Strong comparison principle

In this section, we remove the regularity conditions on $U, V$ in Theorem 4.4.3, by assuming only that $U$ (resp. $V$) is upper-semicontinuous (resp. lower-semicontinuous). The first- and second-order optimality conditions (4.46)-(4.47) cannot be used anymore, and we need other arguments. A key tool is the dedoubling variable technique that we illustrate first in the case of Hamilton-Jacobi equations:

$$-\frac{\partial w}{\partial t} + \beta w - H(t, x, D_x w) = 0, \quad \text{on } [0, T) \times \mathbb{R}^n, \qquad (4.48)$$

with a Hamiltonian $H(t, x, p) = \sup_{a \in A}[b(x, a).p + f(t, x, a)]$ arising typically in deterministic optimal control. We assume that $b$ satisfies a uniform Lipschitz condition in $x$, and $f$ is uniformly continuous in $(t, x)$, uniformly in $a$. The consequence on $H$ is the crucial inequality

$$|H(t, x, p) - H(s, y, p)| \leq \mu(|t - s| + (1 + |p|)|x - y|), \qquad (4.49)$$

for all $(t, s, x, y, p) \in [0, T]^2 \times (\mathbb{R}^n)^2 \times \mathbb{R}^n$, in which $\mu(z)$ converges to zero when $z$ goes to zero.

**Theorem 4.4.4** *Let $U$ (resp. $V$) be a u.s.c. viscosity subsolution (resp. l.s.c. viscosity supersolution) with polynomial growth condition to (4.48). If $U(T, .) \leq V(T, .)$ on $\mathbb{R}^n$, then $U \leq V$ on $[0, T] \times \mathbb{R}^n$.*

**Proof. 1.** By proceeding as in Steps 1 and 2 of Theorem 4.4.3, we may assume w.l.o.g. that $\beta > 0$, and the supremum of the u.s.c. function $U - V$ on $[0, T] \times \mathbb{R}^n$ is attained (up to a penalization) on $[0, T] \times \mathcal{O}$ for some open bounded set $\mathcal{O}$ of $\mathbb{R}^n$. We suppose that $U(T, .) \leq V(T, .)$ on $\mathbb{R}^n$, and let us prove that $U \leq V$ on $[0, T] \times \mathbb{R}^n$. We argue by contradiction, which yields:

$$M := \sup_{[0,T] \times \mathbb{R}^n} (U - V) = \max_{[0,T) \times \mathcal{O}} (U - V) > 0. \tag{4.50}$$

We now use the dedoubling variable technique by considering for any $\varepsilon > 0$, the functions

$$\Phi_\varepsilon(t, s, x, y) = U(t, x) - V(s, y) - \phi_\varepsilon(t, s, x, y),$$
$$\phi_\varepsilon(t, s, x, y) = \frac{1}{\varepsilon} \big[ |t - s|^2 + |x - y|^2 \big].$$

The u.s.c. function $\Phi_\varepsilon$ attains its maximum, denoted by $M_\varepsilon$, on the compact set $[0, T]^2 \times \bar{\mathcal{O}}^2$ at $(t_\varepsilon, s_\varepsilon, x_\varepsilon, y_\varepsilon)$. Let us check that

$$M_\varepsilon \to M, \quad \text{and} \quad \phi_\varepsilon(t_\varepsilon, s_\varepsilon, x_\varepsilon, y_\varepsilon) \to 0, \tag{4.51}$$

as $\varepsilon$ goes to zero. For this, we write that $M \leq M_\varepsilon = \Phi_\varepsilon(t_\varepsilon, s_\varepsilon, x_\varepsilon, y_\varepsilon)$ for all $\varepsilon > 0$, i.e.

$$M \leq M_\varepsilon = U(t_\varepsilon, x_\varepsilon) - V(s_\varepsilon, y_\varepsilon) - \phi_\varepsilon(t_\varepsilon, s_\varepsilon, x_\varepsilon, y_\varepsilon) \tag{4.52}$$
$$\leq U(t_\varepsilon, x_\varepsilon) - V(s_\varepsilon, y_\varepsilon). \tag{4.53}$$

Now, the bounded sequence $(t_\varepsilon, s_\varepsilon, x_\varepsilon, y_\varepsilon)_\varepsilon$ converges, up to a subsequence, to some $(\bar{t}, \bar{s}, \bar{x}, \bar{y}) \in [0, T]^2 \times \bar{\mathcal{O}}^2$. Moreover, since the sequence $(U(t_\varepsilon, x_\varepsilon) - V(s_\varepsilon, y_\varepsilon))_\varepsilon$ is bounded, we see from (4.52) that the sequence $(\phi_\varepsilon(t_\varepsilon, s_\varepsilon, x_\varepsilon, y_\varepsilon))_\varepsilon$ is also bounded, which implies: $\bar{t} = \bar{s}$, $\bar{x} = \bar{y}$. By sending $\varepsilon$ to zero into (4.53), we get $M \leq (U - V)(\bar{t}, \bar{x}) \leq M$, and so $M = (U - V)(\bar{t}, \bar{x})$ with $(\bar{t}, \bar{x}) \in [0, T) \times \mathcal{O}$ by (4.50). By sending again $\varepsilon$ to zero into (4.52)-(4.53), we obtain (4.51).

**2.** Since $(t_\varepsilon, s_\varepsilon, x_\varepsilon, y_\varepsilon)_\varepsilon$ converges to $(\bar{t}, \bar{t}, \bar{x}, \bar{x})$ with $(\bar{t}, \bar{x}) \in [0, T) \times \mathcal{O}$, we may assume that for $\varepsilon$ small enough, $(t_\varepsilon, s_\varepsilon, x_\varepsilon, y_\varepsilon)$ lies in $[0, T)^2 \times \mathcal{O}^2$. Hence, by definition of $(t_\varepsilon, s_\varepsilon, x_\varepsilon, y_\varepsilon)$,

$(t_\varepsilon, x_\varepsilon)$ is a local maximum of $(t, x) \to U(t, x) - \phi_\varepsilon(t, s_\varepsilon, x, y_\varepsilon)$ on $[0, T) \times \mathbb{R}^n$,
$(s_\varepsilon, y_\varepsilon)$ is a local minimum of $(s, y) \to V(s, y) + \phi_\varepsilon(t_\varepsilon, s, x_\varepsilon, y)$ on $[0, T) \times \mathbb{R}^n$.

We can then write the viscosity subsolution property of $U$ applied to the test function $(t, x) \to \phi_\varepsilon(t, s_\varepsilon, x, y_\varepsilon)$ at the point $(t_\varepsilon, x_\varepsilon)$, which gives

$$-\frac{\partial \phi_\varepsilon}{\partial t}(t_\varepsilon, s_\varepsilon, x_\varepsilon, y_\varepsilon) + \beta U(t_\varepsilon, x_\varepsilon) - H(t_\varepsilon, x_\varepsilon, D_x \phi_\varepsilon(t_\varepsilon, s_\varepsilon, x_\varepsilon, y_\varepsilon)) \leq 0,$$

and so

$$-\frac{2}{\varepsilon}(t_\varepsilon - s_\varepsilon) + \beta U(t_\varepsilon, x_\varepsilon) - H\big(t_\varepsilon, x_\varepsilon, \frac{2}{\varepsilon}(x_\varepsilon - y_\varepsilon)\big) \leq 0. \tag{4.54}$$

Similarly, by writing the viscosity supersolution property of $V$ applied to the test function $(s, y) \to -\phi_\varepsilon(t_\varepsilon, s, x_\varepsilon, y)$ at the point $(s_\varepsilon, y_\varepsilon)$, we have

$$-\frac{2}{\varepsilon}(t_\varepsilon - s_\varepsilon) + \beta V(s_\varepsilon, y_\varepsilon) - H\big(s_\varepsilon, y_\varepsilon, \frac{2}{\varepsilon}(x_\varepsilon - y_\varepsilon)\big) \geq 0. \qquad (4.55)$$

By substracting the two inequalities (4.54)-(4.55), and from condition (4.49), we obtain

$$\beta\big[U(t_\varepsilon, x_\varepsilon) - V(s_\varepsilon, y_\varepsilon)\big] \leq H\big(t_\varepsilon, x_\varepsilon, \frac{2}{\varepsilon}(x_\varepsilon - y_\varepsilon)\big) - H\big(s_\varepsilon, y_\varepsilon, \frac{2}{\varepsilon}(x_\varepsilon - y_\varepsilon)\big)$$

$$\leq \mu\big(|t_\varepsilon - s_\varepsilon| + |x_\varepsilon - y_\varepsilon| + \frac{2}{\varepsilon}|x_\varepsilon - y_\varepsilon|^2\big).$$

By sending $\varepsilon$ to zero into this last inequality, and using (4.51), we conclude that $\beta M \leq 0$, a contradiction with (4.50). $\qquad\square$

We turn back to the case of second-order HJB equations (4.42). Let us try to adapt the arguments as in the first-order case of HJ equations (4.48). By keeping the same notations as in the proof of Theorem 4.4.4, the viscosity subsolution (resp. supersolution) property of $U$ (resp. $V$) yields

$$-\frac{2}{\varepsilon}(t_\varepsilon - s_\varepsilon) + \beta U(t_\varepsilon, x_\varepsilon) - H\big(t_\varepsilon, x_\varepsilon, \frac{2}{\varepsilon}(x_\varepsilon - y_\varepsilon), \frac{2}{\varepsilon}\big) \leq 0$$

$$-\frac{2}{\varepsilon}(t_\varepsilon - s_\varepsilon) + \beta V(s_\varepsilon, y_\varepsilon) - H\big(s_\varepsilon, x_\varepsilon, \frac{2}{\varepsilon}(x_\varepsilon - y_\varepsilon), -\frac{2}{\varepsilon}\big) \geq 0.$$

However, in contrast with the first-order PDE case, by substracting the two above inequalities, we cannot get rid of the second-order term $2/\varepsilon$ appearing in the Hamiltonian. We need some additional tools in the proof.

First, we give an equivalent definition of viscosity solutions in terms of superjets and subjets. Notice that if $U$ is u.s.c., $\varphi \in C^{1,2}([0,T) \times \mathbb{R}^n)$, and $(\bar{t}, \bar{x}) \in [0,T) \times \mathbb{R}^n$ is a maximum point of $U - \varphi$, then a second-order Taylor expansion of $\varphi$ yields

$$U(t,x) \leq U(\bar{t}, \bar{x}) + \varphi(t,x) - \varphi(\bar{t}, \bar{x})$$

$$= U(\bar{t}, \bar{x}) + \frac{\partial\varphi}{\partial t}(\bar{t}, \bar{x})(t - \bar{t}) + D_x\varphi(\bar{t}, \bar{x}).(x - \bar{x})$$

$$+ \frac{1}{2}D_x^2\varphi(\bar{t}, \bar{x})(x - \bar{x}).(x - \bar{x}) + o(|t - \bar{t}| + |x - \bar{x}|^2). \qquad (4.56)$$

This naturally leads to the notion of a second-order superjet of a u.s.c. function $U$ at a point $(\bar{t}, \bar{x}) \in [0,T) \times \mathbb{R}^n$, defined as the set $\mathcal{P}^{2,+}U(\bar{t}, \bar{x})$ of elements $(\bar{q}, \bar{p}, \bar{M}) \in \mathbb{R} \times \mathbb{R}^n \times \mathcal{S}_n$ satisfying

$$U(t,x) \leq U(\bar{t}, \bar{x}) + \bar{q}(t - \bar{t}) + \bar{p}.(x - \bar{x}) + \frac{1}{2}\bar{M}(x - \bar{x}).(x - \bar{x}) + o(|t - \bar{t}| + |x - \bar{x}|^2).$$

Similarly, we define the second-order subjet $\mathcal{P}^{2,-}V(\bar{t}, \bar{x})$ of a l.s.c. function $V$ at a point $(\bar{t}, \bar{x}) \in [0,T) \times \mathbb{R}^n$, as the set of elements $(\bar{q}, \bar{p}, \bar{M}) \in \mathbb{R} \times \mathbb{R}^n \times \mathcal{S}_n$ satisfying

$$V(t,x) \geq V(\bar{t}, \bar{x}) + \bar{q}(t - \bar{t}) + \bar{p}.(x - \bar{x}) + \frac{1}{2}\bar{M}(x - \bar{x}).(x - \bar{x}) + o(|t - \bar{t}| + |x - \bar{x}|^2).$$

The inequality (4.56) shows that for a given point $(t,x) \in [0,T) \times \mathbb{R}^n$, if $\varphi \in C^{1,2}([0,T) \times \mathbb{R}^n)$ is such that $(t,x)$ is a maximum of $U - \varphi$, then

$$(q,p,M) = \Big(\frac{\partial\varphi}{\partial t}(t,x), D_x\varphi(t,x), D_x^2\varphi(t,x)\Big) \in \mathcal{P}^{2,+}U(t,x). \qquad (4.57)$$

Actually, the converse property holds true: for any $(q, p, M) \in \mathcal{P}^{2,+} U(t, x)$, there exists $\varphi \in C^{1,2}([0, T) \times \mathbb{R}^n)$ satisfying (4.57). We refer to Lemma 4.1 in [FSo93] for an example of such a construction of $\varphi$. Similarly, we have the following characterization of subjets: given $(t, x) \in [0, T) \times \mathbb{R}^n$, an element $(q, p, M) \in \mathcal{P}^{2,-} V(t, x)$ if and only if there exists $\varphi \in C^{1,2}([0, T) \times \mathbb{R}^n)$ satisfying

$$(q, p, M) = \left( \frac{\partial \varphi}{\partial t}(t, x), D_x \varphi(t, x), D_x^2 \varphi(t, x) \right) \in \mathcal{P}^{2,-} V(t, x),$$

such that $(t, x) \in [0, T) \times \mathbb{R}^n$ is a minimum of $V - \varphi$. For technical reasons related to Ishii's lemma, see below, we also need to consider the limiting superjets and subjets. More precisely, we define $\bar{\mathcal{P}}^{2,+} U(t, x)$ as the set of elements $(q, p, M) \in \mathbb{R} \times \mathbb{R}^n \times \mathcal{S}_n$ for which there exists a sequence $(t_\varepsilon, x_\varepsilon, q_\varepsilon, p_\varepsilon, M_\varepsilon)_\varepsilon$ in $[0, T) \times \mathbb{R}^n \times \mathcal{P}^{2,+} U(t_\varepsilon, x_\varepsilon)$ satisfying $(t_\varepsilon, x_\varepsilon, U(t_\varepsilon, x_\varepsilon), q_\varepsilon, p_\varepsilon, M_\varepsilon) \to (t, x, U(t, x), q, p, M)$. The set $\bar{\mathcal{P}}^{2,-} V(t, x)$ is defined similarly.

We can now state the alternative definition of viscosity solutions for parabolic second-order PDEs:

$$F(t, x, w, \frac{\partial w}{\partial t}, D_x w, D_x^2 w) = 0, \quad \text{on } [0, T) \times \mathbb{R}^n, \tag{4.58}$$

where $F(t, x, r, q, p, M)$ is continuous, and satisfies the parabolicity and ellipticity conditions (4.2)-(4.3).

**Lemma 4.4.5** *A u.s.c. (resp. l.s.c.) function $w$ on $[0, T) \times \mathbb{R}^n$ is a viscosity subsolution (resp. supersolution) of (4.58) on $[0, T) \times \mathbb{R}^n$ if and only if for all $(t, x) \in [0, T) \times \mathbb{R}^n$, and all $(q, p, M) \in \bar{\mathcal{P}}^{2,+} w(t, x)$ (resp. $\bar{\mathcal{P}}^{2,-} w(t, x)$),*

$$F(t, x, w(t, x), q, p, M) \leq (resp. \geq) \ 0.$$

The key tool in the comparison proof for second-order equations in the theory of viscosity solutions is an analysis lemma due to Ishii. We state this lemma without proof, and refer the reader to Theorem 8.3 in the user's guide [CIL92].

**Lemma 4.4.6** *(Ishii's lemma)*
*Let $U$ (resp. $V$) be a u.s.c. (resp. l.s.c.) function on $[0, T) \times \mathbb{R}^n$, $\phi \in C^{1,1,2,2}([0, T)^2 \times \mathbb{R}^n \times \mathbb{R}^n)$, and $(\bar{t}, \bar{s}, \bar{x}, \bar{y}) \in [0, T)^2 \times \mathbb{R}^n \times \mathbb{R}^n$ a local maximum of $U(t, x) - V(t, y) - \phi(t, s, x, y)$. Then, for all $\eta > 0$, there exist $M, N \in \mathcal{S}_n$ satisfying*

$$\left( \frac{\partial \phi}{\partial t}(\bar{t}, \bar{s}, \bar{x}, \bar{y}), D_x \phi(\bar{t}, \bar{s}, \bar{x}, \bar{y}), M \right) \in \bar{\mathcal{P}}^{2,+} U(\bar{t}, \bar{x}),$$

$$\left( -\frac{\partial \phi}{\partial s}(\bar{t}, \bar{s}, \bar{x}, \bar{y}), -D_y \phi(\bar{t}, \bar{s}, \bar{x}, \bar{y}), N \right) \in \bar{\mathcal{P}}^{2,-} V(\bar{s}, \bar{y}),$$

*and*

$$\begin{pmatrix} M & 0 \\ 0 & -N \end{pmatrix} \leq D_{x,y}^2 \phi(\bar{t}, \bar{s}, \bar{x}, \bar{y}) + \eta \big( D_{x,y}^2 \phi(\bar{t}, \bar{s}, \bar{x}, \bar{y}) \big)^2. \tag{4.59}$$

**Remark 4.4.9** We shall use Ishii's lemma with $\phi(t, s, x, y) = \frac{1}{2\varepsilon}[|t-s|^2 + |x-y|^2]$. Then,
$(\frac{\partial \phi}{\partial t}(\bar{t}, \bar{s}, \bar{x}, \bar{y}) = -\frac{\partial \phi}{\partial s}(\bar{t}, \bar{s}, \bar{x}, \bar{y}) = (\bar{t} - \bar{s})/\varepsilon, \ D_x\phi(\bar{t}, \bar{s}, \bar{x}, \bar{y}) = -D_y\phi(\bar{t}, \bar{s}, \bar{x}, \bar{y}) = (\bar{x} - \bar{y})/\varepsilon,$

$$D^2_{x,y}\phi(\bar{t}, \bar{s}, \bar{x}, \bar{y}) = \frac{1}{\varepsilon}\begin{pmatrix} I_n & -I_n \\ -I_n & I_n \end{pmatrix},$$

and $\left(D^2_{x,y}\phi(\bar{t}, \bar{s}, \bar{x}, \bar{y})\right)^2 = \frac{2}{\varepsilon^2}D^2_{x,y}\phi(\bar{t}, \bar{s}, \bar{x}, \bar{y})$. Furthermore, by choosing $\eta = \varepsilon$ in (4.59), we obtain

$$\begin{pmatrix} M & 0 \\ 0 & -N \end{pmatrix} \le \frac{3}{\varepsilon}\begin{pmatrix} I_n & -I_n \\ -I_n & I_n \end{pmatrix}. \tag{4.60}$$

This implies that for any $n \times d$ matrices $C, D$,

$$\text{tr}(CC'M - DD'N) \le \frac{3}{\varepsilon}|C - D|^2. \tag{4.61}$$

Indeed, by noting that the matrix $\Sigma = \begin{pmatrix} CC' & CD' \\ DC' & DD' \end{pmatrix}$ lies in $\mathcal{S}_{2n}$, we get from the inequality (4.60)

$$\text{tr}(CC'M - DD'N) = \text{tr}\left(\Sigma\begin{pmatrix} M & 0 \\ 0 & -N \end{pmatrix}\right)$$
$$\le \frac{3}{\varepsilon}\text{tr}\left(\Sigma\begin{pmatrix} I_n & -I_n \\ -I_n & I_n \end{pmatrix}\right) = \frac{3}{\varepsilon}\text{tr}((C - D)(C - D)').$$

We can finally prove a comparison result for the HJB equation (4.42) with a Hamiltonian given by (4.43). In this Hamiltonian $H$, we assume that the coefficients $b, \sigma$ satisfy a uniform Lipschitz condition in $x$, and $f$ is uniformly continuous in $(t, x)$, uniformly in $a \in A$.

**Theorem 4.4.5** *Let $U$ (resp. $V$) be a u.s.c. viscosity subsolution (resp. l.s.c. viscosity supersolution) with polynomial growth condition to (4.42), such that $U(T, .) \le V(T, .)$ on $\mathbb{R}^n$. Then $U \le V$ on $[0, T] \times \mathbb{R}^n$.*

**Proof.** We proceed similarly as in part **1** of Theorem 4.4.4. We assume w.l.o.g. that $\beta > 0$ in (4.42), and we argue by contradiction by assuming that $M := \sup_{[0,T] \times \mathbb{R}^n}(U - V) > 0$. We consider a bounded sequence $(t_\varepsilon, s_\varepsilon, x_\varepsilon, y_\varepsilon)_\varepsilon$ that maximizes for all $\varepsilon > 0$, the function $\Phi_\varepsilon$ on $[0, T]^2 \times \mathbb{R}^n \times \mathbb{R}^n$ with

$$\Phi_\varepsilon(t, s, x, y) = U(t, x) - V(s, y) - \phi_\varepsilon(t, s, x, y), \quad \phi_\varepsilon(t, s, x, y) = \frac{1}{2\varepsilon}[|t - s|^2 + |x - y|^2].$$

As in (4.51), we have

$$M_\varepsilon = \Phi_\varepsilon(t_\varepsilon, s_\varepsilon, x_\varepsilon, y_\varepsilon) \rightarrow M, \quad \text{and} \quad \phi_\varepsilon(t_\varepsilon, s_\varepsilon, x_\varepsilon, y_\varepsilon) \rightarrow 0, \tag{4.62}$$

as $\varepsilon$ goes to zero. In view of Ishii's lemma 4.4.6 and Remark 4.4.9, there exist $M, N \in \mathcal{S}_n$ satisfying (4.61) and

$$\left(\frac{1}{\varepsilon}(t_\varepsilon - s_\varepsilon), \frac{1}{\varepsilon}(x_\varepsilon - y_\varepsilon), M\right) \in \bar{\mathcal{P}}^{2,+}U(t_\varepsilon, x_\varepsilon),$$

$$\left(\frac{1}{\varepsilon}(t_\varepsilon - s_\varepsilon), \frac{1}{\varepsilon}(x_\varepsilon - y_\varepsilon), N\right) \in \bar{\mathcal{P}}^{2,-}V(s_\varepsilon, y_\varepsilon).$$

From the viscosity subsolution and supersolution characterization of $U$ and $V$ in terms of superjets and subjets, we then have

$$-\frac{1}{\varepsilon}(t_\varepsilon - s_\varepsilon) + \beta U(t_\varepsilon, x_\varepsilon) - H\left(t_\varepsilon, x_\varepsilon, \frac{1}{\varepsilon}(x_\varepsilon - y_\varepsilon), M\right) \le 0,$$

$$-\frac{1}{\varepsilon}(t_\varepsilon - s_\varepsilon) + \beta V(s_\varepsilon, y_\varepsilon) - H\left(s_\varepsilon, y_\varepsilon, \frac{1}{\varepsilon}(x_\varepsilon - y_\varepsilon), N\right) \ge 0.$$

By substracting the above two inequalities, applying inequality (4.61) to $C = \sigma(x_\varepsilon, a)$, $D = \sigma(y_\varepsilon, a)$, and recalling the uniform Lipschitz conditions on $b$, $\sigma$ together with the uniform continuity of $f$, we get

$$\beta\left[U(t_\varepsilon, x_\varepsilon) - V(s_\varepsilon, y_\varepsilon)\right] \le H\left(t_\varepsilon, x_\varepsilon, \frac{1}{\varepsilon}(x_\varepsilon - y_\varepsilon), M\right) - H\left(s_\varepsilon, y_\varepsilon, \frac{1}{\varepsilon}(x_\varepsilon - y_\varepsilon), N\right)$$

$$\le \mu\left(|t_\varepsilon - s_\varepsilon| + |x_\varepsilon - y_\varepsilon| + \frac{2}{\varepsilon}|x_\varepsilon - y_\varepsilon|^2\right),$$

where $\mu(z)$ converges to zero as $z$ goes to zero. By sending $\varepsilon$ to zero in this last inequality, and using (4.62), we conclude that $\beta M \le 0$, a contradiction.    □

In this section, we stated comparison principles for regular HJB equations of type (4.42). For HJB variational inequalities in the form (4.17), one can also prove comparison results under suitable conditions on the function $G$. There are no general results, and the proof should be adapted depending on the form of the function $G$. We shall give in Theorem 5.3.3 an example of such comparison results for variational inequalities.

## 4.5 An irreversible investment model

### 4.5.1 Problem

We consider a firm producing some output (electricity, oil, etc.). The firm can increase its production capacity $X_t$ by transferring capital from another production activity. The controlled dynamics of the production capacity evolves according to

$$dX_t = X_t(-\delta dt + \sigma dW_t) + l_t dt.$$

$\delta \ge 0$ is the depreciation rate, $\sigma > 0$ is the volatility related to random fluctuations of the capacity, $l_t dt$ is the number of capital units purchased by the company at a cost $\lambda l_t dt$, $\lambda > 0$ is interpreted as a factor of conversion from one production activity to another. The control $l_t$ is nonnegative, unbounded valued in $\mathbb{R}_+$, and it is suitable to set $L_t = \int_0^t l_s ds$: more generally, we consider as control any càd-làg adapted process $(L_t)_{t \ge 0}$, nondecreasing with $L_{0^-} = 0$, and we write $L \in \mathcal{A}$: $L_t$ represents the cumulated amount of capital up to time $t$. Given $x \ge 0$ and $L \in \mathcal{A}$, we denote by $X^x$ the solution to

$$dX_t = X_t(-\delta dt + \sigma dW_t) + dL_t, \quad X_{0^-} = x. \tag{4.63}$$

The profit function of the firm is a continuous function $\Pi$ from $\mathbb{R}_+$ into $\mathbb{R}$, concave, nondecreasing, $C^1$ on $(0,\infty)$, with $\Pi(0) = 0$, and satisfying the usual Inada conditions in 0:

$$\Pi'(0^+) := \lim_{x\downarrow 0} \Pi'(x) = \infty \quad \text{and} \quad \Pi'(\infty) := \lim_{x\to\infty} \Pi'(x) = 0. \tag{4.64}$$

We introduce the Fenchel-Legendre transform of $\Pi$, which is finite on $\mathbb{R}_+$:

$$\tilde{\Pi}(y) := \sup_{x\geq 0}[\Pi(x) - xy] < \infty, \quad \forall y > 0. \tag{4.65}$$

The objective of the firm is defined by

$$v(x) = \sup_{L\in\mathcal{A}} E\left[\int_0^\infty e^{-\beta t}(\Pi(X_t^x)dt - \lambda dL_t)\right], \quad x \geq 0. \tag{4.66}$$

The Hamiltonian of this singular stochastic control problem is

$$H(x,p,M) = \begin{cases} -\delta xp + \frac{1}{2}\sigma^2 x^2 M + \Pi(x) & \text{if } \lambda - p \geq 0 \\ \infty & \text{if } \lambda - p < 0. \end{cases}$$

The associated HJB variational inequality is then written as

$$\min[\beta v - \mathcal{L}v - \Pi , \lambda - v'] = 0, \tag{4.67}$$

with

$$\mathcal{L}v = -\delta xv' + \frac{1}{2}\sigma^2 x^2 v''.$$

### 4.5.2 Regularity and construction of the value function

We state some useful properties on the value function.

**Lemma 4.5.7** *(a) $v$ is finite on $\mathbb{R}_+$, and satisfies for all $\mu \in [0,\lambda]$:*

$$0 \leq v(x) \leq \frac{\tilde{\Pi}((\beta+\delta)\mu)}{\beta} + \mu x, \quad x \geq 0. \tag{4.68}$$

*(b) $v$ is concave and continuous on $(0,\infty)$.*

**Proof.** (a) By considering the null control $L = 0$, we clearly see that $v \geq 0$. Let us consider for $\mu \in [0,\lambda]$ the positive function

$$\varphi(x) = \mu x + \frac{\tilde{\Pi}((\beta+\delta)\mu)}{\beta}.$$

Then $\varphi' \leq \lambda$, and we have for all $x \geq 0$,

$$\beta\varphi - \mathcal{L}\varphi - \Pi(x) = \tilde{\Pi}((\beta+\delta)\mu) + (\beta+\delta)\mu x - \Pi(x) \geq 0,$$

by definition of $\tilde{\Pi}$ in (4.65). Given $L \in \mathcal{A}$, we apply Itô's formula to $e^{-\beta t}\varphi(X_t^x)$ between 0 and $\tau_n$ where $\tau_n$ is the bounded stopping time: $\tau_n = \inf\{t \geq 0 : X_t^x \geq n\} \wedge n$, $n \in \mathbb{N}$.

By taking the expectation, and since the stopped stochastic integral is a martingale, we have

$$E\left[e^{-\beta\tau_n}\varphi(X_{\tau_n}^x)\right] = \varphi(x) + E\left[\int_0^{\tau_n} e^{-\beta t}\left(-\beta\varphi + \mathcal{L}\varphi\right)(X_t^x)dt\right] \tag{4.69}$$

$$+ E\left[\int_0^{\tau_n} e^{-\beta t}\varphi'(X_t^x)dL_t^c\right] + E\left[\sum_{0\leq t\leq\tau_n} e^{-\beta t}\left[\varphi(X_t^x) - \varphi(X_{t-}^x)\right]\right],$$

where $L^c$ is the continuous part of $L$. Since $\varphi' \leq \lambda$ and $X_t^x - X_{t-}^x = L_t - L_{t-}$, the mean-value theorem implies

$$\varphi(X_t^x) - \varphi(X_{t-}^x) \leq \lambda(L_t - L_{t-}).$$

By using again the inequality $\varphi' \leq \lambda$ in the integrals in $dL^c$ in (4.69), and recalling that $-\beta\varphi + \mathcal{L}\varphi \leq -\Pi$, we obtain:

$$E\left[e^{-\beta\tau_n}\varphi(X_{\tau_n}^x)\right] \leq \varphi(x) - E\left[\int_0^{\tau_n} e^{-\beta t}\Pi(X_t^x)dt\right]$$

$$+ E\left[\int_0^{\tau_n} e^{-\beta t}\lambda dL_t^c\right] + E\left[\sum_{0\leq t\leq\tau_n} e^{-\beta t}\lambda(L_t - L_{t-})\right]$$

$$= \varphi(x) - E\left[\int_0^{\tau_n} e^{-\beta t}\Pi(X_t^x)dt\right] + E\left[\int_0^{\tau_n} e^{-\beta t}\lambda dL_t\right],$$

and so

$$E\left[\int_0^{\tau_n} e^{-\beta t}\left(\Pi(X_t^x)dt - \lambda dL_t\right)\right] + E\left[e^{-\beta\tau_n}\varphi(X_{\tau_n}^x)\right] \leq \varphi(x).$$

Since $\varphi \geq 0$, we thus derive:

$$\varphi(x) \geq E\left[\int_0^{\tau_n} e^{-\beta t}\Pi(X_t^x)dt\right] - E\left[\int_0^{\infty} e^{-\beta t}\lambda dL_t\right].$$

We can apply Fatou's lemma, and send $n$ to infinity to get:

$$E\left[\int_0^{\infty} e^{-\beta t}\left(\Pi(X_t^x)dt - \lambda dL_t\right)\right] \leq \varphi(x),$$

and finally $v(x) \leq \varphi(x)$ since $L$ was arbitrary.

(b) The proof of the concavity of $v$ is standard: it is derived (as in Section 3.6.1) by considering convex combinations of initial states and controls, and by relying on the linearity of the dynamics of $X$, and the concavity of $\Pi$. The continuity of $v$ on $\mathbb{R}_+$ is then a consequence of the general continuity property of a concave function on the interior of its domain.    □

Since $v$ is concave on $(0, \infty)$, it admits a right-derivative $v'_+(x)$ and a left-derivative $v'_-(x)$ at any point $x > 0$, with $v'_+(x) \leq v'_-(x)$.

**Lemma 4.5.8**

$$v'_-(x) \leq \lambda, \quad x > 0. \tag{4.70}$$

**Proof.** For any $x > 0$, $0 < l < x$, $L \in \mathcal{A}$, let us consider the control $\tilde{L}$ defined by $\tilde{L}_{0-} = 0$, and $\tilde{L}_t = L_t + l$, for $t \geq 0$. Denote by $\tilde{X}$ the controlled diffusion with $\tilde{L}$ and initial condition $\tilde{X}_{0-} = x - l$. Then, $\tilde{X}_t^{x-l} = X_t^x$ for all $t \geq 0$, and we have

$$v(x - l) \geq E\left[ \int_0^\infty e^{-\beta t} \left( \Pi(\tilde{X}_t^{x-l})dt - \lambda d\tilde{L}_t \right) \right]$$

$$= E\left[ \int_0^\infty e^{-\beta t} \left( \Pi(X_t^x)dt - \lambda dL_t \right) \right] - \lambda l.$$

Since $L$ is arbitrary, this yields

$$v(x - l) \geq -\lambda l + v(x), \quad x > 0.$$

We obtain the required result by dividing by $l$, and sending $l$ to zero.    □

**Lemma 4.5.9** *There exists $x_b \in [0, \infty]$ such that*

$$\mathcal{NT} := \{x > 0 : v_-'(x) < \lambda\} = (x_b, \infty), \tag{4.71}$$

*Furthermore, $v$ is differentiable on $(0, x_b)$, and*

$$v'(x) = \lambda, \quad \text{on } \mathcal{B} = (0, x_b). \tag{4.72}$$

**Proof.** We set $x_b = \inf\{x \geq 0 : v_+'(x) < \lambda\}$. Then, $\lambda \leq v_+'(x) \leq v_-'(x)$, for all $x < x_b$. By combining with (4.70), this proves (4.72). Finally, the concavity of $v$ shows the relation (4.71).    □

We know from the general results of Section 4.3 (see Theorem 4.3.1) that the value function $v$ is a viscosity solution to the HJB variational inequality (4.67). By exploiting the concavity of $v$ for this unidimensional problem, we can show that the value function is actually a smooth $C^2$ solution.

**Theorem 4.5.6** *The value function $v$ is a classical solution $C^2$ on $(0, \infty)$ to*

$$\min\left\{\beta v - \mathcal{L}v - \Pi(x), \ -v'(x) + \lambda\right\} = 0, \quad x > 0.$$

**Proof.** *Step 1.* We first prove that $v$ is $C^1$ on $(0, \infty)$. Since $v$ is concave, its left and right derivatives $v_-'(x)$ and $v_+'(x)$ exist for all $x > 0$ with $v_+'(x) \leq v_-'(x)$. We argue by contradiction by assuming that $v_+'(x_0) < v_-'(x_0)$ for some $x_0 > 0$. Let us then fix $\mu \in (v_+'(x_0), v_-'(x_0))$, and consider the smooth test function

$$\varphi_\varepsilon(x) = v(x_0) + \mu(x - x_0) - \frac{1}{2\varepsilon}(x - x_0)^2,$$

with $\varepsilon > 0$. Then $x_0$ is a local maximum of $(v - \varphi_\varepsilon)$ with $\varphi_\varepsilon(x_0) = v(x_0)$. Since $\varphi_\varepsilon'(x_0) = \mu < \lambda$ by (4.70), and $\varphi_\varepsilon''(x_0) = -1/\varepsilon$, the subsolution property of $v$ implies that

$$\beta v(x_0) + \delta x_0 \mu + \frac{1}{2\varepsilon}\sigma^2 x_0^2 - \Pi(x_0) \leq 0. \tag{4.73}$$

By choosing $\varepsilon$ sufficiently small, we get the required contradiction, which shows that $v_+'(x_0) = v_-'(x_0)$.

*Step 2.* From Lemma 4.5.9, $v$ is $C^2$ on $(0, x_b)$, and satisfies $v'(x) = \lambda$, $x \in (0, x_b)$. By Step 1, we have: $\mathcal{NT} = (x_b, \infty) = \{x > 0 : v'(x) < \lambda\}$. Let us check that $v$ is a viscosity solution to

$$\beta v - \mathcal{L}v - \Pi = 0, \quad \text{on } (x_b, \infty). \tag{4.74}$$

Let $x_0 \in (x_b, \infty)$ and $\varphi$ a function $C^2$ on $(x_b, \infty)$ such that $x_0$ is a local maximum of $v - \varphi$, with $(v - \varphi)(x_0) = 0$. Since $\varphi'(x_0) = v'(x_0) < \lambda$, the viscosity subsolution property of $v$ to (4.67) yields

$$\beta \varphi(x_0) - \mathcal{L}\varphi(x_0) - \Pi(x_0) \leq 0.$$

This shows that $v$ is a viscosity subsolution to (4.74) on $(x_b, \infty)$. The proof of the viscosity supersolution property to (4.74) is similar. Let us consider for arbitrary $x_1 \leq x_2$ in $(x_b, \infty)$, the Dirichlet problem

$$\beta V - \mathcal{L}V - \Pi(x) = 0, \quad \text{on } (x_1, x_2) \tag{4.75}$$

$$V(x_1) = v(x_1), \quad V(x_2) = v(x_2). \tag{4.76}$$

Classical results provide the existence and uniqueness of a smooth $C^2$ solution $V$ to (4.75)-(4.76). In particular, this smooth function $V$ is a viscosity solution to (4.74) on $(x_1, x_2)$. From comparison results for viscosity solutions (see the previous section) for linear PDEs in bounded domain, we deduce that $v = V$ on $(x_1, x_2)$. Since $x_1$ and $x_2$ are arbitrary in $(x_b, \infty)$, this proves that $v$ is $C^2$ on $(x_b, \infty)$, and satisfies the PDE (4.74) in a classical sense.

*Step 3.* It remains to prove the $C^2$ condition on $x_b$ in the case where $0 < x_b < \infty$. Let $x \in (0, x_b)$. Since $v$ is $C^2$ on $(0, x_b)$, the viscosity supersolution property of $v$ applied to the point $x$ and the test function $\varphi = v$ implies that $v$ satisfies in a classical sense

$$\beta v(x) - \mathcal{L}v(x) - \Pi(x) \geq 0, \quad 0 < x < x_b.$$

Since the derivative of $v$ is constant, equal to $\lambda$ on $(0, x_b)$, we have

$$\beta v(x) + \delta x \lambda - \Pi(x) \geq 0, \quad 0 < x < x_b,$$

and so

$$\beta v(x_b) + \delta x_b \lambda - \Pi(x_b) \geq 0. \tag{4.77}$$

On the other hand, from the $C^1$ condition of $v$ at $x_b$, we obtain by sending $x$ to $x_b$ in (4.74)

$$\beta v(x_b) + \delta x_b \lambda - \Pi(x_b) = \frac{1}{2}\sigma^2 x_b^2 v''(x_b^+). \tag{4.78}$$

From the concavity of $v$, the right-hand term of (4.78) is nonpositive, which, together with (4.77), implies that $v''(x_b^+) = 0$. This proves that $v$ is $C^2$ at $x_b$ with $v''(x_b) = 0$. $\square$

We can now give an explicit form for the value function. Let us consider the ordinary differential equation (ODE) arising in the HJB equation:

$$\beta v - \mathcal{L}v - \Pi = 0. \tag{4.79}$$

We recall that the general solution to (4.79) (with $\Pi = 0$) is given by

$$\hat{V}(x) = Ax^m + Bx^n,$$

where

$$m = \frac{\delta}{\sigma^2} + \frac{1}{2} - \sqrt{\left(\frac{\delta}{\sigma^2} + \frac{1}{2}\right)^2 + \frac{2\beta}{\sigma^2}} < 0$$

$$n = \frac{\delta}{\sigma^2} + \frac{1}{2} + \sqrt{\left(\frac{\delta}{\sigma^2} + \frac{1}{2}\right)^2 + \frac{2\beta}{\sigma^2}} > 1.$$

Moreover, the ODE (4.79) admits a particular solution given by

$$\hat{V}_0(x) = E\left[\int_0^\infty e^{-\beta t} \Pi(\hat{X}_t^x)\right]$$

$$= \frac{2}{\sigma^2(n-m)}\left[x^n \int_x^\infty s^{-n-1}\Pi(s)ds + x^m \int_0^x s^{-m-1}\Pi(s)ds\right], \quad x > 0,$$

where $\hat{X}^x$ is the solution to (4.63) with $L = 0$. We easily check (exercise left to the reader) that under the Inada condition (4.64):

$$\hat{V}_0'(0^+) = \infty \quad \text{and} \quad \hat{V}_0'(\infty) = 0.$$

**Lemma 4.5.10** *The boundary $x_b$ lies in $(0, \infty)$.*

**Proof.** We first check that $x_b > 0$. On the contrary, the region $\mathcal{B}$ is empty, and we would get from Lemma 4.5.9 and Theorem 4.5.6

$$\beta v - \mathcal{L}v - \Pi = 0, \quad x > 0.$$

Thus, $v$ should take the form

$$v(x) = Ax^m + Bx^n + \hat{V}_0(x), \quad x > 0.$$

Since $m < 0$ and $|v(0^+)| < \infty$, this implies $A = 0$. Moreover, since $n > 1$, we have $v'(0^+)$ $= \hat{V}_0'(0^+) = \infty$, which is in contradiction with the fact that $v'(x) \leq \lambda$ for all $x > 0$.

We immediately show that $x_b < \infty$. Otherwise, $v$ satisfies $v' = \lambda$ on $(0, \infty)$, which is in contradiction with the growth condition (4.68). □

**Theorem 4.5.7** *The value function has the explicit form*

$$v(x) = \begin{cases} \lambda x + v(0^+) & x \leq x_b \\ Ax^m + \hat{V}_0(x) & x_b < x, \end{cases} \tag{4.80}$$

*where the three parameters $v(0^+)$, $A$ and $x_b$ are determined by the continuity, $C^1$ and $C^2$ conditions of $v$ at $x_b$:*

$$Ax_b^m + \hat{V}_0(x_b) = \lambda x_b + v(0^+), \tag{4.81}$$

$$mAx_b^{m-1} + \hat{V}_0'(x_b) = \lambda, \tag{4.82}$$

$$m(m-1)Ax_b^{m-2} + \hat{V}_0''(x_b) = 0. \tag{4.83}$$

**Proof.** We already know from Lemma 4.5.9 that on $(0, x_b)$, which is nonempty by Lemma 4.5.10, $v$ has the structure described in (4.80). Moreover, on $(x_b, \infty)$, we have $v' < \lambda$ by Lemma 4.5.9. From Theorem 4.5.6, we then deduce that $v$ satisfies $\beta v - \mathcal{L}v - \Pi = 0$, and so is in the form:

$$v(x) = Ax^m + Bx^n + \hat{V}_0(x), \quad x > x_b.$$

Since $m < 0$, $n > 1$, $\hat{V}_0'(x)$ converges to 0 when $x$ goes to infinity, and $0 \le v'(x) \le \lambda$, we must have $B = 0$. Thus, $v$ has the form described in (4.80). Finally, the three conditions (4.81), (4.82) and (4.83), which follow from the $C^2$ condition of $v$ at $x_b$, determine the three constants $A$, $x_b$ and $v(0^+)$. □

### 4.5.3 Optimal strategy

We recall the Skorohod lemma proved for example in P.L. Lions and Snitzman [LS84].

**Lemma 4.5.11** *Given any initial state $x \ge 0$ and any boundary $x_b \ge 0$, there exists a unique adapted process $X^*$, and a nondecreasing process $L^*$, right-continuous, satisfying the Skorohod problem $\mathcal{S}(x, x_b)$:*

$$dX_t^* = X_t^* \left( -\delta dt + \sigma dW_t \right) + dL_t^*, \ t \ge 0, \qquad X_{0-}^* = x, \qquad (4.84)$$

$$X_t^* \in [x_b, \infty) \quad a.s., \ t \ge 0, \qquad (4.85)$$

$$\int_0^\infty 1_{X_u^* > x_b} dL_u^* = 0. \qquad (4.86)$$

*Moreover, if $x \ge x_b$, then $L^*$ is continuous. When $x < x_b$, $L_0^* = x_b - x$, and $X_0^* = x_b$.*

The solution $X^*$ to the above equations is a reflected diffusion on the boundary $x_b$, and the process $L^*$ is the local time of $X^*$ at $x_b$. The condition (4.86) means that $L^*$ increases only when $X^*$ reaches the boundary $x_b$. The $\beta$-potential of $L^*$ is finite, i.e. $E[\int_0^\infty e^{-\beta t} dL_t^*]$, see Chapter X in Revuz and Yor [ReY91], which implies that

$$E\left[ \int_0^\infty e^{-\beta t} X_t^* dt \right] < \infty. \qquad (4.87)$$

**Theorem 4.5.8** *Given $x \ge 0$, let $(X^*, L^*)$ be the solution to the Skorohod problem $\mathcal{S}(x, x_b)$. We then have*

$$v(x) = E\left[ \int_0^\infty e^{-\beta t} \left( \Pi(X_t^*) dt - \lambda dL_t^* \right) \right].$$

**Proof.** (1) We first consider the case where $x \ge x_b$. Then the processes $X^*$ and $L^*$ are continuous. From (4.85) and Theorem 4.5.6, we have

$$\beta v(X_t^*) - \mathcal{L}v(X_t^*) - \Pi(X_t^*) = 0, \ a.s. \ t \ge 0.$$

By applying Itô's formula to $e^{-\beta t} v(X_t^*)$ between 0 and $T$, we then obtain

$$E\left[ e^{-\beta T} v(X_T^*) \right] = v(x) - E\left[ \int_0^T e^{-\beta t} \Pi(X_t^*) dt \right] + E\left[ \int_0^T e^{-\beta t} v'(X_t^*) dL_t^* \right]. \quad (4.88)$$

(The stochastic integral appearing in Itô's formula is vanishing in expectation by (4.87)). Thus, from (4.86), we have

$$E\Big[\int_0^T e^{-\beta t} v'(X_t^*) dL_t^*\Big] = E\Big[\int_0^T e^{-\beta t} v'(X_t^*) 1_{X_t^*=x_b} dL_t^*\Big]$$
$$= E\Big[\int_0^T e^{-\beta t} \lambda dL_t^*\Big],$$

since $v'(x_b) = \lambda$. By substituting into (4.88), we get

$$v(x) = E\big[e^{-\beta T} v(X_T^*)\big] + E\Big[\int_0^T e^{-\beta t} \Pi(X_t^*) dt\Big] - E\Big[\int_0^T e^{-\beta t} \lambda dL_t^*\Big]. \quad (4.89)$$

From (4.87), we have $\lim_{T\to\infty} E[e^{-\beta T} X_T^*] = 0$, and so by the linear growth condition on $v$

$$\lim_{T\to\infty} E[e^{-\beta T} v(X_T^*)] = 0.$$

By sending $T$ to infinity into (4.89), we get the required equality.

(2) When $x < x_b$, and since $L_0^* = x - x_b$, we have

$$E\Big[\int_0^\infty e^{-\beta t} \left(\Pi(X_t^x) - \lambda dL_t^*\right)\Big] = E\Big[\int_0^\infty e^{-\beta t} \left(\Pi(X_t^{x_b}) - \lambda dL_t^*\right)\Big] - \lambda(x - x_b)$$
$$= v(x_b) - \lambda(x - x_b) = v(x),$$

by recalling that $v' = \lambda$ on $(0, x_b)$.    □

In conclusion, Theorems 4.5.7 and 4.5.8 provide an explicit solution to this irreversible investment problem. They validate the economic intuition that a firm should invest in the augmentation of capital to maintain its production capacity above some determined level.

## 4.6 Superreplication cost in uncertain volatility model

We consider the controlled diffusion

$$dX_s = \alpha_s X_s dW_s, \quad t \le s \le T,$$

valued in $(0, \infty)$ (for an initial condition $x > 0$), and where the control process $\alpha \in \mathcal{A}$ is valued in $A = [\underline{a}, \bar{a}]$, with $0 \le \underline{a} \le \bar{a} \le \infty$. To avoid trivial cases, we assume that $\bar{a} > 0$ and $\underline{a} \ne \infty$. In finance, $\alpha$ represents the uncertain volatility process of the stock price $X$. Given a continuous function $g$ with linear growth condition, representing the payoff of an European option, we want to calculate its superreplication cost, which is given by

$$v(t, x) = \sup_{\alpha \in \mathcal{A}} E[g(X_T^{t,x})], \quad (t, x) \in [0, T] \times (0, \infty). \quad (4.90)$$

Since the positive process $\{X_s^{t,x}, t \le s \le T\}$ is a supermartingale for all $\alpha \in \mathcal{A}$, it is easy to see that $v$ inherits from $g$ the linear growth condition, and is in particular locally bounded.

The Hamiltonian of this stochastic control problem is

$$H(x, M) = \sup_{a \in [\underline{a}, \bar{a}]} \left[ \frac{1}{2} a^2 x^2 M \right], \quad (x, M) \in (0, \infty) \times \mathbb{R}.$$

We shall then distinguish two cases according to the finiteness of the upper bound volatility $\bar{a}$.

### 4.6.1 Bounded volatility

We suppose that

$$\bar{a} < \infty.$$

In this regular case, the Hamiltonian $H$ is finite on the whole domain $(0, \infty) \times \mathbb{R}$, and is explicitly given by

$$H(x, M) = \frac{1}{2} \hat{a}^2(M) x^2 M,$$

with

$$\hat{a}(M) = \begin{cases} \bar{a} & \text{if } M \geq 0 \\ \underline{a} & \text{if } M < 0. \end{cases}$$

According to the general results of the previous sections, we have the following characterization on the superreplication cost.

**Theorem 4.6.9** *Suppose $\bar{a} < \infty$. Then $v$ is continuous on $[0, T] \times (0, \infty)$, and is the unique viscosity solution with linear growth condition to the so-called Black-Scholes-Barenblatt equation*

$$-\frac{\partial v}{\partial t} - \frac{1}{2} \hat{a}^2 \left( \frac{\partial^2 v}{\partial x^2} \right) x^2 \frac{\partial^2 v}{\partial x^2} = 0, \quad (t, x) \in [0, T) \times (0, \infty), \tag{4.91}$$

*satisfying the terminal condition*

$$v(T, x) = g(x), \quad x \in (0, \infty). \tag{4.92}$$

**Proof.** We know from Theorem 4.3.1 (and Remark 4.3.4) that $v$ is a viscosity solution to (4.91). Moreover, Theorem 4.3.2 and Remark 4.3.5 show that $v$ is continuous on $T$ with $v(T^-, .) = v(T, .) = g$. Finally, we obtain the uniqueness and continuity of $v$ from the strong comparison principle in Theorem 4.4.5. $\qquad\square$

**Remark 4.6.10** When $\underline{a} > 0$, there is existence and uniqueness of a smooth solution to the Black-Scholes-Barenblatt (4.91) together with the terminal condition (4.92), and so $v$ is this smooth solution. Indeed, in this case, after the change of variable $x \to \ln x$, we have a uniform ellipticity condition on the second-order term of the PDE, and we may apply classical results of Friedman [Fr75].

**Remark 4.6.11** When $g$ is convex, the function

$$w(t,x) = E[g(\hat{X}_T^{t,x})], \quad (t,x) \in [0,T] \times (0,\infty),$$

where $\{\hat{X}_s^{t,x}, t \leq s \leq T\}$ is the geometric Brownian motion, the solution to

$$d\hat{X}_s = \bar{a}\hat{X}_s dW_s, \quad t \leq s \leq T, \quad \hat{X}_t = x,$$

is also convex in $x$. Note that $w$ is the Black-Scholes price of the option payoff $g$ with volatility $\bar{a}$. Moreover, $w$ is continuous on $[0,T] \times (0,\infty)$ and is a classical solution to the Black-Scholes equation

$$-\frac{\partial w}{\partial t} - \frac{1}{2}\bar{a}^2 x^2 \frac{\partial^2 w}{\partial x^2} = 0, \quad (t,x) \in [0,T) \times (0,\infty),$$

and satisfies the terminal condition $w(T,x) = g(x)$. Since $\hat{a}\left(\dfrac{\partial^2 w}{\partial x^2}\right) = \bar{a}$, this implies that $w$ is also a solution to (4.91). By uniqueness, we conclude that $w = v$. Similarly, when $g$ is concave, the function $v$ is equal to the Black-Scholes price of the option payoff $g$ with volatility $\underline{a}$.

### 4.6.2 Unbounded volatility

In this section, we suppose that

$$\bar{a} = \infty.$$

In this singular case, the Hamiltonian is given by

$$H(x,M) = \begin{cases} \frac{1}{2}\underline{a}^2 x^2 M & \text{if } -M \geq 0 \\ \infty & \text{if } -M < 0. \end{cases}$$

According to Theorem 4.3.1, the function $v$ is then a viscosity solution to

$$\min\left[-\frac{\partial v}{\partial t} - \frac{1}{2}\underline{a}^2 x^2 \frac{\partial^2 v}{\partial x^2}, -\frac{\partial^2 v}{\partial x^2}\right] = 0, \quad (t,x) \in [0,T) \times (0,\infty). \tag{4.93}$$

Moreover, from Theoerem 4.3.2, the terminal condition is determined by the equation (in the viscosity sense)

$$\min\left[v(T^-,.) - g, -D_x^2 v(T^-,.)\right] = 0 \quad \text{on } (0,\infty). \tag{4.94}$$

In view of this terminal condition (see also Remark 4.3.5), we introduce the upper concave envelope of $g$, denoted by $\hat{g}$, which is the smallest concave function above $g$. We can then explicitly characterize the superreplication cost $v$.

**Theorem 4.6.10** *Suppose $\bar{a} = \infty$. Then $v = w$ on $[0,T) \times (0,\infty)$ where $w$ is the Black-Scholes price for the payoff function $\hat{g}(x)$:*

$$w(t,x) = E\left[\hat{g}\left(\hat{X}_T^{t,x}\right)\right], \quad \forall (t,x) \in [0,T] \times (0,\infty), \tag{4.95}$$

*in a Black-Scholes model with lower volatility $\underline{a}$, i.e. $\{\hat{X}_s^{t,x}, t \leq s \leq T\}$ is the solution to*

$$d\hat{X}_s = \underline{a}\hat{X}_s dW_s, \quad t \leq s \leq T, \quad \hat{X}_t = x.$$

**Proof. 1.** We know from Theorem 4.3.2 that $v_*(T,.)$ is a viscosity supersolution to (4.94), i.e. $v_*(T,.) \geq g$, and $v_*(T,.)$ is a viscosity supersolution to $-D_x^2 v_*(T,.) \geq 0$. In the classical case, this last property means that $v_*(T,.)$ is concave. This result is still true with the notion of viscosity solution. To see this, let $x_0 < x_1$ be two arbitrary points in $(0,\infty)$. Since $v_*(T,.)$ is lower-bounded on $[x_0, x_1]$, we may assume, by adding eventually a constant to $v_*(T,.)$ (which does not change its viscosity supersolution property), that $v_*(T,.) \geq 0$ on $[x_0, x_1]$. We thus clearly see that $v_*(T,.)$ is a viscosity supersolution to $\varepsilon^2 u - D^2 u = 0$ on $(x_0, x_1)$ for all $\varepsilon > 0$. Let us consider the equation

$$\varepsilon u - D^2 u = 0 \quad \text{on } (x_0, x_1),$$

with the boundary conditions

$$u(x_0) = v_*(T, x_0), \ u(x_1) = v_*(T, x_1).$$

The solution to this linear Dirichlet problem is smooth on $[x_0, x_1]$, and given by

$$u_\varepsilon(x) = \frac{v_*(T, x_0)\left[e^{\varepsilon(x_1-x)} - 1\right] + v_*(T, x_1)\left[e^{\varepsilon(x-x_0)} - 1\right]}{e^{\varepsilon(x_1-x_0)} - 1}.$$

From the comparison principle for this Dirichlet problem, we deduce that $v_*(T, x) \geq u_\varepsilon(x)$ for all $x \in [x_0, x_1]$. By sending $\varepsilon$ to zero, this yields

$$\frac{v_*(T, x) - v_*(T, x_0)}{x - x_0} \geq \frac{v_*(T, x_1) - v_*(T, x_0)}{x_1 - x_0}, \quad \forall x \in (x_0, x_1),$$

which proves the concavity of $v_*(T,.)$. By definition of $\hat{g}$, and since $v_*(T,.) \geq g$, we obtain

$$v_*(T,.) \geq \hat{g}, \quad \text{on } (0,\infty). \tag{4.96}$$

**2.** We know from Theorem 4.3.2 that $v^*(T,.)$ is a viscosity subsolution to (4.94). One could invoke a comparison principle for (4.94) to deduce that $v^*(T,.) \leq \hat{g}$, which is a viscosity solution to this equation. We provide here an alternative direct argument. Denote by $\mathcal{A}_b$ the subset of bounded controls in $\mathcal{A}$. We then have $v(t, x) \geq \sup_{\alpha \in \mathcal{A}_b} E[g(X_T^{t,x})]$. Conversely, given an arbitrary $\hat{\alpha} \in \mathcal{A}$, we set for $n \in \mathbb{N}$, $\alpha^n := \hat{\alpha} 1_{|\hat{\alpha}| \leq n} \in \mathcal{A}_b$, and we denote by $\hat{X}_s^{t,x}$ (resp. $X_s^n$), $t \leq s \leq T$, the diffusion process controlled by $\hat{\alpha}$ (resp. $\alpha^n$). i.e.

$$\hat{X}_s^{t,x} = x \exp\left(\int_t^s \hat{\alpha}_u dW_u - \frac{1}{2}\int_t^s |\hat{\alpha}_u|^2 du\right),$$

$$X_s^n = x \exp\left(\int_t^s \alpha_u^n dW_u - \frac{1}{2}\int_t^s |\alpha_u^n|^2 du\right).$$

Then $X_T^n$ converges a.s. to $\hat{X}_T^{t,x}$ when $n$ goes to infinity, and by Fatou's lemma, we get

$$\sup_{\alpha \in \mathcal{A}_b} E[g(X_T^{t,x})] \geq \liminf_{n \to +\infty} E[g(X_T^n)] \geq E[g(\hat{X}_T^{t,x})].$$

Since $\hat{\alpha} \in \mathcal{A}$ is arbitrary, we obtain the converse inequality, and so

$$v(t, x) = \sup_{\alpha \in \mathcal{A}_b} E[g(X_T^{t,x})]. \tag{4.97}$$

By noting that for any $\alpha \in \mathcal{A}_b$, the process $X_s^{t,x}$, $t \leq s \leq T$, is a martingale, Jensen's inequality yields

$$v(t,x) \leq \sup_{\alpha \in \mathcal{A}_b} E[\hat{g}(X_T^{t,x})]$$

$$\leq \sup_{\alpha \in \mathcal{A}_b} \hat{g}\left(E[X_T^{t,x}]\right) = \hat{g}(x). \tag{4.98}$$

This shows in particular that $v^*(T,.) \leq g$, and together with (4.96) that

$$v_*(T,x) = v^*(T,x) = \hat{g}(x), \quad x > 0. \tag{4.99}$$

**3.** The Black-Scholes price $w$ in (4.95) is a continuous function on $[0,T] \times (0,\infty)$, and is a viscosity solution with linear growth condition to the Black-Scholes PDE

$$-\frac{\partial w}{\partial t} - \frac{1}{2}\underline{a}^2 x^2 \frac{\partial^2 w}{\partial x^2} = 0, \quad (t,x) \in [0,T) \times (0,\infty), \tag{4.100}$$

together with the terminal condition

$$w(T,x) = \hat{g}(x), \quad x \in (0,\infty). \tag{4.101}$$

From (4.93), we know that $v$ is a viscosity supersolution to (4.100), and by (4.96): $v_*(T,.) \geq w(T,.)$. From the comparison principle for the linear PDE (4.100), we deduce that

$$v_* \geq w \quad \text{on } [0,T] \times (0,\infty). \tag{4.102}$$

**4.** Since $\hat{g}$ is concave and it is well-known that Black-Scholes price inherits concavity from its payoff, we deduce that

$$-\frac{\partial^2 w}{\partial x^2}(t,x) \geq 0 \quad \text{on } [0,T) \times (0,\infty).$$

This characterization of concavity is obviously true when $w$ is smooth, and is still valid in the viscosity sense (exercise left to the reader). Together with the previous equality (4.100), this proves that $w$ is a viscosity solution to

$$\min\left[-\frac{\partial w}{\partial t} - \frac{1}{2}\underline{a}^2 x^2 \frac{\partial^2 w}{\partial x^2}, \ -\frac{\partial^2 w}{\partial x^2}\right] = 0, \quad (t,x) \in [0,T) \times (0,\infty).$$

We could invoke a comparison principle for the above variational inequality to conclude that $v = w$ on $[0,T) \times (0,\infty)$. Here, we provide a direct alternative argument. When $\underline{a} = 0$, we have $w = \hat{g} \geq v$ by (4.98), and so the required equality with (4.102). When $\underline{a} > 0$, the function $w$ lies in $C^{1,2}([0,T) \times (0,\infty))$. By applying Itô's formula to $w(s, X_s^{t,x})$ given an arbitrary $\alpha \in \mathcal{A}$ (after localization for removing in expectation the stochastic integral term), we obtain

$$E[\hat{g}(X_T^{t,x})] = w(t,x) + E\left[\int_t^T \frac{\partial w}{\partial t}(s, X_s^{t,x}) + (\alpha_s)^2(X_s^{t,x})^2 \frac{\partial^2 w}{\partial x^2}(s,, X_s^{t,x}) \, ds\right].$$

Since $w$ is concave, and $\alpha_s \geq \underline{a}$, we deduce that

$$E[g(X_T^{t,x})] \le E[\hat{g}(X_T^{t,x})]$$

$$\le w(t,x) + E\Big[\int_t^T \frac{\partial w}{\partial t}(s, X_s^{t,x}) + \underline{a}^2(X_s^{t,x})^2 \frac{\partial^2 w}{\partial x^2}(s,, X_s^{t,x})\, ds\Big] = w(t,x),$$

since $w$ is solution to (4.100). From the arbitrariness of $\alpha \in \mathcal{A}$, we conclude that $v \le w$, and finally with (4.102) that $v = w$ on $[0,T) \times (0,\infty)$. $\qquad\square$

**Remark 4.6.12** The previous theorem shows that when $\underline{a} = 0$, then $v(t,.) = \hat{g}$ for all $t \in [0,T)$. This also holds true for any $\underline{a} \ge 0$ when we further assume that $g$ is convex. Indeed, we already know from (4.98) that $v \le \hat{g}$. Moreover, if $g$ is convex, then by (4.97) and Jensen's inequality, we have $v(t,x) \ge g(x)$ for all $(t,x) \in [0,T] \times (0,\infty)$. By concavity of $v(t,.)$ for $t \in [0,T)$, we conclude that $v(t,x) \ge \hat{g}(x)$, and so $v(t,x) = \hat{g}(x)$ for all $(t,x) \in [0,T) \times (0,\infty)$.

## 4.7 Bibliographical remarks

Viscosity solutions for second-order equations of HJB type related to stochastic control problems were introduced by P.L. Lions [Lio83]. Comparison results for viscosity solutions were extensively studied in the literature, and we cite the seminal papers by Jensen [Je88], Ishii [Ish89], see also the user's guide by Crandall, Ishii and P.L. Lions [CIL92]. Viscosity solutions in finance was pioneered in the thesis of Zariphopoulou [Zar88].

The application of the viscosity solutions approach to an irreversible investment model in Section 4.5 is a simplified case of a reversible investment model studied in Guo and Pham [GP05]. Other examples of applications of viscosity solutions to singular stochastic control problems in finance are considered in Shreve and Soner [ShSo94] for models with proportional transaction costs, or in Choulli, Taksar and Zhou [CTZ03] for models of optimal dividends. The application to the calculation of superreplication cost in uncertain volatility model is inspired from Cvitanic, Pham and Touzi [CPT99a]. An extension to the multidimensional case for risky assets is dealt with in Gozzi and Vargiolu [GV02]. Other variations and extensions to financial models with portfolio constraints or transaction costs are studied in Soner and Touzi [ST00], Cvitanic, Pham and Touzi [CPT99b] or Bentahar and Bouchard [BeBo07].

# 5

# Optimal switching and free boundary problems

## 5.1 Introduction

The theory of optimal stopping and its generalization as optimal switching is an important and classical field of stochastic control, which knows a renewed increasing interest due to its numerous and various applications in economy and finance, in particular for real options. Actually, it provides a suitable modeling framework for the evaluation of optimal investment decisions in capital for firms. Hence, it permits to capture the value of managerial flexibility to adapt decisions in response to unexpected markets developments, which is a key element in the modern theory of real options.

In this chapter, we present the tools of optimal switching, revisited under the approach of viscosity solutions, and we illustrate its applications through several examples related to real options. We essentially focus on infinite horizon problems. Section 5.2 is devoted to the optimal stopping problem, which appears classically in finance in the computation of American option prices. We derive the corresponding free boundary problem in the viscosity sense, and discuss the smooth-fit principle, i.e. the continuous differentiability of the value function. We give examples of application to the problem of selling an asset at an optimal date, and to the valuation of natural resources. In Section 5.3, we develop the optimal switching problem, where the controller may intervene successively on the system, whose coefficients may take different values, called regimes. This is an extension of the optimal stopping problem, and we show how the dynamic programming principle leads to a system of variational inequalities for the value functions that we characterize in the viscosity sense. We also state a smooth fit property in this context, and give an explicit solution in the two-regime case.

## 5.2 Optimal stopping

We consider a diffusion process on $\mathbb{R}^n$ driven by the SDE

$$dX_t = b(X_t)dt + \sigma(X_t)dW_t, \tag{5.1}$$

where $W$ is a $d$-dimensional standard Brownian motion on a probability space $(\Omega, \mathcal{F}, P)$ equipped with a filtration $\mathbb{F} = (\mathcal{F}_t)_{t \geq 0}$ satisfying the usual conditions. The coefficients

H. Pham, *Continuous-time Stochastic Control and Optimization with Financial Applications*, Stochastic Modelling and Applied Probability 61, DOI 10.1007/978-3-540-89500-8_5, © Springer-Verlag Berlin Heidelberg 2009

$b$ and $\sigma$ satisfy the usual Lipschitz conditions ensuring the existence and uniqueness of a solution given an initial condition: we denote by $\{X_t^x, t \geq 0\}$ the solution to (5.1) starting from $x$ at time $t = 0$, and $\{X_s^{t,x}, s \geq t\}$ the solution to (5.1) starting from $x$ at $t$. Given two continuous functions $f, g : \mathbb{R}^n \to \mathbb{R}$, satisfying a linear growth condition, and a positive discount factor $\beta > 0$, we consider the infinite horizon optimal stopping problem

$$v(x) = \sup_{\tau \in \mathcal{T}} E\left[ \int_0^\tau e^{-\beta t} f(X_t^x) dt + e^{-\beta \tau} g(X_\tau^x) \right], \quad x \in \mathbb{R}^n. \tag{5.2}$$

Here $\mathcal{T}$ denotes the set of stopping times valued in $[0, \infty]$, and we use the convention that $e^{-\beta \tau} = 0$ when $\tau = \infty$. For simplicity (see also Remark 5.2.1), we assume that $f$ and $g$ are Lipschitz. The finite horizon optimal stopping problem is formulated as

$$v(t, x) = \sup_{\tau \in \mathcal{T}_{t,T}} E\left[ \int_0^\tau e^{-\beta(s-t)} f(X_s^{t,x}) ds + e^{-\beta(\tau-t)} g(X_\tau^{t,x}) \right], \quad (t, x) \in [0, T] \times \mathbb{R}^n,$$

where $\mathcal{T}_{t,T}$ denotes the set of stopping times valued in $[0, T]$.

### 5.2.1 Dynamic programming and viscosity property

The aim of this section is to relate the value function $v$ in (5.2) to a PDE in variational form

$$\min\left[ \beta v - \mathcal{L}v - f , \; v - g \right] = 0, \tag{5.3}$$

in the viscosity sense. Here $\mathcal{L}$ is the second-order operator associated to the diffusion $X$:

$$\mathcal{L}v = b(x).D_x v + \frac{1}{2}\text{tr}(\sigma\sigma'(x)D_x^2 v).$$

Formally, this variational inequality (5.3) means that the PDE: $\beta v - \mathcal{L}v - f = 0$ is satisfied in the domain $\mathcal{C} = \{x \in \mathbb{R}^n : v(x) > g(x)\}$ with the boundary $v = g$ on the boundary of $\mathcal{C}$. Since this boundary is unknown, we also call this problem a *free boundary* problem.

We first prove the continuity of the value function.

**Lemma 5.2.1** *There exist positive constants $\beta_0, C$ such that for all $\beta > \beta_0$, $x, y \in \mathbb{R}^n$*

$$|v(x) - v(y)| \leq C|x - y|.$$

**Proof.** We recall from the estimation (1.19), that there exists some positive constant $\beta_0$ (depending on the Lipschitz constants of $b$ and $\sigma$) s.t. $E[\sup_{0 \leq u \leq t} |X_u^x - X_u^y|] \leq e^{\beta_0 t}|x-y|$. For $\beta > \beta_0$, we can prove similarly that $E[\sup_{t \geq 0} e^{-\beta t}|X_t^x - X_t^y|] \leq |x - y|$. Therefore, from the Lipschitz conditions on $f$ and $g$, we deduce that

$$|v(x) - v(y)| \leq \sup_{\tau \in \mathcal{T}} E\left[ \int_0^\tau e^{-\beta t}|f(X_t^x) - f(X_t^y)|dt + e^{-\beta \tau}|g(X_\tau^x) - g(X_\tau^y)| \right]$$

$$\leq CE\left[ \int_0^\infty e^{-\beta t}|X_t^x - X_t^y|dt \right] + CE\left[ \sup_{t \geq 0} e^{-\beta t}|X_t^x - X_t^y| \right]$$

$$\leq C|x - y|.$$

$\square$

In the sequel, we assume that $\beta$ is large enough, e.g. $\beta > \beta_0$, which ensures that $v$ is finite, satisfying a linear growth condition, and is continuous. As in the previous chapters, we shall appeal to the dynamic programming principle, which takes the following form.

DYNAMIC PROGRAMMING.
Fix $x \in \mathbb{R}^n$. For all stopping time $\theta \in \mathcal{T}$, we have

$$v(x) = \sup_{\tau \in \mathcal{T}} E\Big[ \int_0^{\tau \wedge \theta} e^{-\beta t} f(X_t^x)dt + e^{-\beta \tau} g(X_\tau^x)1_{\tau < \theta} + e^{-\beta \theta} v(X_\theta^x)1_{\theta \leq \tau}\Big]. \quad (5.4)$$

This principle means formally that at any time $\theta$, we may either decide to stop the process and receive the gain, or decide to continue by expecting to get a better reward: we then choose the best of these two possible decisions. A rigorous statement of this principle can be found e.g. in [Elk81] or [Kry80]. We now investigate the analytic implications of the dynamic programming principle.

**Theorem 5.2.1** *The value function $v$ in (5.2) is the unique viscosity solution to (5.3) on $\mathbb{R}^n$ satisfying a linear growth function.*

**Proof.** *Viscosity property.* We first check the viscosity supersolution property. By definition of $v$, it is clear that $v \geq g$ (take $\tau = 0$ in (5.2)). Moreover, for all $\theta \in \mathcal{T}$, we have from (5.4) (take $\tau = \theta$)

$$v(x) \geq E\Big[ \int_0^\theta e^{-\beta t} f(X_t^x)dt + e^{-\beta \theta} v(X_\theta^x)\Big].$$

Therefore, for any test function $\varphi \in C^2(\mathbb{R}^n)$ s.t. $0 = (v - \varphi)(x) = \min(v - \varphi)$, and by applying Itô's formula to $e^{-\beta t}\varphi(X_t^x)$ between $t = 0$ and $\theta \wedge h$, with $h > 0$, and $\theta$ the first exit time of $X^x$ outside some ball around $x$, we get

$$E\Big[\frac{1}{h} \int_0^{\theta \wedge h} e^{-\beta t}\big(\beta \varphi - \mathcal{L}\varphi - f\big)(X_t^x)dt\Big] \geq 0.$$

By sending $h$ to zero, and from the mean-value theorem, we obtain the other required supersolution inequality:

$$(\beta \varphi - \mathcal{L}\varphi - f)(x) \geq 0.$$

For the viscosity subsolution property, we take $\bar{x} \in \mathbb{R}^n$ and a test function $\varphi \in C^2(\mathbb{R}^n)$ s.t. $0 = (v - \varphi)(\bar{x}) = \max(v - \varphi)$, and we argue by contradiction by assuming that

$$(\beta \varphi - \mathcal{L}\varphi - f)(\bar{x}) > 0 \quad \text{and} \quad (v - g)(\bar{x}) > 0.$$

By continuity of $v$, $\mathcal{L}\varphi$ and $f$, we can find $\delta > 0$, and a ball $B(\bar{x}, \delta)$, s.t.

$$(\beta \varphi - \mathcal{L}\varphi - f)(X_t^{\bar{x}}) \geq \delta \quad \text{and} \quad (v - g)(X_t^{\bar{x}}) \geq \delta, \quad 0 \leq t \leq \theta, \quad (5.5)$$

where $\theta$ is the first exit time of $X^{\bar{x}}$ outside $B(\bar{x}, \delta)$. For any $\tau \in \mathcal{T}$, we now apply Itô's formula to $e^{-\beta t}\varphi(X_t^x)$ between $0$ and $\theta \wedge \tau$:

$$v(\bar{x}) \;=\; \varphi(\bar{x}) = E\!\left[\int_0^{\theta\wedge\tau} e^{-\beta t}(\beta\varphi - \mathcal{L}\varphi)(X_t^x)dt + e^{-\beta(\theta\wedge\tau)}\varphi(X_{\theta\wedge\tau}^x)\right]$$

$$\geq E\!\left[\int_0^{\theta\wedge\tau} e^{-\beta t}f(X_t^x)dt + e^{-\beta\tau}g(X_\tau^x)\mathbf{1}_{\tau<\theta} + e^{-\beta\theta}v(X_\theta^x)\mathbf{1}_{\theta\leq\tau}\right]$$

$$+\,\delta E\!\left[\int_0^{\theta\wedge\tau} e^{-\beta t}dt + e^{-\beta\tau}\mathbf{1}_{\tau<\theta}\right], \tag{5.6}$$

where we used the fact that $\varphi \geq v$ and (5.5). We now claim that there exists some positive constant $c_0 > 0$ s.t.

$$E\!\left[\int_0^{\theta\wedge\tau} e^{-\beta t}dt + e^{-\beta\tau}\mathbf{1}_{\tau<\theta}\right] \geq c_0, \quad \forall \tau \in \mathcal{T}. \tag{5.7}$$

For this, we construct a smooth function $w$ s.t.

$$\max\Big\{\beta w(x) - \mathcal{L}w(x) - 1\,,\; w(x) - 1\Big\} \leq 0,\; \forall x \in B(\bar{x},\delta) \tag{5.8}$$

$$w(x) = 0,\; \forall x \in \partial B(\bar{x},\delta) \tag{5.9}$$

$$w(\bar{x}) > 0. \tag{5.10}$$

For instance, we can take the function $w(x) = c_0\left(1 - \frac{|x-\bar{x}|^2}{\delta^2}\right)$, with

$$0 < c_0 \leq \min\left\{\left(\beta + \frac{2}{\delta}\sup_{x\in B(\bar{x},\delta)}|b_i(x)| + \frac{1}{\delta^2}\sup_{x\in B(\bar{x},\delta)}\mathrm{tr}(\sigma\sigma'(x))\right)^{-1},\; 1\right\}.$$

Then, by applying Itô's formula to $e^{-\beta t}w(X_t^{\bar{x}})$ between $0$ and $\theta\wedge\tau$, we have

$$0 < c_0 \;=\; w(\bar{x}) = E\!\left[\int_0^{\theta\wedge\tau} e^{-\beta t}(\beta w - \mathcal{L}w)(X_t^{\bar{x}})dt + e^{-\beta(\theta\wedge\tau)}w(X_{\theta\wedge\tau}^{\bar{x}})\right]$$

$$\leq E\!\left[\int_0^{\theta\wedge\tau} e^{-\beta t}dt + e^{-\beta\tau}\mathbf{1}_{\tau<\theta}\right],$$

from (5.8), (5.9) and (5.10). By plugging this last relation into (5.6), and taking the supremum over $\tau \in \mathcal{T}$, we get a contradiction with (5.4).

*Uniqueness property.* This is a consequence of the following general comparison result: let $U$ (resp. $V$) be a u.s.c. viscosity subsolution (resp. l.s.c. viscosity supersolution) of (5.3), satisfying a linear growth condition. Then $U \leq V$ on $\mathbb{R}^n$.

We explain how to adapt the arguments in Section 4.4, and sketch the proof. We argue by contradiction by assuming that $M := \sup(U - V) > 0$. Up to a quadratic penalization term, we may assume that $U-V$ attains its maximum on $\mathbb{R}^n$. Let us consider the bounded sequence $(x_m, y_m)_m$ that attains the maximum of the functions

$$\Phi_m(x,y) = U(x) - V(y) - \phi_m(x,y), \quad \phi_m(x,y) = m|x-y|^2, \quad m \in \mathbb{N}.$$

By the same arguments as in the proof of Theorem 4.4.4, we have

$$M_m = \max\Phi_m(x,y) = \Phi_m(x_m, y_m) \to M, \quad \text{and}\quad \phi_m(x_m, y_m) \to 0. \tag{5.11}$$

Moreover, from Ishii's lemma, there exist $M$ and $N \in \mathcal{S}_n$ s.t.

$$(2m(x_m - y_m), M) \in \bar{\mathcal{P}}^{2,+}U(x_m), \quad (2m(x_m - y_m), N) \in \bar{\mathcal{P}}^{2,-}V(y_m),$$

and

$$\text{tr}\big(\sigma\sigma'(x_m)M - \sigma\sigma'(y_m)N\big) \leq 3m|\sigma(x_m) - \sigma(y_m)|^2. \qquad (5.12)$$

From the viscosity subsolution (resp. supersolution) property of $U$ (resp. $V$) at $x_m$ (resp. $y_m$), we then have

$$\min\Big[\beta U(x_m) - b(x_m).2m(x_m - y_m) - \frac{1}{2}\text{tr}(\sigma\sigma'(x_m)M) - f(x_m),$$
$$U(x_m) - g(x_m)\Big] \leq 0, \qquad (5.13)$$

$$\min\Big[\beta V(y_m) - b(y_m).2m(x_m - y_m) - \frac{1}{2}\text{tr}(\sigma\sigma'(y_m)N) - f(y_m),$$
$$V(y_m) - g(y_m)\Big] \geq 0. \qquad (5.14)$$

If $U(x_m) - g(x_m) > 0$ for $m$ large enough, then the first term of the l.h.s. of (5.13) is nonpositive, and by substracting it with the nonnegative first term of the l.h.s. of (5.14), we get

$$\beta\big(U(x_m) - V(y_m)\big) \leq 2m\big(b(x_m) - b(y_m)\big).(x_m - y_m)$$
$$+ \frac{1}{2}\text{tr}\big(\sigma\sigma'(x_m)M - \sigma\sigma'(y_m)N\big) + f(x_m) - f(y_m).$$

From the Lipschitz condition on $b$, $\sigma$, continuity of $f$, (5.11), and (5.12), we get the contradiction by sending $m$ to infinity: $\beta M \leq 0$. Otherwise, up to a subsequence, $U(x_m) - g(x_m) \leq 0$, for all $m$, and since $V(y_m) - g(y_m) \geq 0$ by (5.14), we get $U(x_m) - V(y_m) \leq g(x_m) - g(y_m)$. By sending $m$ to infinity, and from the continuity of $g$, we also get the required contradiction: $M \leq 0$. $\qquad\square$

**Remark 5.2.1** Theorem 5.2.1 and continuity of the value function also hold true when we remove the Lipschitz condition on $f$, $g$, by assuming only that these reward functions are continuous and satisfy a linear growth condition. Actually, in this case, we only have a priori the linear growth condition on $v$, but we can prove by similar arguments as in Theorem 5.2.1 that $v$ is a (discontinuous) viscosity solution to (5.3). Then, from the (strong) comparison principle for (5.3), which holds under continuity and linear growth properties of $f$, $g$, we conclude in addition to the uniqueness, that $v$ is continuous.

### 5.2.2 Smooth-fit principle

In view of Theorem 5.2.1, we introduce the open set

$$\mathcal{C} = \big\{x \in \mathbb{R}^n : v(x) > g(x)\big\},$$

and its complement set

$$\mathcal{S} = \big\{x \in \mathbb{R}^n : v(x) = g(x)\big\}.$$

The set $\mathcal{S}$ is called the stopping (or exercise) region, since it corresponds to the state values, where it is optimal to stop the process, and then receive the profit $g$. The complement set $\mathcal{C}$ is called the continuation region, since it corresponds to the state values where it is optimal to let the diffusion continue. A probabilistic interpretation of these results will be given in the next section. The analytic interpretation of the continuation region is justified by the following result.

**Lemma 5.2.2** *The value function $v$ in (5.2) is a viscosity solution to*

$$\beta v - \mathcal{L}v - f = 0 \quad on \quad \mathcal{C}. \tag{5.15}$$

*Moreover, if the function $\sigma$ is uniformly elliptic, i.e. there exists $\varepsilon > 0$ s.t. $x'\sigma\sigma'(x)x \geq \varepsilon|x|^2$ for all $x \in \mathbb{R}^n$, then $v$ is $C^2$ on $\mathcal{C}$.*

**Proof.** First, we notice that the supersolution property of $v$ to (5.15) is immediate from the viscosity property of $v$ to (5.3). On the other hand, let $\bar{x} \in \mathcal{C}$, and $\varphi$ a $C^2$ test function s.t. $\bar{x}$ is a maximum of $v - \varphi$ with $v(\bar{x}) = \varphi(\bar{x})$. Now, by definition of $\mathcal{C}$, we have $v(\bar{x}) > g(\bar{x})$, so that from the viscosity subsolution property of $v$ to (5.3), we have

$$\beta v(\bar{x}) - \mathcal{L}\varphi(\bar{x}) - f(\bar{x}) \leq 0,$$

which implies the viscosity subsolution, and then the viscosity solution property of $v$ to (5.15).

Given an arbitrary bounded open domain $\mathcal{O}$ in $\mathcal{C}$, let us consider the linear Dirichlet boundary problem:

$$\beta w - \mathcal{L}w - f = 0 \quad on \quad \mathcal{O}, \quad w = v \quad on \quad \partial\mathcal{O}. \tag{5.16}$$

Under the uniform ellipticity condition on $\sigma$, classical results (see e.g. [Fr75]) provide the existence and uniqueness of a smooth $C^2$ solution $w$ on $\mathcal{O}$ to (5.16). In particular, this smooth solution $w$ is a viscosity solution to (5.15) on $\mathcal{O}$. From standard uniqueness results on viscosity solutions (here for linear PDE on a bounded domain), we deduce that $v = w$ on $\mathcal{O}$. From the arbitrariness of $\mathcal{O} \subset \mathcal{C}$, this proves that $v$ is smooth on $\mathcal{C}$, and so satisfies (5.15) in a classical sense. $\square$

The boundary $\partial\mathcal{C}$ (also called free boundary or exercise boundary) of the set $\mathcal{C}$ is included in the stopping region $\mathcal{D}$, and we have

$$v = g \quad on \quad \partial\mathcal{C}.$$

The smooth-fit principle for optimal stopping problems states that the value function $v$ is smooth $C^1$ through the free boundary, once the diffusion is uniformly elliptic, and $g$ is $C^1$ on the free boundary. This general classical result is proved in [Sh78], [Ja93], or [PeSh06]. We give here a simple proof based on viscosity solution arguments in the one-dimensional case for the diffusion.

**Proposition 5.2.1** *Assume that $X$ is one-dimensional, $\sigma$ is uniformly elliptic, and $g$ is $C^1$ on $\mathcal{S}$. Then, $v$ is $C^1$ on $\partial\mathcal{C}$.*

**Proof. 1.** We first check that $v$ admits a left and right derivative $v'_-(\bar{x})$ and $v'_+(\bar{x})$ at any point $\bar{x} \in \partial \mathcal{C}$. Let $\bar{x} \in \partial \mathcal{C}$, and suppose that $\bar{x}$ lies in the right boundary of $\partial \mathcal{C}$, i.e. there exists $\varepsilon > 0$ s.t. $(\bar{x} - \varepsilon, \bar{x}) \subset \mathcal{C}$. (The other possible case where $\bar{x}$ lies in the left boundary of $\mathcal{C}$ is dealt with similarly.) Recalling from Lemma 5.2.2, that $v$ satisfies $\beta v - \mathcal{L}v - f = 0$ on $\mathcal{C}$, we deduce that $v = w$ on $(\bar{x} - \varepsilon, \bar{x})$, where $w$ is the unique smooth $C^2$ solution on $[\bar{x} - \delta, \bar{x}]$ to $\beta w - \mathcal{L}w - f = 0$ with the boundary data: $w(\bar{x} - \delta) = v(\bar{x} - \delta)$, $w(\bar{x}) = v(\bar{x})$. This shows that $v$ admits a left derivative at $\bar{x}$ with $v'_-(\bar{x}) = w'(\bar{x})$. For the right derivative, we distinguish two cases: (i) If there exists $\delta > 0$ s.t. $(\bar{x}, \bar{x} + \delta) \subset \mathcal{S}$, then $v = g$ is $C^1$ on $(\bar{x}, \bar{x} + \delta)$, and so admits a right derivative at $\bar{x}$: $v'_+(\bar{x}) = g'(\bar{x})$. (ii) Otherwise, one can find a sequence $(x_n)$ in $\mathcal{C}$, $x_n > x$, converging to $\bar{x}$. Since $\mathcal{C}$ is open, there would exist $\delta' > 0$ s.t. $(x, x + \delta') \subset \mathcal{C}$. In this case, by same arguments as for the left derivative, we deduce that $v$ admits a right derivative at $\bar{x}$.

**2.** Let $\bar{x} \in \partial \mathcal{C}$. Since $v(\bar{x}) = g(\bar{x})$, and $v \geq g$ on $\mathbb{R}$, we have

$$\frac{v(x) - v(\bar{x})}{x - \bar{x}} \leq \frac{g(x) - g(\bar{x})}{x - \bar{x}}, \quad \forall x < \bar{x},$$

$$\frac{v(x) - v(\bar{x})}{x - \bar{x}} \geq \frac{g(x) - g(\bar{x})}{x - \bar{x}}, \quad \forall x > \bar{x},$$

and so $v'_-(\bar{x}) \leq g'(\bar{x}) \leq v'_+(\bar{x})$. We argue by contradiction and suppose that $v$ is not differentiable at $\bar{x}$. Then, in view of the above inequality, one can find some $p \in (v'_-(\bar{x}), v'_+(\bar{x}))$. Consider, for $\varepsilon > 0$, the smooth $C^2$ function:

$$\varphi_\varepsilon(x) = v(\bar{x}) + p(x - \bar{x}) + \frac{1}{2\varepsilon}(x - \bar{x})^2.$$

Then, we see that $v$ dominates locally in a neighborhood of $\bar{x}$ the function $\varphi_\varepsilon$, i.e. $\bar{x}$ is a local minimum of $v - \varphi_\varepsilon$. From the supersolution viscosity property of $v$ to (5.3), this yields

$$\beta v(\bar{x}) - \mathcal{L}\varphi_\varepsilon(\bar{x}) - f(\bar{x}) \geq 0,$$

which is written as

$$\beta v(\bar{x}) - b(\bar{x})p - f(\bar{x}) - \frac{1}{2\varepsilon}\sigma^2(\bar{x}) \geq 0.$$

Sending $\varepsilon$ to zero provides the required contradiction since $\sigma^2(\bar{x}) > 0$. We have then proved that for $\bar{x} \in \partial \mathcal{C}$, $v'(\bar{x}) = g'(\bar{x})$. $\qquad\square$

### 5.2.3 Optimal strategy

We first discuss the probabilistic interpretation of the optimal stopping problem (5.2). Let us define the process

$$Z_t = e^{-\beta t}v(X_t) + \int_0^t e^{-\beta s}f(X_s)ds,$$

and the first exit time of the continuation region

$$\tau^* = \inf\{t \geq 0 : v(X_t) = g(X_t)\}.$$

From the dynamic programming principle (which implies $\beta v - \mathcal{L}v - f \geq 0$), we have

$$Z_t \geq E[Z_\theta | \mathcal{F}_t], \quad \text{for any stopping time } \theta \text{ valued in } [t, \infty),$$

which means that $Z$ is a supermartingale. Moreover, since $(\beta v - \mathcal{L}v - f)(s, X_s) = 0$ for $0 \leq s < \theta^*$, the process $Z$ is a martingale on $[0, \tau^*]$, and so

$$v(X_0) = E\Big[ \int_0^{\tau^*} e^{-\beta t} f(X_t) dt + e^{-\beta \tau^*} g(X_{\tau^*}) \Big],$$

which implies that $\tau^*$ is an optimal stopping strategy. We refer to [Elk81] for a general analysis of optimal stopping problems by probabilistic methods.

We shall now provide some results on the optimal stopping strategy. Let us introduce the function

$$\hat{V}(x) = E\Big[ \int_0^\infty e^{-\beta t} f(X_t^x) dt \Big],$$

which corresponds to the value function of the total expected profit when we never stop the process $X$. We give sufficient conditions ensuring that it is optimal to never stop the process, i.e. $v = \hat{V}$.

**Lemma 5.2.3** *We have the following implications:*

$$\mathcal{S} = \emptyset \Longrightarrow \hat{V} \geq g \Longrightarrow v = \hat{V}.$$

**Proof.** Assume that $\mathcal{S} = \emptyset$. Then $v$ is a solution to the linear PDE

$$\beta v - \mathcal{L}v - f = 0 \quad \text{on } \mathbb{R}^n.$$

Since, by the Feynman-Kac formula, $\hat{V}$ is also solution to the same linear PDE, and satisfies a linear growth condition, we deduce by uniqueness that $v = \hat{V}$. However, we know that $v \geq g$, and so $\hat{V} \geq g$. Assume now that $\hat{V} \geq g$. Then, $\hat{V}$ is a solution to the same PDE variational inequality as $v$, i.e.

$$\min \big[ \beta \hat{V} - \mathcal{L}\hat{V} - f , \ \hat{V} - g \big] = 0, \quad \text{on } \mathbb{R}^n.$$

By uniqueness, we conclude that $v = \hat{V}$.    $\square$

**Remark 5.2.2** A trivial sufficient condition ensuring that $\hat{V} \geq g$ is: $\inf f \geq \beta \sup g$.

We end this section by providing some partial information on the stopping region.

**Lemma 5.2.4** *Assume that $g$ is $C^2$ on some open set $\mathcal{O}$ of $\mathbb{R}^n$ on which $\mathcal{S} \subset \mathcal{O}$. Then*

$$\mathcal{S} \subset \mathcal{D} := \big\{ x \in \mathcal{O} : \beta g(x) - \mathcal{L}g(x) - f(x) \geq 0 \big\}.$$

**Proof.** Let $x \in \mathcal{S}$. Since $v(x) = g(x)$ and $v \geq g$, this implies $x$ is a local mimimun of $v - \varphi$ with the smooth test function $\varphi = g$ on $\mathcal{O}$. From the viscosity supersolution property of $v$ to (5.3), this yields

$$\beta g(x) - \mathcal{L}g(x) - f(x) \geq 0,$$

which proves the result.    $\square$

The set $\mathcal{D}$ may be usually explicitly computed, and we shall show in the one-dimensional case how this can be used to determine the stopping regions.

### 5.2.4 Methods of solution in the one-dimensional case

We now focus on the one-dimensional diffusion process $X$ valued in $(0, \infty)$, and we assume that the real-valued coefficients $b$ and $\sigma$ of $X$ can be extended to Lipschitz continuous functions on $\mathbb{R}_+$, and for all $t \geq 0$,

$$X_t^x \to X_t^0 = 0, \quad a.s. \quad \text{as } x \text{ goes to zero.} \tag{5.17}$$

The typical example is the geometric Brownian motion. The function $f$ is continuous on $\mathbb{R}_+$, satisfying a linear growth condition, and $g$ is Lipschitz continuous functions on $\mathbb{R}_+$. The state domain on which $v$ satisfies the PDE variational inequality is then $(0, \infty)$, and we shall need to specify a boundary condition for $v(x)$ when $x$ approaches $0$ in order to get a uniqueness result.

**Lemma 5.2.5** *We have* $v(0^+) := \lim_{x \downarrow 0} v(x) = \max\left(\frac{f(0)}{\beta}, g(0)\right)$.

**Proof.** By definition of $v$, we have

$$v(x) \geq \max(\hat{V}(x), g(x)). \tag{5.18}$$

By continuity and the linear growth condition on $f$, we deduce by (5.17) and the dominated convergence theorem that $\hat{V}(0^+) = E[\int_0^\infty e^{-\beta t} f(X_t^0) dt] = \frac{f(0)}{\beta}$. Together with (5.18), we get $\liminf_{x \downarrow 0} v(x) > \max\left(\frac{f(0)}{\beta}, g(0)\right)$. To prove the converse inequality, let us define $\tilde{f} = f - f(0)$, $\tilde{g} = g - f(0)/\beta$, $\tilde{f}_+ = \max(\tilde{f}, 0)$, and $\tilde{g}_+ = \max(g, 0)$. Then, we have

$$v(x) = \frac{f(0)}{\beta} + \sup_{\tau \in \mathcal{T}} E\left[\int_0^\tau e^{-\beta t} \tilde{f}(X_t^x) dt + e^{-\beta \tau} \tilde{g}(X_\tau^x)\right]$$

$$\leq \frac{f(0)}{\beta} + E\left[\int_0^\infty e^{-\beta t} \tilde{f}_+(X_t^x) dt\right] + E[\sup_{t \geq 0} e^{-\beta t} \tilde{g}_+(X_t^x)].$$

Similarly as for $\hat{V}$, we see that that $E[\int_0^\infty e^{-\beta t} \tilde{f}_+(X_t^x) dt]$ goes to $\frac{\tilde{f}(0)}{\beta} = 0$ as $x$ goes to zero. From the Lipschitz conditions on $b$, $\sigma$ on $\mathbb{R}_+$, and recalling that $X_t^0 = 0$, we easily deduce by standard arguments that

$$E\left[\sup_{t \geq 0} e^{-\beta t} X_t^x\right] = E\left[\sup_{t \geq 0} e^{-\beta t} |X_t^x - X_t^0|\right] \leq x, \quad \forall x \in \mathbb{R}_+. \tag{5.19}$$

Moreover, by (5.19), and the Lipschitz property of $\tilde{g}_+$, the function $E[\sup_{t \geq 0} e^{-\beta t} \tilde{g}_+(X_t^x)]$ goes to $E[\sup_{t \geq 0} e^{-\beta t} \tilde{g}_+(0)] = \tilde{g}_+(0)$ as $x$ goes to zero. This proves that $\limsup_{x \downarrow 0} v(x) \leq \frac{f(0)}{\beta} + \tilde{g}_+(0) = \max\left(\frac{f(0)}{\beta}, g(0)\right)$, which ends the proof. $\square$

In the sequel, we shall assume that

$$\hat{V}(x_0) < g(x_0) \quad \text{for some } x_0 > 0, \tag{5.20}$$

so that the stopping region

$$\mathcal{S} = \{x \in (0, \infty) : v(x) = g(x)\},$$

is nonempty. We shall assume that $\mathcal{S}$ is included in some open set $\mathcal{O}$ of $(0, \infty)$ on which $g$ is $C^2$, and we know from Lemma 5.2.4 that

$$\mathcal{S} \subset \mathcal{D} = \{x \in \mathcal{O} : \beta g(x) - \mathcal{L}g(x) - f(x) \geq 0\}.$$

**Lemma 5.2.6** *Assume* (5.20) *holds.*

*(1) If* $\mathcal{D} = [a, \infty)$ *for some* $a > 0$, *then* $\mathcal{S}$ *is of the form* $[x^*, \infty)$ *for some* $x^* \in (0, \infty)$.
*(2) If* $g(0) \geq \frac{f(0)}{\beta}$ *and* $\mathcal{D} = (0, a]$ *for some* $a > 0$, *then* $\mathcal{S}$ *is of the form* $(0, x^*]$ *for some* $x^* \in (0, \infty)$.

**Proof.** (1) We set $x^* = \inf \mathcal{S}$, and since $\mathcal{S}$ is nonempty and included in $\mathcal{D}$, we know that $x^* \in [a, \infty)$. In order to prove that $\mathcal{S} = [x^*, \infty)$, let us consider the function $w = g$ on $[x^*, \infty)$. Since $[x^*, \infty) \subset \mathcal{D}$, and by definition of $\mathcal{D}$, we obtain that $w$ satisfies $\beta w - \mathcal{L}w - f \geq 0$, and so $\min[\beta w - \mathcal{L}w - f, w - g] = 0$ on $(\bar{x}, \infty)$. Moreover, since $w(x^*) = g(x^*) = v(x^*)$, and $w$, $v$ satisfy a linear growth condition, we deduce by uniqueness that $w = v$ on $[x^*, \infty)$. This proves that $\mathcal{S} = [x^*, \infty)$.

2) We set $x^* = \sup \mathcal{S}$, and since $\mathcal{S}$ is nonempty and included in $\mathcal{D}$, we know that $x^* \in (0, a]$. Let us consider the function $w = g$ on $(0, x^*]$. Since $(0, x^*] \subset \mathcal{D}$, we obtain that $w$ satisfies $\beta w - \mathcal{L}w - f \geq 0$, and so $\min[\beta w - \mathcal{L}w - f, w - g] = 0$ on $(0, x^*)$. Moreover, since $w(0^+) = g(0) = v(0^+)$ by Lemma 5.2.5 and the condition on $g(0)$, and $w(x^*) = g(x^*) = v(x^*)$, we deduce by uniqueness that $w = v$ on $(0, x^*)$, and so $\mathcal{S} = (0, x^*]$. □

### 5.2.5 Examples of applications

We consider a geometric Brownian motion (GBM) for $X$:

$$dX_t = \rho X_t dt + \gamma X_t dW_t,$$

where $\rho$ and $\gamma > 0$ are constants, and we shall solve various examples of optimal stopping problems arising from financial motivations.

**Perpetual American Put options**

We consider an American Put option of payoff on the underlying asset price $X$, which can be exercised at any time, and whose price is given by:

$$v(x) = \sup_{\tau \in \mathcal{T}} E[e^{-\beta \tau}(K - X_\tau^x)_+].$$

Here $\beta = \rho$ is the constant interest rate, and $K > 0$ is the strike. With the notations of the previous section, we have $f = 0$, $\hat{V} = 0$, $g(x) = (K - x)_+$. By convexity of $g$, and the linearity of the GBM $X^x$ in function of the initial condition $x$, we easily see that $v$ is convex. It is also clear that $v(x) > 0$ for all $x > 0$. We deduce that the exercise region is of the form $\mathcal{S} = (0, x^*]$ for some $x^* \in (0, K]$. Moreover, on the continuation region $(x^*, \infty)$, $v$ satisfies the ODE

$$\rho v(x) - \rho x v'(x) - \frac{1}{2}\gamma^2 x^2 v''(x) = 0, \quad x > x^*.$$

The general solution to the above ODE is of the form

$$v(x) = Ax^m + Bx^n,$$

for some constants $A$ and $B$, where $m$, $n$ are the roots of the second-degree equation $\frac{1}{2}\gamma^2 m^2 + (\rho - \frac{1}{2}\gamma^2)m - \rho = 0$, and are given by

$$m = -2\frac{\rho}{\gamma^2}, \quad n = 1.$$

Since $v$ is bounded, we should have $B = 0$ so that

$$v(x) = Ax^m, \quad x > x^*.$$

By writing that $v$ satisfies the smooth-fit condition at $x^*$, i.e. $A(x^*)^m = (K - x^*)$, and $Am(x^*)^{m-1} = -1$, we determine the parameters $A$ and $x^*$:

$$x^* = K\frac{m}{m - 1} = \frac{K}{1 + \frac{\gamma^2}{2\rho}}, \quad A = -\frac{1}{m}(x^*)^{1-m}.$$

We conclude that

$$v(x) = \begin{cases} K - x, & x \leq x^* = \frac{K}{1 + \frac{\gamma^2}{2\rho}}, \\ -\frac{x^*}{m}\left(\frac{x}{x^*}\right)^m, & x > x^*, \end{cases}$$

and it is optimal to exercise the option whenever the asset price is below $x^*$.

### When is it optimal to sell?

We consider an owner of an asset or firm with value process $X$. The owner may sell this asset at any time, but has to pay a fixed fee $K > 0$. He is looking for the optimal time to sell his asset, and his objective is then to solve the optimal stopping problem:

$$v(x) = \sup_{\tau \in \mathcal{T}} E\left[e^{-\beta\tau}(X_\tau^x - K)\right],$$

where $\beta > 0$ is a discount factor. With the notations of the previous section, we have $\hat{V} = f = 0$, $g(x) = x - K$, which is $C^2$ on $(0, \infty)$, and the set $\mathcal{D}$ is equal to

$$\mathcal{D} = \{x \in (0, \infty) : (\beta - \rho)x \geq \beta K\}.$$

We are then led to distinguish the following cases:

(i) $\beta < \rho$. It is then easy to see that $v(x) = \infty$ for all $x > 0$. Indeed, in this case, from the explicit expression of the geometric Brownian motion, we have for any $T > 0$,

$$v(x) \geq E\left[e^{-\beta T}(X_T^x - K)\right] = xe^{(\rho-\beta)T}E\left[e^{\gamma W_T - \frac{\rho^2}{2}T}\right] - Ke^{-\beta T}$$
$$= xe^{(\rho-\beta)T} - Ke^{-\beta T},$$

and by sending $T$ to infinity, we get the announced result.

(ii) $\beta = \rho$. In this case, we have $v(x) = x$. Indeed, for any stopping time $\tau$, for any $n \in \mathbb{N}$, we have by the optional sampling theorem for martingales

$$E\left[e^{-\beta(\tau \wedge n)}(X_{\tau \wedge n}^x - K)\right] = x - KE[e^{-\beta(\tau \wedge n)}] \tag{5.21}$$
$$\leq x,$$

By sending $n$ to infinity, and from Fatou's lemma, we then get for any $\tau \in \mathcal{T}$

$$E\big[e^{-\beta\tau}(X_\tau^x - K)\big] \le x,$$

and so $v(x) \le x$. Conversely, by taking $\tau = n$, we have from (5.21): $v(x) \ge x - Ke^{-\beta n}$, and so by sending $n$ to infinity: $v(x) \ge x$.

(iii) $\beta > \rho$. In this case, $\mathcal{D} = [a, \infty)$ for $a = \beta K/(\beta - \rho) > 0$. Moreover, we obviously have $\hat{V} = 0 < g(x) = x - K$ for $x > K$. We deduce from Lemma 5.2.6 that $\mathcal{S}$ is of the form $\mathcal{S} = [x^*, \infty)$ for some $x^* \in (0, \infty)$. Moreover, on the continuation region $(0, x^*)$, $v$ satisfies the ODE

$$\beta v - \rho x v'(x) - \frac{1}{2}\gamma^2 x^2 v''(x) = 0, \quad x < x^*,$$

whose general solution is of the form

$$v(x) = Ax^m + Bx^n,$$

for some constants $A$ and $B$, and $m$, $n$ are the roots of the second-degree equation $\frac{1}{2}\gamma^2 m^2 + (\rho - \frac{1}{2}\gamma^2)m - \beta = 0$, and are given by

$$m = \frac{1}{2} - \frac{\rho}{\gamma^2} - \sqrt{(\frac{1}{2} - \frac{\rho}{\gamma^2})^2 + \frac{2\beta}{\gamma^2}} < 0 \qquad (5.22)$$

$$n = \frac{1}{2} - \frac{\rho}{\gamma^2} + \sqrt{(\frac{1}{2} - \frac{\rho}{\gamma^2})^2 + \frac{2\beta}{\gamma^2}} > 1. \qquad (5.23)$$

Since $v(0^+) = \max(-K, 0) = 0$, we should have $A = 0$. Finally, by writing the smooth-fit condition at $x^*$, we determine the coefficients $B$ and $x^*$ by

$$x^* = K\frac{n}{n-1}, \quad B = \frac{1}{n}\frac{1}{(x^*)^{n-1}}.$$

We conclude that

$$v(x) = \begin{cases} \frac{x^*}{n}\left(\frac{x}{x^*}\right)^n, & x < x^*, \\ x - K, & x \ge x^*, \end{cases}$$

and it is optimal to sell the asset once it is above $x^*$.

**Valuation of natural resources**

We consider a firm producing some natural resource (oil, gas, etc.) with price process $X$. The running profit of this production is given by a nondecreasing function $f$ depending on the price, and the firm may decide at any time to stop the production at a fixed constant cost $-K$. The real options value of the firm is then given by the optimal stopping problem:

$$v(x) = \sup_{\tau \in \mathcal{T}} E\Big[\int_0^\tau e^{-\beta t} f(X_t^x)dt + e^{-\beta\tau}K\Big].$$

We assume that $f$ satisfies a linear growth condition, and we notice that for $\beta > \rho$, there exists some $C > 0$ s.t. $E[\int_0^\infty e^{-\beta t}|f(X_t^x)|dt] \le C(1 + x)$ for all $x \in \mathbb{R}_+$, so that $v$ also satisfies a linear growth condition. With the notations of the previous section, the set $\mathcal{D}$ is equal to

$$\mathcal{D} = \{x \in (0, \infty) : f(x) \leq \beta K\}.$$

Since $f$ is nondecreasing, we are led to consider the following three cases:

(i) $f(\infty) := \lim_{x \nearrow \infty} f(x) \leq \beta K$. In this case, $\mathcal{D} = (0, \infty)$, which means that $g = K$ is solution to the variational PDE $\min[\beta g - \mathcal{L}g - f, g - g] = 0$. Since $v(0^+) = \max(f(0)/\beta, K)$ $= K = g$, it follows by uniqueness that $v = K$, and the optimal strategy is to stop immediately.

(ii) $f(\infty) > \beta K$ and $f(0) = \inf f(x) \geq \beta K$. In this case $\hat{V}(x) = E[\int_0^\infty e^{-\beta t} f(X_t^x) dt] \geq K = g$, and so by Lemma 5.2.3, $v = \hat{V}$, and it is optimal to never stop the production.

(iii) $f(\infty) > \beta K$ and $f(0) < \beta K$. In this case, $\mathcal{D} = (0, a)$ for some $a > 0$. Moreover, there exists some $x > 0$ s.t. $\hat{V}(x) < g(x) = K$. Otherwise, we would have $\hat{V}(x) \geq K$ for all $x > 0$, and by sending $x$ to zero, this would imply $\hat{V}(0^+) = f(0)/\beta \geq K$, a contradiction. By Lemma 5.2.6, the stopping region is in the form $\mathcal{S} = (0, x^*]$ for some $x^* > 0$. Moreover, on the continuation region $(x^*, \infty)$, $v$ satisfies the ODE

$$\beta v - \rho x v'(x) - \frac{1}{2} \gamma^2 x^2 v''(x) - f(x) = 0, \quad x > x^*,$$

whose general solution is of the form

$$v(x) = A x^m + B x^n + \hat{V}(x),$$

for some constants $A$ and $B$, and $m$, $n$ are given in (5.22)-(5.23). Since $v$ satisfies a linear growth condition, we should have $B = 0$. Finally, the coefficients $A$ and $x^*$ are determined by the smooth-fit condition at $x^*$:

$$K = A(x^*)^m + \hat{V}(x^*), \quad 0 = m A(x^*)^{m-1} + \hat{V}'(x^*).$$

We conclude that

$$v(x) = \begin{cases} K, & x \leq x^* \\ A x^m + \hat{V}(x), & x > x^*, \end{cases}$$

and it is optimal to stop the production once the price is below $x^*$.

## 5.3 Optimal switching

In this section, we consider a diffusion $X$ and a profit function, whose coefficients may take different values, called regimes, depending on the value taken by an underlying control. The goal of the controller is to manage the different regimes in order to maximize an expected total profit. Such problems appear in the real options literature. For example, in the firm's investment problem under uncertainty, a company (oil tanker, eletricity station, etc.) manages several production activities operating in different modes or regimes representing a number of different economic outlooks (e.g. state of economic growth, open or closed production activity). The process $X$ is the price of input or output goods of the firm and its dynamics may differ according to the regimes. The firm's project yields a running profit payoff that depends on the commodity price and on the regime choice. The transition from one regime to another is realized sequentially at time decisions and incurs costs. The problem is to find the strategy that maximizes the expected value of profits resulting from the project.

### 5.3.1 Problem formulation

We formulate optimal switching problem on infinite horizon. We fix some filtered probability space $(\Omega, \mathcal{F}, \mathbb{F} = (\mathcal{F}_t)_{t \geq 0}, P)$ satisfying the usual conditions. We first define a set of possible regimes $\mathbb{I}_m = \{1, \ldots, m\}$. A switching control is a double sequence $\alpha = (\tau_n, \iota_n)_{n \geq 1}$, where $(\tau_n)$ is an increasing sequence of stopping times, $\tau_n \in \mathcal{T}$, $\tau_n \to \infty$, representing the decision on "when to switch", and $\iota_n$ are $\mathcal{F}_{\tau_n}$-measurable valued in $\mathbb{I}_m$ representing the new value of the regime at time $\tau_n$ until time $\tau_{n+1}$ or the decision on "where to switch". We denote by $\mathcal{A}$ the set of switching controls. Given an initial regime value $i \in \mathbb{I}_m$, and a control $\alpha = (\tau_n, \iota_n)_{n \geq 1} \in \mathcal{A}$, we define

$$I_t^i = \sum_{n \geq 0} \iota_n 1_{[\tau_n, \tau_{n+1})}(t), \quad t \geq 0, \quad I_{0-}^i = i,$$

which is the piecewise constant process indicating the regime value at any time $t$. Here, we set $\tau_0 = 0$ and $\iota_0 = i$. We notice that $I^i$ is a càd-làg process, possibly with a jump at time 0 if $\tau_1 = 0$ and so $I_0^i = \iota_1$. Given an initial state-regime $(x, i) \in \mathbb{R}^d \times \mathbb{I}_m$, and a switching control $\alpha \in \mathcal{A}$, the controlled process $X^{x,i}$ is the solution to

$$dX_t = b(X_t, I_t^i)dt + \sigma(X_t, I_t^i)dW_t, \quad t \geq 0, \quad X_0 = x,$$

where $W$ is a standard Brownian motion on $(\Omega, \mathcal{F}, \mathbb{F} = (\mathcal{F}_t)_{t \geq 0}, P)$, and $b_i(.) := b(., i)$, $\sigma_i(.) := \sigma(., i)$, $i \in \mathbb{I}_m$, satisfy the Lipschitz condition.

The operational regimes are characterized by their running reward functions $f : \mathbb{R}^d \times \mathbb{I}_m \to \mathbb{R}$, and we set $f_i(.) = f(., i)$, $i \in \mathbb{I}_m$, which are continuous and satisfy a linear growth condition. For simplicity, we shall assume that $f_i$ is Lipschitz (see Remark 5.3.3). Switching from regime $i$ to $j$ incurs an instantaneous cost, denoted by $g_{ij}$, with the convention $g_{ii} = 0$. The following triangular condition is reasonable:

$$g_{ik} < g_{ij} + g_{jk}, \quad j \neq i, k, \tag{5.24}$$

which means that it is less expensive to switch directly in one step from regime $i$ to $k$ than in two steps via an intermediate regime $j$. Notice that a switching cost $g_{ij}$ may be negative, and condition (5.24) for $i = k$ prevents an arbitrage by simply switching back and forth, i.e.

$$g_{ij} + g_{ji} > 0, \quad i \neq j \in \mathbb{I}_m. \tag{5.25}$$

Here, for simplicity, we assume that the switching costs $g_{ij}$ are constants.

The expected total profit of running the system when the initial state is $(x, i)$ and using the impulse control $\alpha = (\tau_n, \iota_n)_{n \geq 1} \in \mathcal{A}$ is

$$J(x, i, \alpha) = E\left[\int_0^\infty e^{-\beta t} f(X_t^{x,i}, I_t^i) dt - \sum_{n=1}^\infty e^{-\beta \tau_n} g_{\iota_{n-1}, \iota_n}\right].$$

Here $\beta > 0$ is a positive discount factor, and we use the convention that $e^{-\beta \tau_n(\omega)} = 0$ when $\tau_n(\omega) = \infty$. The objective is to maximize this expected total profit over $\mathcal{A}$. Accordingly, we define the value functions

$$v(x, i) = \sup_{\alpha \in \mathcal{A}} J_i(x, \alpha), \quad x \in \mathbb{R}^d, \ i \in \mathbb{I}_m. \tag{5.26}$$

We write $J_i(x, \alpha) = J(x, i, \alpha)$ and $v_i(.) = v(., i)$. We shall see later that for $\beta$ large enough, the expectation defining $J_i(x, \alpha)$ is well-defined and the value function $v_i$ is finite.

### 5.3.2 Dynamic programming and system of variational inequalities

We first state the finiteness and Lipschitz continuity of the value functions.

**Lemma 5.3.7** *There exists some positive constant $\rho$ such that for $\beta > \rho$, the value functions $v_i$, $i \in \mathbb{I}_m$, are finite on $\mathbb{R}^d$. In this case, the value functions $v_i$, $i \in \mathbb{I}_m$, are Lipschitz continuous:*

$$|v_i(x) - v_i(y)| \le C|x - y|, \quad \forall x, y \in \mathbb{R}^d,$$

*for some positive constant $C$.*

**Proof.** First, we prove by induction that for all $N \ge 1$, $\tau_1 \le \ldots \le \tau_N$, $\iota_0 = i$, $\iota_n \in \mathbb{I}_m$, $n = 1, \ldots, N$

$$-\sum_{n=1}^N e^{-\beta \tau_n} g_{\iota_{n-1}, \iota_n} \le \max_{j \in \mathbb{I}_m}(-g_{ij}), \quad a.s. \tag{5.27}$$

Indeed, the above assertion is obviously true for $N = 1$. Suppose now it holds true at step $N$. Then, at step $N + 1$, we distinguish two cases: If $g_{\iota_N, \iota_{N+1}} \ge 0$, then we have $-\sum_{n=1}^{N+1} e^{-\beta \tau_n} g_{\iota_{n-1}, \iota_n} \le -\sum_{n=1}^N e^{-\beta \tau_n} g_{\iota_{n-1}, \iota_n}$ and we conclude by the induction hypothesis at step $N$. If $g_{\iota_N, \iota_{N+1}} < 0$, then by (5.24), and since $\tau_N \le \tau_{N+1}$, we have $-e^{-\beta \tau_N} g_{\iota_{N-1}, \iota_N} - e^{-\beta \tau_{N+1}} g_{\iota_N, \iota_{N+1}} \le -e^{-\beta \tau_N} g_{\iota_{N-1}, \iota_{N+1}}$, and so $-\sum_{n=1}^{N+1} e^{-\beta \tau_n} g_{\iota_{n-1}, \iota_n} \le -\sum_{n=1}^N e^{-\beta \tau_n} g_{\tilde{\iota}_{n-1}, \tilde{\iota}_n}$, with $\tilde{\iota}_n = \iota_n$ for $n = 1, \ldots, N-1$, $\tilde{\iota}_N = \iota_{N+1}$. We then conclude by the induction hypothesis at step $N$.

By definition, using the latter inequality and the growth condition on $f$, we have for all $i \in \mathbb{I}_m$, $x \in \mathbb{R}^d$, and $\alpha \in \mathcal{A}$

$$J_i(x, \alpha) \le E\left[\int_0^\infty e^{-\beta t}|f(X_t^{x,i}, I_t^i)|dt + \max_{j \in \mathbb{I}_m}(-g_{ij})\right]$$

$$\le E\left[\int_0^\infty e^{-\beta t} C(1 + |X_t^{x,i}|)dt + \max_{j \in \mathbb{I}_m}(-g_{ij})\right]. \tag{5.28}$$

Now, a standard estimate on the process $(X_t^{x,i})_{t \ge 0}$, based on Itô's formula and Gronwall's lemma, yields

$$E\left[|X_t^{x,i}|\right] \le e^{\rho t}(1+|x|),$$

for some positive constant $\rho$ (independent of $t$ and $x$). Plugging the above inequality into (5.28), and from the arbitrariness of $\alpha \in \mathcal{A}$, we get

$$|v_i(x)| \le \frac{C}{\beta} + C(1+|x|) \int_0^\infty e^{(\rho-\beta)t} dt + \max_{j\in\mathbb{I}_m}(-g_{ij}).$$

We therefore have the finiteness of the value functions if $\beta > \rho$, in which case the value functions satisfy the linear growth condition.

Moreover, by a standard estimate for the SDE applying Itô's formula to $|X_t^{x,i} - X_t^{y,i}|^2$ and using Gronwall's lemma, we then obtain from the Lipschitz condition on $b$, $\sigma$, the following inequality uniformly in $\alpha \in \mathcal{A}$:

$$E\left|X_t^{x,i} - X_t^{y,i}\right| \le e^{\rho t}|x-y|, \quad \forall x,y \in \mathbb{R}^d, \; t \ge 0.$$

From the Lipschitz condition on $f$, we deduce

$$
\begin{aligned}
|v_i(x) - v_i(y)| &\le \sup_{\alpha\in\mathcal{A}} E\left[\int_0^\infty e^{-\beta t}\left|f(X_t^{x,i}, I_t^i) - f(X_t^{y,i}, I_t^i)\right| dt\right] \\
&\le C \sup_{\alpha\in\mathcal{A}} E\left[\int_0^\infty e^{-\beta t}\left|X_t^{x,i} - X_t^{y,i}\right| dt\right] \\
&\le C \int_0^\infty e^{-\beta t} e^{\rho t}|x-y| dt \; \le \; C|x-y|,
\end{aligned}
$$

for $\beta > \rho$. This ends the proof.    □

In the sequel, we shall assume that $\beta$ is large enough, which ensures that $v_i$ is Lipschitz continuous, and in order to derive a PDE characterization of $v_i$, we shall appeal to the dynamic programming principle, which takes the following form

DYNAMIC PROGRAMMING PRINCIPLE: For any $(x,i) \in \mathbb{R}^d \times \mathbb{I}_m$, we have

$$
\begin{aligned}
v(x,i) = \sup_{(\tau_n, \iota_n)_n \in \mathcal{A}} E\Big[ &\int_0^\theta e^{-\beta t} f(X_t^{x,i}, I_t^i) dt - \sum_{\tau_n \le \theta} e^{-\beta \tau_n} g_{\iota_{n-1}, \iota_n} \\
&+ e^{-\beta\theta} v(X_\theta^{x,i}, I_\theta^i)\Big],
\end{aligned}
\tag{5.29}
$$

where $\theta$ is any stopping time, possibly depending on $\alpha \in \mathcal{A}$ in (5.29).

The aim of this section is to relate via the dynamic programming principle the value functions $v_i$, $i \in \mathbb{I}_m$, to the system of variational inequalities

$$\min\left[\beta v_i - \mathcal{L}_i v_i - f_i \,,\; v_i - \max_{j\neq i}(v_j - g_{ij})\right] = 0, \quad x \in \mathbb{R}^d, \; i \in \mathbb{I}_m, \tag{5.30}$$

where $\mathcal{L}_i$ is the generator of the diffusion $X$ in regime $i$:

$$\mathcal{L}_i \varphi = b_i.D_x\varphi + \frac{1}{2}\mathrm{tr}(\sigma_i\sigma_i'(x)D_x^2\varphi)$$

**Theorem 5.3.2** *For each $i \in \mathbb{I}_m$, the value function $v_i$ is a viscosity solution to (5.30).*

**Proof.** (1) We first prove the viscosity supersolution property. Fix $i \in \mathbb{I}_m$, and let $\bar{x} \in \mathbb{R}^d$, $\varphi \in C^2(\mathbb{R}^d)$ s.t. $\bar{x}$ is a minimum of $v_i - \varphi$ with $v_i(\bar{x}) = \varphi(\bar{x})$. By taking the immediate switching control $\tau_1 = 0$, $\iota_1 = j \neq i$, $\tau_n = \infty$, $n \geq 2$, and $\theta = 0$ in the relation (5.29), we obtain

$$v_i(\bar{x}) \geq v_j(\bar{x}) - g_{ij}, \quad \forall j \neq i. \tag{5.31}$$

On the other hand, by taking the no-switching control $\tau_n = \infty$, $n \geq 1$, i.e. $I_t^i = i$, $t \geq 0$, $X^{\bar{x},i}$ stays in regime $i$ with diffusion coefficients $b_i$ and $\sigma_i$, and $\theta = \tau_\varepsilon \wedge h$, with $h > 0$ and $\tau_\varepsilon = \inf\{t \geq 0 : X_t^{\bar{x},i} \notin B_\varepsilon(\bar{x})\}$, $\varepsilon > 0$, we get from (5.29)

$$\varphi(\bar{x}) \; = \; v_i(\bar{x}) \geq E\Big[ \int_0^\theta e^{-\beta t} f_i(X_t^{\bar{x},i}) dt + e^{-\beta\theta} v_i(X_\theta^{\bar{x},i}) \Big]$$

$$\geq E\Big[ \int_0^\theta e^{-\beta t} f_i(X_t^{\bar{x},i}) dt + e^{-\beta\theta} \varphi(X_\theta^{\bar{x},i}) \Big]$$

By applying Itô's formula to $e^{-\beta t}\varphi(X_t^{\bar{x},i})$ between $0$ and $\theta = \tau_\varepsilon \wedge h$ and plugging into the last inequality, we obtain

$$\frac{1}{h}E\Big[ \int_0^{\tau_\varepsilon \wedge h} e^{-\beta t}(\beta\varphi - \mathcal{L}_i\varphi - f_i)(X_t^{\bar{x},i}) \Big] \geq 0.$$

From the dominated convergence theorem and the mean-value theorem, this yields by sending $h$ to zero

$$(\beta\varphi - \mathcal{L}_i\varphi - f_i)(\bar{x}) \geq 0.$$

By combining with (5.31), we obtain the required viscosity supersolution inequality.

(2) We next prove the viscosity subsolution property. Fix $i \in \mathbb{I}_m$, and consider any $\bar{x} \in \mathbb{R}^d$, $\varphi \in C^2(\mathbb{R}^d)$ s.t. $\bar{x}$ is a maximum of $v_i - \varphi$ with $v_i(\bar{x}) = \varphi(\bar{x})$. We argue by contradiction by assuming on the contrary that the subsolution inequality does not hold so that by continuity of $v_i$, $v_j$, $j \neq i$, $\varphi$ and its derivatives, there exists some $\delta > 0$ s.t.

$$(\beta\varphi - \mathcal{L}_i\varphi - f_i)(x) \geq \delta, \quad \forall x \in B_\delta(\bar{x}) = (x - \delta, x + \delta) \tag{5.32}$$

$$v_i(x) - \max_{j \neq i}(v_j - g_{ij})(x) \geq \delta, \quad \forall x \in B_\delta(\bar{x}). \tag{5.33}$$

For any $\alpha = (\tau_n, \iota_n)_{n \geq 1} \in \mathcal{A}$, consider the exit time $\tau_\delta = \inf\{t \geq 0 : X_t^{\bar{x},i} \notin B_\delta(\bar{x})\}$. By applying Itô's formula to $e^{-\beta t}\varphi(X_t^{\bar{x},i})$ between $0$ and $\theta = \tau_1 \wedge \tau_\delta$, we have by noting that before $\theta$, $X^{\bar{x},i}$ stays in regime $i$ and in the ball $B_\delta(\bar{x})$

$$v_i(\bar{x}) \; = \; \varphi(\bar{x}) = E\Big[ \int_0^\theta e^{-\beta t}(\beta\varphi - \mathcal{L}_i\varphi)(X_t^{\bar{x},i}) dt + e^{-\beta\theta}\varphi(X_\theta^{\bar{x},i}) \Big]$$

$$\geq E\Big[ \int_0^\theta e^{-\beta t}(\beta\varphi - \mathcal{L}_i\varphi)(X_t^{\bar{x},i}) dt + e^{-\beta\theta} v_i(X_\theta^{\bar{x},i}) \Big]. \tag{5.34}$$

Now, since $\theta = \tau_\delta \wedge \tau_1$, we have

$$e^{-\beta\theta}v(X_\theta^{\bar{x},i}, I_\theta^i) - \sum_{\tau_n \leq \theta} e^{-\beta\tau_n} g_{\iota_{n-1},\iota_n}$$

$$= e^{-\beta\tau_1} \left(v(X_{\tau_1}^{\bar{x},i}, \iota_1) - g_{i\iota_1}\right) 1_{\tau_1 \leq \tau_\delta} + e^{-\beta\tau_\delta} v_i(X_{\tau_\delta}^{\bar{x},i}) 1_{\tau_\delta < \tau_1}$$

$$\leq e^{-\beta\tau_1} \left(v_i(X_{\tau_1}^{\bar{x},i}) - \delta\right) 1_{\tau_1 \leq \tau_\delta} + e^{-\beta\tau_\delta} v_i(X_{\tau_\delta}^{\bar{x},i}) 1_{\tau_\delta < \tau_1}$$

$$= e^{-\beta\theta} v_i(X_\theta^{\bar{x},i}) - \delta e^{-\beta\tau_1} 1_{\tau_1 \leq \tau_\delta},$$

where the inequality follows from (5.33). By plugging into (5.34) and using (5.32), we get

$$v_i(\bar{x}) \geq E\Big[ \int_0^\theta e^{-\beta t} f_i(X_t^{\bar{x},i}) dt + e^{-\beta\theta} v(X_\theta^{\bar{x},i}, I_\theta^i) - \sum_{\tau_n \leq \theta} g_{\iota_{n-1},\iota_n} \Big]$$

$$+ \delta E\Big[ \int_0^\theta e^{-\beta t} dt + e^{-\beta\tau_1} 1_{\tau_1 \leq \tau_\delta} \Big]. \tag{5.35}$$

Now, by the same arguments as in (5.7), there exists some positive constant $c_0 > 0$ s.t.

$$E\Big[ \int_0^\theta e^{-\beta t} dt + e^{-\beta\tau_1} 1_{\tau_1 \leq \tau_\delta} \Big] \geq c_0, \quad \forall \alpha \in \mathcal{A}.$$

By plugging this last inequality (uniform in $\alpha$) into (5.35), we then obtain

$$v_i(\bar{x}) \geq \sup_{\alpha \in \mathcal{A}} E\Big[ \int_0^\theta e^{-\beta t} f_i(X_t^{\bar{x},i}) dt + e^{-\beta\theta} v(X_\theta^{\bar{x},i}, I_\theta^i) - \sum_{\tau_n \leq \theta} g_{\iota_{n-1},\iota_n} \Big] + \delta c_0,$$

which is in contradiction with the dynamic programming principle (5.29). $\qquad\square$

We complete the characterization of the value functions with the following comparison principle for the system of variational inequalities (5.30).

**Theorem 5.3.3** *Let $U_i$ (resp. $V_i$), $i \in \mathbb{I}_m$ a family of u.s.c. viscosity subsolutions (resp. l.s.c. viscosity supersolutions) to (5.30), and satisfying a linear growth condition. Then, $U_i \leq V_i$ on $\mathbb{R}^d$ for all $i \in \mathbb{I}_m$.*

**Proof.** *Step 1.* We first construct strict supersolutions to the system (5.30) with suitable perturbations of $V_i$, $i \in \mathbb{I}_m$. For $i \in \mathbb{I}_m$, we set $\alpha_i = \min_{j \neq i} g_{ji}$, and $\psi_i(x) = C(1 + |x|^2) + \alpha_i$, $x \in \mathbb{R}^d$, where $C$ is a positive constant to be determined later. From the linear growth conditions on $b_i$, $\sigma_i$ and $f_i$, there exist some positive constants $\rho$ and $C_1$ s.t. for all $x \in \mathbb{R}^d$, $i \in \mathbb{I}_m$,

$$\beta\psi_i(x) - \mathcal{L}_i\psi_i(x) - f_i(x) \geq C(\beta - \rho)(1 + |x|^2) + \beta\alpha_i - C_1(1 + |x|^2).$$

Thus, by choosing $C > 0$ s.t. $C(\beta - \rho) + \beta \min_i \alpha_i - C_1 \geq 1$, we see that

$$\beta\psi_i(x) - \mathcal{L}_i\psi_i(x) - f_i(x) \geq 1, \quad \forall x \in \mathbb{R}^d, \ i \in \mathbb{I}_m. \tag{5.36}$$

We now define for all $\lambda \in (0,1)$, the u.s.c. functions on $\mathbb{R}^d$ by

$$V_i^\lambda = (1 - \lambda)V_i + \lambda\psi_i, \quad i \in \mathbb{I}_m.$$

We then see that for all $\lambda \in (0,1)$, $i \in \mathbb{I}_m$

$$V_i^\lambda - \max_{j \neq i}(V_j^\lambda - g_{ij}) = (1-\lambda)V_i + \lambda\alpha_i - \max_{j \neq i}[(1-\lambda)(V_j - g_{ij}) + \lambda\alpha_j - \lambda g_{ij}]$$

$$\geq (1-\lambda)[V_i - \max_{j \neq i}(V_j - g_{ij})] + \lambda\big(\alpha_i + \min_{j \neq i}(g_{ij} - \alpha_j)\big)$$

$$\geq \lambda \min_{i \in \mathbb{I}_m}\big(\alpha_i + \min_{j \neq i}(g_{ij} - \alpha_j)\big) := \lambda\underline{\nu}. \tag{5.37}$$

We now check that $\underline{\nu} > 0$, i.e. $\nu_i := \alpha_i + \min_{j \neq i}(g_{ij} - \alpha_j) > 0$, $\forall i \in \mathbb{I}_m$. Indeed, fix $i \in \mathbb{I}_m$, and let $k \in \mathbb{I}_m$ such that $\min_{j \neq i}(g_{ij} - \alpha_j) = g_{ik} - \alpha_k$ and set $\underline{i}$ such that $\alpha_i = \min_{j \neq i} g_{ji} = g_{\underline{i}i}$. We then have

$$\nu_i = g_{\underline{i}i} + g_{ik} - \min_{j \neq k} g_{jk} > g_{\underline{i}k} - \min_{j \neq k} g_{jk} \geq 0,$$

by (5.24) and thus $\underline{\nu} > 0$. From (5.37) and (5.36), we then deduce that for all $i \in \mathbb{I}_m$, $\lambda \in (0,1)$, $V_i^\lambda$ is a supersolution to

$$\min\left\{\beta V_i^\lambda - \mathcal{L}_i V_i^\lambda - f_i, V_i^\lambda - \max_{j \neq i}(V_j^\lambda - g_{ij})\right\} \geq \lambda\delta, \quad \text{on } \mathbb{R}^d, \tag{5.38}$$

where $\delta = \underline{\nu} \wedge 1 > 0$.

*Step 2.* In order to prove the comparison principle, it suffices to show that for all $\lambda \in (0,1)$

$$\max_{j \in \mathbb{I}_m} \sup(U_j - V_j^\lambda) \leq 0,$$

since the required result is obtained by letting $\lambda$ to 0. We argue by contradiction and suppose that there exists some $\lambda \in (0,1)$ and $i \in \mathbb{I}_m$ s.t.

$$M := \max_{j \in \mathbb{I}_m} \sup(U_j - V_j^\lambda) = \sup(U_i - V_i^\lambda) > 0. \tag{5.39}$$

From the linear growth condition on $U_i$, $V_i$, and the quadratic growth condition on $\psi_i$, we observe that $U_i(x) - V_i^\lambda(x)$ goes to $-\infty$ when $x$ goes to infinity. Hence, the u.s.c. function $U_i - V_i^\lambda$ attains its maximum $M$. Let us consider the family of u.s.c. functions

$$\Phi_\varepsilon(x,y) = U_i(x) - V_i^\lambda(y) - \phi_\varepsilon(x,y), \quad \phi_\varepsilon(x,y) = \frac{1}{2\varepsilon}|x-y|^2, \quad \varepsilon > 0,$$

and the bounded family $(x_\varepsilon, y_\varepsilon)$ that attains the maximum of $\Phi_\varepsilon$. By standard arguments, we have

$$M_\varepsilon = \max \Phi_\varepsilon = \Phi_\varepsilon(x_\varepsilon, y_\varepsilon) \rightarrow M, \quad \text{and} \quad \phi_\varepsilon(x_\varepsilon, y_\varepsilon) \rightarrow 0, \tag{5.40}$$

as $\varepsilon$ goes to zero. From Ishii's lemma, we get the existence of $M_\varepsilon$, $N_\varepsilon \in \mathcal{S}_d$ s.t.

$$\left(\frac{x_\varepsilon - y_\varepsilon}{\varepsilon}, M_\varepsilon\right) \in \bar{\mathcal{P}}^{2,+}U_i(x_\varepsilon), \quad \left(\frac{x_\varepsilon - y_\varepsilon}{\varepsilon}, N_\varepsilon\right) \in \bar{\mathcal{P}}^{2,-}V_i^\lambda(y_\varepsilon),$$

and

$$\text{tr}\big(\sigma\sigma'(x_\varepsilon)M_\varepsilon - \sigma\sigma'(y_\varepsilon)N_\varepsilon\big) \leq \frac{3}{\varepsilon}|\sigma(x_\varepsilon) - \sigma(y_\varepsilon)|^2.$$

By writing the viscosity subsolution property (5.30) of $U_i$ and the viscosity strict super-solution property (5.38) of $V_i^\lambda$, we have the following inequalities:

$$\min\left[\beta U_i(x_\varepsilon) - \frac{1}{\varepsilon}(x_\varepsilon - y_\varepsilon).b_i(x_\varepsilon) - \frac{1}{2}\mathrm{tr}(\sigma_i\sigma_i'(x_\varepsilon)M_\varepsilon) - f_i(x_\varepsilon),\right.$$
$$\left. U_i(x_\varepsilon) - \max_{j\neq i}(U_j - g_{ij})(x_\varepsilon)\right] \leq 0 \qquad (5.41)$$

$$\min\left[\beta V_i^\lambda(y_\varepsilon) - \frac{1}{\varepsilon}(x_\varepsilon - y_\varepsilon).b_i(y_\varepsilon) - \frac{1}{2}\mathrm{tr}(\sigma_i\sigma_i'(y_\varepsilon)N_\varepsilon) - f_i(y_\varepsilon),\right.$$
$$\left. V_i^\lambda(y_\varepsilon) - \max_{j\neq i}(V_j^\lambda - g_{ij})(y_\varepsilon)\right] \geq \lambda\delta. \qquad (5.42)$$

If $U_i(x_\varepsilon) - \max_{j\neq i}(U_j - g_{ij})(x_\varepsilon) \leq 0$ in (5.41), then, since $V_i^\lambda(y_\varepsilon) - \max_{j\neq i}(V_j^\lambda - g_{ij})(y_\varepsilon)$ $\geq \lambda\delta$ by (5.42), we obtain

$$U_i(x_\varepsilon) - V_i^\lambda(y_\varepsilon) \leq -\lambda\delta + \max_{j\neq i}(U_j - g_{ij})(x_\varepsilon) - \max_{j\neq i}(V_j^\lambda - g_{ij})(y_\varepsilon)$$
$$\leq -\lambda\delta + \max_{j\neq i}(U_j(x_\varepsilon) - V_j^\lambda(y_\varepsilon)).$$

By sending $\varepsilon$ to zero, and from (5.40), we get the contradiction $M \leq -\lambda\delta + M$. Otherwise, if $U_i(x_\varepsilon) - \max_{j\neq i}(U_j - g_{ij})(x_\varepsilon) > 0$ in (5.41), then the first term of the l.h.s. of (5.41) is nonpositive, and by substracting with the nonnegative first term of the l.h.s. of (5.42), we conclude as in the proof of Theorem 5.2.1.    □

**Remark 5.3.3** As in the case of the optimal stopping problem, the continuity and the unique characterization of the value functions $v_i$ to the system of variational inequalities (5.30) also hold true when we remove the Lipschitz condition on $f_i$, by assuming only that they are continuous and satisfy a linear growth condition. Indeed, the viscosity property is proved similarly by means of discontinuous viscosity solutions, and the strong comparison principle in Theorem 5.3.3 (which only requires continuity and linear growth condition on $f_i$) implies, in addition to the uniqueness, the continuity of the value fuctions.

### 5.3.3 Switching regions

For any regime $i \in \mathbb{I}_m$, we introduce the closed set

$$\mathcal{S}_i = \left\{x \in \mathbb{R}^d : v_i(x) = \max_{j\neq i}(v_j - g_{ij})(x)\right\}. \qquad (5.43)$$

$\mathcal{S}_i$ is called the switching region since it corresponds to the region where it is optimal for the controller to change the regime. The complement set $\mathcal{C}_i$ of $\mathcal{S}_i$ in $\mathbb{R}^d$ is the so-called continuation region:

$$\mathcal{C}_i = \left\{x \in \mathbb{R}^d : v_i(x) > \max_{j\neq i}(v_j - g_{ij})(x)\right\},$$

where it is optimal to stay in regime $i$. Similarly as in Lemma 5.2.2 for optimal stopping problem, we show that the value function $v_i$ is a viscosity solution to

$$\beta v_i - \mathcal{L}_i v_i - f_i = 0 \quad \text{on } \mathcal{C}_i.$$

Moreover, if the function $\sigma_i$ is uniformly elliptic, then $v_i$ is $C^2$ on $\mathcal{C}_i$.

Let us introduce the functions

$$\hat{V}_i(x) = E\left[\int_0^\infty e^{-\beta t} f_i(\hat{X}_t^{x,i}) dt\right], \quad x \in \mathbb{R}^d, \ i \in \mathbb{I}_m, \tag{5.44}$$

where $\hat{X}^{x,i}$ is the solution to the diffusion in regime $i$ starting from $x$ at time 0, i.e. the solution to

$$dX_t = b_i(X_t)dt + \sigma_i(X_t)dW_t, \quad X_0 = x.$$

$\hat{V}_i$ corresponds to the expected profit $J_i(x, \alpha)$ where $\alpha$ is the strategy of never switching. In particular, we obviously have $v_i \geq \hat{V}_i$.

**Lemma 5.3.8** *Let $i \in \mathbb{I}_m$. If $\mathcal{S}_i = \emptyset$, then $v_i = \hat{V}_i$, and so*

$$\sup_{x \in \mathbb{R}^d} \max_{j \neq i} (\hat{V}_j - \hat{V}_i - g_{ij}) \leq 0. \tag{5.45}$$

**Proof.** If $\mathcal{S}_i = \emptyset$, i.e. $\mathcal{C}_i = \mathbb{R}^d$, then $v_i$ is a viscosity solution to

$$\beta v_i - \mathcal{L}_i v_i - f_i = 0 \quad \text{on } \mathbb{R}^d.$$

Since $\hat{V}_i$ is a viscosity solution to the same equation, and satisfies as well as $v_i$ a linear growth condition, we deduce by uniqueness that $v_i = \hat{V}_i$. Moreover, recalling that $v_i \geq v_j - g_{ij}$, and $v_j \geq \hat{V}_j$, for all $j \neq i$, we obtain (5.45).    □

**Remark 5.3.4** We shall use this lemma by contradiction: if relation (5.45) is violated, i.e. there exist $x_0 \in \mathbb{R}^d$, $j \neq i$, s.t. $(\hat{V}_j - \hat{V}_i - g_{ij})(x_0) > 0$, then $\mathcal{S}_i$ is nonempty.

From the definition (5.43) of the switching regions, we have the elementary decomposition property

$$\mathcal{S}_i = \cup_{j \neq i} \mathcal{S}_{ij}, \quad i \in \mathbb{I}_m, \tag{5.46}$$

where

$$\mathcal{S}_{ij} = \left\{x \in \mathbb{R}^d : v_i(x) = (v_j - g_{ij})(x)\right\}$$

is the switching region from regime $i$ to regime $j$. Moreover, from the triangular condition (5.24), when one switches from regime $i$ to regime $j$, one does not switch immediately to another regime, i.e. one stays for a while in the continuation region of regime $j$. In other words,

$$\mathcal{S}_{ij} \subset \mathcal{C}_j, \quad j \neq i \in \mathbb{I}_m.$$

The following useful lemma gives some partial information about the structure of the switching regions.

**Lemma 5.3.9** *Let* $i \neq j$ *in* $\mathbb{I}_m$, *and assume that* $\sigma_j$ *is uniformly elliptic. Then, we have*

$$\mathcal{S}_{ij} \subset \mathcal{Q}_{ij} := \{x \in \mathcal{C}_j : (\mathcal{L}_j - \mathcal{L}_i)v_j(x) + (f_j - f_i)(x) - \beta g_{ij} \geq 0\}.$$

**Proof.** Let $x \in \mathcal{S}_{ij}$. By setting $\varphi_j = v_j - g_{ij}$, it follows that $x$ is a minimum of $v_i - \varphi_j$ with $v_i(x) = \varphi_j(x)$. Moreover, since $x$ lies in the open set $\mathcal{C}_j$ where $v_j$ is smooth under the condition that $\sigma_j$ is uniformly elliptic , we have that $\varphi_j$ is $C^2$ in a neighborhood of $x$. By the supersolution viscosity property of $v_i$ to the PDE (5.30), this yields

$$\beta\varphi_j(x) - \mathcal{L}_i\varphi_j(x) - f_i(x) \geq 0. \tag{5.47}$$

Now recall that for $x \in \mathcal{C}_j$, we have

$$\beta v_j(x) - \mathcal{L}_j v_j(x) - f_j(x) = 0,$$

so by substituting into (5.47), we obtain

$$(\mathcal{L}_j - \mathcal{L}_i)v_j(x) + (f_j - f_i)(x) - \beta g_{ij} \geq 0,$$

which is the required result. □

### 5.3.4 The one-dimensional case

In this section, we consider the case where the state process $X$ is real-valued. Similarly as for optimal stopping problem, we prove the smooth-fit property, i.e. the continuous differentiability of the value functions.

**Proposition 5.3.2** *Assume that* $X$ *is one-dimensional and* $\sigma_i$ *is uniformly elliptic, for all* $i \in \mathbb{I}_m$. *Then, the value functions* $v_i$, $i \in \mathbb{I}_m$, *are continuously differentiable on* $\mathbb{R}$. *Moreover, at* $x \in \mathcal{S}_{ij}$, *we have* $v_i'(x) = v_j'(x)$, $i \neq j \in \mathbb{I}_m$.

**Proof.** We already know that $v_i$ is smooth $C^2$ on the open set $\mathcal{C}_i$ for all $i \in \mathbb{I}_m$. We have to prove the $C^1$ property of $v_i$ at any point of the closed set $\mathcal{S}_i$. We denote for all $j \in \mathbb{I}_m$, $j \neq i$, $h_j = v_j - g_{ij}$ and we notice that $h_j$ is smooth $C^1$ (actually even $C^2$) on $\mathcal{C}_j$.

**1.** We first check that $v_i$ admits a left and right derivative $v_{i,-}'(x_0)$ and $v_{i,+}'(x_0)$ at any point $x_0$ in $\mathcal{S}_i = \cup_{j \neq i}\mathcal{S}_{ij}$. We distinguish the two following cases:

• *(a)* $x_0$ lies in the interior $\text{Int}(\mathcal{S}_i)$ of $\mathcal{S}_i$. Then, we have two subcases:

  ⋆ $x_0 \in \text{Int}(\mathcal{S}_{ij})$ for some $j \neq i$, i.e. there exists some $\delta > 0$ s.t. $[x_0 - \delta, x_0 + \delta] \subset \mathcal{S}_{ij}$. By definition of $\mathcal{S}_{ij}$, we then have $v_i = h_j$ on $[x_0 - \delta, x_0 + \delta] \subset \mathcal{C}_j$, and so $v_i$ is differentiable at $x_0$ with $v_i'(x_0) = h_j'(x_0)$.

  ⋆ There exists $j \neq k \neq i$ in $\mathbb{I}_m$ and $\delta > 0$ s.t. $[x_0 - \delta, x_0] \subset \mathcal{S}_{ij}$ and $[x_0, x_0 + \delta] \subset \mathcal{S}_{ik}$. We then have $v_i = h_j$ on $[x_0 - \delta, x_0] \subset \mathcal{C}_j$ and $v_i = h_k$ on $[x_0, x_0 + \delta] \subset \mathcal{C}_k$. Thus, $v_i$ admits a left and right derivative at $x_0$ with $v_{i,-}'(x_0) = h_j'(x_0)$ and $v_{i,+}'(x_0) = h_k'(x_0)$.

• *(b)* $x_0$ lies in the boundary $\partial\mathcal{S}_i = \mathcal{S}_i \setminus \text{Int}(\mathcal{S}_i)$ of $\mathcal{S}_i$. We assume that $x_0$ lies in the left-boundary of $\mathcal{S}_i$, i.e. there exists $\delta > 0$ s.t. $[x_0 - \delta, x_0) \subset \mathcal{C}_i$ (the other case where $x_0$ lies in the right-boundary is dealt with similarly). Recalling that on $\mathcal{C}_i$, $v_i$ is solution to $\beta v_i - \mathcal{L}_i v_i - f_i = 0$, we deduce that on $[x_0 - \delta, x_0)$, $v_i$ is equal to $w_i$, the unique smooth

$C^2$ solution to the ODE $\beta w_i - \mathcal{L}_i w_i - f_i = 0$ with the boundary conditions $w_i(x_0 - \delta)$ $= v_i(x_0 - \delta)$, $w_i(x_0) = v_i(x_0)$. Therefore, $v_i$ admits a left derivative at $x_0$ with $v'_{i,-}(x_0)$ $= w'_i(x_0)$. In order to prove that $v_i$ admits a right derivative, we distinguish the two subcases:

$\star$ There exists $j \neq i$ in $\mathbb{I}_m$ and $\delta' > 0$ s.t. $[x_0, x_0 + \delta'] \subset \mathcal{S}_{ij}$. Then, on $[x_0, x_0 + \delta']$, $v_i$ is equal to $h_j$. Hence $v_i$ admits a right derivative at $x_0$ with $v'_{i,+}(x_0) = h'_j(x_0)$.

$\star$ Otherwise, for all $j \neq i$, we can find a sequence $(x_n^j)$ s.t. $x_n^j \geq x_0$, $x_n^j \notin \mathcal{S}_{ij}$ and $x_n^j$ $\to x_0$. By a diagonalization procedure, we then construct a sequence $(x_n)$ s.t. $x^n \geq x_0$, $x_n \notin \mathcal{S}_{ij}$ for all $j \neq i$, i.e. $x_n \in \mathcal{C}_i$, and $x_n \to x_0$. Since $\mathcal{C}_i$ is open, there exists then $\delta''$ $> 0$ s.t. $[x_0, x_0 + \delta''] \subset \mathcal{C}_i$. We deduce that on $[x_0, x_0 + \delta'']$, $v_i$ is equal to $\hat{w}_i$ the unique smooth $C^2$ solution to the o.d.e. $\beta \hat{w}_i - \mathcal{L}_i \hat{w}_i - f_i = 0$ with the boundary conditions $\hat{w}_i(x_0) = v_i(x_0)$, $\hat{w}_i(x_0 + \delta'') = v_i(x_0 + \delta'')$. In particular, $v_i$ admits a right derivative at $x_0$ with $v'_{i,+}(x_0) = \hat{w}'_i(x_0)$.

**2.** Consider now some point in $\mathcal{S}_i$ eventually on its boundary. We recall from (5.46) that there exists some $j \neq i$ s.t. $x_0 \in \mathcal{S}_{ij}$: $v_i(x_0) = h_j(x_0)$, and $h_j$ is smooth $C^1$ on $x_0$ in $\mathcal{C}_j$. Since $v_j \geq h_j$, we deduce that

$$\frac{v_i(x) - v_i(x_0)}{x - x_0} \leq \frac{h_j(x) - h_j(x_0)}{x - x_0}, \quad \forall x < x_0$$
$$\frac{v_i(x) - v_i(x_0)}{x - x_0} > \frac{h_j(x) - h_j(x_0)}{x - x_0}, \quad \forall x > x_0,$$

and so:

$$v'_{i,-}(x_0) \leq h'_j(x_0) \leq v'_{i,+}(x_0).$$

We argue by contradiction and suppose that $v_i$ is not differentiable at $x_0$. Then, in view of the above inequality, one can find some $p \in (v'_{i,-}(x_0), v'_{i,+}(x_0))$. Consider, for $\varepsilon > 0$, the smooth $C^2$ function

$$\varphi_\varepsilon(x) = v_i(x_0) + p(x - x_0) + \frac{1}{2\varepsilon}(x - x_0)^2.$$

Then, we see that $v_i$ dominates locally in a neighborhood of $x_0$ the function $\varphi_\varepsilon$, i.e $x_0$ is a local minimum of $v_i - \varphi_\varepsilon$. From the supersolution viscosity property of $v_i$ to the PDE (5.30), this yields

$$\beta \varphi_\varepsilon(x_0) - \mathcal{L}_i \varphi_\varepsilon(x_0) - f_i(x_0) \geq 0,$$

which is written as

$$\beta v_i(x_0) - b_i(x_0)p - f_i(x_0) - \frac{1}{2\varepsilon}\sigma_i^2(x_0) \geq 0.$$

Sending $\varepsilon$ to zero provides the required contradiction since $\sigma(x_0) > 0$. We have then proved that for $x_0 \in \mathcal{S}_{ij}$, $v'_i(x_0) = h'_j(x_0) = v'_j(x_0)$. $\square$

We now focus on the case where $X$ is valued in $(0, \infty)$, and we assume that the real-valued coefficients $b_i$ and $\sigma_i$, $i \in \mathbb{I}_m$, of $X$ can be extended to Lipschitz continuous functions on $\mathbb{R}_+$, and $x = 0$ is an absorbing state in the sense that for all $t \geq 0$, $\alpha \in \mathcal{A}$,

$X_t^{0,i} = 0$. The functions $f_i$ are continuous on $\mathbb{R}_+$, and satisfy a linear growth condition. The state domain on which $v_i$ satisfy the PDE variational inequalities is then $(0, \infty)$, and we shall need to specify a boundary condition for $v_i(x)$ when $x$ approaches 0 in order to get a uniqueness result.

**Lemma 5.3.10** *For all $i \in \mathbb{I}_m$, we have*

$$v_i(0^+) = \max_{j \in \mathbb{I}_m} \left( \frac{f_j(0)}{\beta} - g_{ij} \right).$$

**Proof.** By considering the particular strategy $\alpha = (\tau_n, \iota_n)$ of immediately switching from the initial state $(x, i)$ to state $(x, j)$, $j \in \mathbb{I}_d$, at cost $g_{ij}$ and then doing nothing, i.e. $\tau_1 = 0$, $\iota_1 = j$, $\tau_n = \infty$, $\iota_n = j$ for all $n \geq 2$, we have

$$v_i(x) \geq J_i(x, \alpha) = E\left[ \int_0^\infty e^{-\beta t} f_j(\hat{X}_t^{x,j}) dt - g_{ij} \right] = \hat{V}_j(x) - g_{ij},$$

where $\hat{X}^{x,j}$ denotes the controlled process in regime $j$ starting from $x$ at time 0. Since $\hat{X}_t^{x,j}$ goes to $\hat{X}_t^{0,j} = 0$ as $x$ goes to zero, for all $t \geq 0$, we deduce by continuity of $f_j$, and dominated convergence theorem that $\hat{V}_j(0^+) = f_j(0)/\beta$. From the arbitrariness of $j \in \mathbb{I}_m$, this implies $\liminf_{x \downarrow 0} v_i(x) \geq \max_j (f_j(0)/\beta - g_{ij})$.

Conversely, let us define $\bar{f}_i = f_i - f_i(0)$, $i \in \mathbb{I}_m$, and observe that for any $x > 0$, $\alpha \in \mathcal{A}$, we have

$$
\begin{aligned}
J_i(x, \alpha) &= E\left[ \sum_{n=1}^\infty \int_{\tau_{n-1}}^{\tau_n} e^{-\beta t} f(X_t^{x,i}, \iota_{n-1}) dt - \sum_{n=1}^\infty e^{-\beta \tau_n} g_{\iota_{n-1},\iota_n} \right] \\
&= E\left[ \sum_{n=1}^\infty \int_{\tau_{n-1}}^{\tau_n} e^{-\beta t} \big( \bar{f}(X_t^{x,i}, \iota_{n-1}) + f_{\iota_{n-1}}(0) \big) dt - \sum_{n=1}^\infty e^{-\beta \tau_n} g_{\iota_{n-1},\iota_n} \right] \\
&= E\left[ \sum_{n=1}^\infty \int_{\tau_{n-1}}^{\tau_n} e^{-\beta t} \bar{f}(X_t^{x,i}, \iota_{n-1}) dt + \frac{f_{\iota_0}(0)}{\beta} \right. \\
&\qquad \left. - \sum_{n=1}^\infty e^{-\beta \tau_n} \Big( g_{\iota_{n-1},\iota_n} + \frac{f_{\iota_{n-1}}(0) - f_{\iota_n}(0)}{\beta} \Big) \right] \\
&= \frac{f_i(0)}{\beta} + E\left[ \int_0^\infty e^{-\beta t} \bar{f}(X_t^{x,i}, I_t^i) dt - \sum_{n=1}^\infty e^{-\beta \tau_n} \bar{g}_{\iota_{n-1},\iota_n} \right],
\end{aligned}
\tag{5.48}
$$

where

$$\bar{g}_{ij} = g_{ij} + \frac{f_i(0) - f_j(0)}{\beta}, \tag{5.49}$$

also satisfy the triangular condition $\bar{g}_{ik} < \bar{g}_{ij} + \bar{g}_{jk}$, $j \neq i, k$. As in (5.27) in Lemma 5.3.7, we have for any $\alpha \in \mathcal{A}$,

$$-\sum_{n=1}^\infty e^{-\beta \tau_n} \bar{g}_{\iota_{n-1},\iota_n} \leq \max_{j \in \mathbb{I}_m} (-\bar{g}_{ij}).$$

Define $\bar{f}_i^*(y) = \sup_{x>0} [\bar{f}_i(x) - xy]$, $i \in \mathbb{I}_m$, and $\bar{f}^*(y) = \max_i \bar{f}_i^*(y)$, $y > 0$, and notice from the linear growth conditions on $\bar{f}_i$ that $\bar{f}_i^*(y)$ is finite for $y > C$ large enough, with $\bar{f}_i^*(\infty) = \bar{f}_i(0) = 0$, and so $\bar{f}^*(\infty) = 0$. From (5.48), we have for all $y > C$, $\alpha \in \mathcal{A}$,

$$J_i(x, \alpha) \leq \frac{f_i(0)}{\beta} + E\left[\int_0^\infty e^{-\beta t}\left(y X_t^{x,i} + \bar{f}^*(y)\right) dt\right] + \max_{j \in \mathbb{I}_m}(-\bar{g}_{ij}). \tag{5.50}$$

Now, from the Lipschitz condition on $b_i$, $\sigma_i$ on $\mathbb{R}_+$, and by standard arguments based on Itô's formula and Gronwall's lemma, we have for $i \in \mathbb{I}_m$, $\alpha \in \mathcal{A}$, $x > 0$

$$E[X_t^{x,i}] = E[|X_t^{x,i} - X_t^{0,i}|] \leq e^{\rho t}|x - 0| = e^{\rho t}x,$$

where $\rho$ is a positive constant depending on the Lipschitz coefficients of $b_i$, $\sigma_i$, $i \in \mathbb{I}_m$. We deduce from (5.50) that for $\beta > \rho$,

$$v_i(x) \leq \frac{f_i(0)}{\beta} + \frac{xy}{\beta - \rho} + \frac{\bar{f}^*(y)}{\beta} + \max_{j \in \mathbb{I}_m}(-\bar{g}_{ij}), \quad \forall x > 0, y > C.$$

By sending $x$ to zero, and then $y$ to infinity, we conclude that

$$\limsup_{x \downarrow 0} v_i(x) \leq \frac{f_i(0)}{\beta} + \max_{j \in \mathbb{I}_m}(-\bar{g}_{ij}) = \max_{j \in \mathbb{I}_m}\left(\frac{f_j(0)}{\beta} - g_{ij}\right),$$

which ends the proof. $\qquad\square$

**Remark 5.3.5** From the relations (5.48), we may assume without loss of generality that $f_i(0) = 0$, for all $i \in \mathbb{I}_m$, by modifying switching costs as in (5.49), to take into account the possibly different initial values of the profit functions.

### 5.3.5 Explicit solution in the two-regime case

In this section, we consider the case where the number of regimes is $m = 2$ for a one-dimensional state process in $(0, \infty)$ as in the previous section. Without loss of generality (see Remark 5.3.5), we shall assume that $f_i(0) = 0$, $i = 1, 2$. We then know that the value functions $v_i$, $i = 1, 2$, are the unique continuous viscosity solutions with linear growth condition on $(0, \infty)$, and boundary conditions $v_i(0^+) = (-g_{ij})_+ := \max(-g_{ij}, 0)$, $j \neq i$, to the system

$$\min\left\{\beta v_1 - \mathcal{L}_1 v_1 - f_1, v_1 - (v_2 - g_{12})\right\} = 0 \tag{5.51}$$
$$\min\left\{\beta v_2 - \mathcal{L}_2 v_2 - f_2, v_2 - (v_1 - g_{21})\right\} = 0. \tag{5.52}$$

Moreover, the switching regions are

$$\mathcal{S}_i = \mathcal{S}_{ij} = \{x > 0 : v_i(x) = v_j(x) - g_{ij}\}, \quad i, j = 1, 2, \ i \neq j.$$

We set

$$\underline{x}_i^* = \inf \mathcal{S}_i \in [0, \infty], \quad \bar{x}_i^* = \sup \mathcal{S}_i \in [0, \infty],$$

with the usual convention that $\inf \emptyset = \infty$.

We shall provide explicit solutions in the two following situations:

- Different diffusion regimes with identical profit functions
- Different profit functions with identical diffusion regimes

We also consider the cases for which both switching costs are positive, and for which one of the two is negative, the other then being positive according to (5.25). This last case is interesting in applications where a firm chooses between an open or closed activity, and may regain a fraction of its opening costs when it decides to close.

**Different diffusion regimes with identical profit functions**

We suppose that the running profit functions are identical in the form

$$f_1(x) \;=\; f_2(x) = x^p, \quad x > 0, \quad \text{for some } 0 < p < 1.$$

The diffusion regimes correspond to two different geometric Brownian motions, i.e.

$$b_i(x) \;=\; \rho_i x, \; \sigma_i(x) \;=\; \gamma_i x, \quad \text{for some constants } \rho_i, \; \gamma_i > 0, \; i = 1, 2.$$

We assume that $\beta > \max(0, \rho_1, \rho_2)$, which ensures that the value functions $v_i$ are finite and satisfy a linear growth condition. In this case, a straightforward calculation shows that the value functions $\hat{V}_i$ in (5.44) corresponding to the problem without regime switching are given by

$$\hat{V}_i(x) = K_i x^p \quad \text{with} \quad K_i = \frac{1}{\beta - \rho_i p + \frac{1}{2}\gamma_i^2 p(1-p)} > 0, \; i = 1, 2.$$

Recall that $\hat{V}_i$ is a particular solution to the second-order ODE

$$\beta w - \mathcal{L}_i w - f_i = 0, \tag{5.53}$$

whose general solution (without second member $f_i$) is of the form

$$w(x) = A x^{m_i^+} + B x^{m_i^-},$$

for some constants $A$, $B$, and where

$$m_i^- = -\frac{\rho_i}{\gamma_i^2} + \frac{1}{2} - \sqrt{\left(-\frac{\rho_i}{\gamma_i^2} + \frac{1}{2}\right)^2 + \frac{2\beta}{\gamma_i^2}} < 0,$$

$$m_i^+ = -\frac{\rho_i}{\gamma_i^2} + \frac{1}{2} + \sqrt{\left(-\frac{\rho_i}{\gamma_i^2} + \frac{1}{2}\right)^2 + \frac{2\beta}{\gamma_i^2}} > 1.$$

We show that the structure of the switching regions depends actually only on the sign of $K_2 - K_1$, and of the sign of the switching costs $g_{12}$ and $g_{21}$. More precisely, we have the following explicit result.

**Theorem 5.3.4** *Let $i, j = 1, 2, i \neq j$.*
(1) *If $K_i = K_j$, then*

$$v_i(x) = \hat{V}_i(x) + (-g_{ij})_+, \quad x \in (0, \infty),$$

$$\mathcal{S}_i = \begin{cases} \emptyset & \text{if } g_{ij} > 0, \\ (0, \infty) & \text{if } g_{ij} \leq 0. \end{cases}$$

*It is always optimal to switch from regime $i$ to $j$ if the corresponding switching cost is nonpositive, and never optimal to switch otherwise.*
(2) *If $K_j > K_i$, then we have the following situations depending on the switching costs:*

(a) *$g_{ij} \leq 0$: We have $\mathcal{S}_i = (0, \infty)$, $\mathcal{S}_j = \emptyset$, and*

$$v_i = \hat{V}_j - g_{ij}, \quad v_j = \hat{V}_j.$$

(b) $g_{ij} > 0$:

- If $g_{ji} \geq 0$, then $\mathcal{S}_i = [\underline{x}_i^*, \infty)$ with $\underline{x}_i^* \in (0, \infty)$, $\mathcal{S}_j = \emptyset$, and

$$
v_i(x) = \begin{cases} Ax^{m_i^+} + \hat{V}_i(x), & x < \underline{x}_i^*, \\ v_j(x) - g_{ij}, & x \geq \underline{x}_i^*, \end{cases} \tag{5.54}
$$

$$
v_j(x) = \hat{V}_j(x), \quad x \in (0, \infty), \tag{5.55}
$$

where the constants $A$ and $\underline{x}_i^*$ are determined by the continuity and smooth-fit conditions of $v_i$ at $\underline{x}_i^*$, and explicitly given by

$$
\underline{x}_i^* = \left( \frac{m_i^+}{m_i^+ - p} \frac{g_{ij}}{K_j - K_i} \right)^{\frac{1}{p}}, \tag{5.56}
$$

$$
A = (K_j - K_i) \frac{p}{m_i^+} (\underline{x}_i^*)^{p - m_i^+}. \tag{5.57}
$$

When we are in regime $i$, it is optimal to switch to regime $j$ whenever the state process $X$ exceeds the threshold $\underline{x}_i^*$, while when we are in regime $j$, it is optimal to never switch.

- If $g_{ji} < 0$, then $\mathcal{S}_i = [\underline{x}_i^*, \infty)$ with $\underline{x}_i^* \in (0, \infty)$, $\mathcal{S}_j - (0, \bar{x}_j^*]$, and

$$
v_i(x) = \begin{cases} Ax^{m_i^+} + \hat{V}_i(x), & x < \underline{x}_i^*, \\ v_j(x) - g_{ij}, & x \geq \underline{x}_i^*, \end{cases} \tag{5.58}
$$

$$
v_j(x) = \begin{cases} v_i(x) - g_{ji}, & x \leq \bar{x}_j^*, \\ Bx^{m_j^-} + \hat{V}_j(x), & x > \bar{x}_i^*, \end{cases} \tag{5.59}
$$

where the constants $A$, $B$ and $\bar{x}_j^* < \underline{x}_i^*$ are determined by the continuity and smooth-fit conditions of $v_i$ and $v_j$ at $\underline{x}_i^*$ and $\bar{x}_j^*$, and explicitly given by

$$
\bar{x}_j^* = \left[ \frac{-m_j^-(g_{ji} + g_{ij} y^{m_i^+})}{(K_i - K_j)(p - m_j^-)(1 - y^{m_i^+ - p})} \right]^{\frac{1}{p}},
$$

$$
\underline{x}_i^* = \frac{\bar{x}_j^*}{y},
$$

$$
B = \frac{(K_i - K_j)(m_i^+ - p)\underline{x}_i^{*p - m_j^-} + m_i^+ g_{ij} \underline{x}_i^{*-m_j^-}}{m_i^+ - m_j^-},
$$

$$
A = B\underline{x}_i^{*m_j^- - m_i^+} - (K_i - K_j)\underline{x}_i^{*p - m_i^+} - g_{ij}\underline{x}_i^{*-m_i^+},
$$

with $y$ solution in

$$
\left( 0, \left( -\frac{g_{ji}}{g_{ij}} \right)^{\frac{1}{m_i^+}} \right)
$$

to the equation

$$
m_i^+(p - m_j^-)(1 - y^{m_i^+ - p})(g_{ij} y^{m_j^-} + g_{ji})
$$
$$
+ m_j^-(m_i^+ - p)(1 - y^{m_j^- - p})(g_{ij} y^{m_i^+} + g_{ji}) = 0.
$$

*When we are in regime i, it is optimal to switch to regime j whenever the state process X exceeds the threshold $\underline{x}_i^*$, while when we are in regime j, it is optimal to switch to regime i for values of the state process X under the threshold $\bar{x}_j^*$.*

### Economic interpretation

In the particular case where $\gamma_1 = \gamma_2$, the condition $K_2 - K_1 > 0$ means that regime 2 provides a higher expected return $\rho_2$ than $\rho_1$ of regime 1 for the same volatility coefficient $\gamma_i$. Moreover, if the switching cost $g_{21}$ from regime 2 to regime 1 is nonnegative, it is intuitively clear that it is in our best interest to always stay in regime 2, which is formalized by the property that $\mathcal{S}_2 = \emptyset$. However, if one receives some gain compensation to switch from regime 2 to regime 1, i.e., the corresponding cost $g_{21}$ is negative, then it is in our best interest to change regime for small values of the current state. This is formalized by the property that $\mathcal{S}_2 = (0, \bar{x}_2^*]$. On the other hand, in regime 1, our best interest is to switch to regime 2, for all current values of the state if the corresponding switching cost $g_{12}$ is nonpositive, or from a certain threshold $\underline{x}_1^*$ if the switching cost $g_{12}$ is positive. A similar interpretation holds when $\rho_1 = \rho_2$, and $K_2 - K_1 > 0$, i.e., $\gamma_2 < \gamma_1$. Theorem 5.3.4 extends these results for general coefficients $\rho_i$ and $\gamma_i$, and shows that the critical parameter value determining the form of the optimal strategy is given by the sign of $K_2 - K_1$ and the switching costs. The different optimal strategy structures are depicted in Figure I.

### Proof of Theorem 5.3.4.

(1) If $K_i = K_j$, then $\hat{V}_i = \hat{V}_j$. We consider the smooth functions $w_i = \hat{V}_i + (-g_{ij})_+$ for $i, j = 1, 2$ and $j \neq i$. Since $\hat{V}_i$ are solution to (5.53), we see that $w_i$ satisfy

$$\beta w_i - \mathcal{L} w_i - f_i = \beta(-g_{ij})_+ \tag{5.60}$$

$$w_i - (w_j - g_{ij}) = g_{ij} + (-g_{ij})_+ - (-g_{ji})_+. \tag{5.61}$$

Notice that the l.h.s. of (5.60) and (5.61) are both nonnegative by (5.25). Moreover, if $g_{ij} > 0$, then the l.h.s. of (5.60) is zero, and if $g_{ij} \leq 0$, then $g_{ji} > 0$ and the l.h.s. of (5.61) is zero. Therefore, $w_i$, $i = 1, 2$ is solution to the system

$$\min\{\beta w_i - \mathcal{L}_i w_i - f_i, w_i - (w_j - g_{ij})\} = 0.$$

Since $\hat{V}_i(0^+) = 0$, we have $w_i(0^+) = (-g_{ij})_+ = v_i(0^+)$ by Lemma 5.3.10. Moreover, $w_i$ satisfy like $\hat{V}_i$ a linear growth condition. Therefore, by uniqueness to the system (5.51)-(5.52), we deduce that $v_i = w_i$. As observed above, if $g_{ij} \leq 0$, then the l.h.s. of (5.61) is zero, and so $\mathcal{S}_i = (0, \infty)$. Finally, if $g_{ij} > 0$, then the l.h.s. of (5.61) is positive, and so $\mathcal{S}_i = \emptyset$.

(2) We now suppose w.l.o.g. that $K_2 > K_1$.

(a) Consider first the case where $g_{12} \leq 0$, and so $g_{21} > 0$. We set $w_1 = \hat{V}_2 - g_{12}$ and $w_2 = \hat{V}_2$. Then, by construction, we have $w_1 = w_2 - g_{12}$ on $(0, \infty)$, and by definition of $\hat{V}_1$ and $\hat{V}_2$

$$\beta w_1(x) - \mathcal{L}_1 w_1(x) - f_1(x) = \frac{K_2 - K_1}{K_1} x^p - \beta g_{12} > 0, \quad \forall x > 0.$$

On the other hand, we also have $\beta w_2 - \mathcal{L}_2 w_2 - f_2 = 0$ on $(0, \infty)$, and $w_2 > w_1 - g_{21}$ since $g_{12} + g_{21} > 0$. Hence, $w_1$ and $w_2$ are smooth (hence viscosity) solutions to the

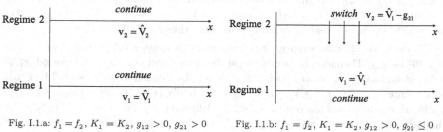

Fig. I.1.a: $f_1 = f_2$, $K_1 = K_2$, $g_{12} > 0$, $g_{21} > 0$    Fig. I.1.b: $f_1 = f_2$, $K_1 = K_2$, $g_{12} > 0$, $g_{21} \leq 0$

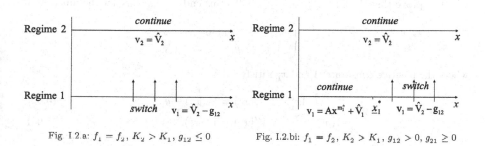

Fig. I.2.a: $f_1 = f_2$, $K_2 > K_1$, $g_{12} \leq 0$    Fig. I.2.bi: $f_1 = f_2$, $K_2 > K_1$, $g_{12} > 0$, $g_{21} \geq 0$

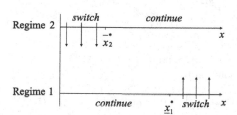

Fig. I.2.bii: $f_1 = f_2$, $K_2 > K_1$, $g_{12} > 0$, $g_{21} < 0$

Fig. I

system (5.51)-(5.52), with linear growth conditions and boundary conditions $w_1(0^+)$ $= V_1(0^+) - g_{12} = (-g_{12})_+ = v_1(0^+)$, $w_2(0^+) = \hat{V}_2(0^+) = 0 = (-g_{21})_+ = v_2(0^+)$. By uniqueness, we deduce that $v_1 = w_1$, $v_2 = w_2$, and thus $\mathcal{S}_1 = (0, \infty)$, $\mathcal{S}_2 = \emptyset$.

(b) Consider now the case where $g_{12} > 0$. From (5.25), we have $v_1(0^+) = 0 > (-g_{21})_+ - g_{12}$ $= v_2(0^+) - g_{12}$. Therefore, by continuity of the value functions on $(0, \infty)$, we get $\underline{x}_1^* > 0$. We claim that $\underline{x}_1^* < \infty$. Otherwise, $\mathcal{S}_1 = \emptyset$, and by Lemma 5.3.8, $v_1$ should be equal to $\hat{V}_1$. Since $v_1 \geq v_2 - g_{12} \geq \hat{V}_2 - g_{12}$, this would imply $(\hat{V}_2 - \hat{V}_1)(x) = (K_2 - K_1)x^p \leq g_{12}$ for all $x > 0$, an obvious contradiction. By definition of $\underline{x}_1^*$, we have $(0, \underline{x}_1^*) \subset \mathcal{C}_1$. We shall actually prove the equality $(0, \underline{x}_1^*) = \mathcal{C}_1$, i.e. $\mathcal{S}_1 = [\underline{x}_1^*, \infty)$. On the other hand, the form of $\mathcal{S}_2$ will depend on the sign of $g_{21}$.

• *Case:* $g_{21} \geq 0$.

We shall prove that $\mathcal{C}_2 = (0, \infty)$, i.e. $\mathcal{S}_2 = \emptyset$. To this end, let us consider the function

$$w_1(x) = \begin{cases} Ax^{m_1^+} + \hat{V}_1(x), & 0 < x < x_1 \\ \hat{V}_2(x) - g_{12}, & x \geq x_1, \end{cases}$$

where the positive constants $A$ and $x_1$ satisfy

$$Ax_1^{m_1^+} + \hat{V}_1(x_1) = \hat{V}_2(x_1) - g_{12} \tag{5.62}$$

$$Am_1^+ x_1^{m_1^+ - 1} + \hat{V}_1'(x_1) = \hat{V}_2'(x_1), \tag{5.63}$$

and are explicitly determined by

$$(K_2 - K_1)x_1^p = \frac{m_1^+}{m_1^+ - p} g_{12} \tag{5.64}$$

$$A = (K_2 - K_1)\frac{p}{m_1^+} x_1^{p - m_1^+}. \tag{5.65}$$

Notice that by construction, $w_1$ is $C^2$ on $(0, x_1) \cup (x_1, \infty)$, and $C^1$ on $x_1$.

⋆ Let us now show that $w_1$ is a viscosity solution to

$$\min\left\{ \beta w_1 - \mathcal{L}_1 w_1 - f_1, w_1 - (\hat{V}_2 - g_{12}) \right\} = 0, \quad \text{on } (0, \infty). \tag{5.66}$$

We first check that

$$w_1(x) \geq \hat{V}_2(x) - g_{12}, \quad \forall\, 0 < x < x_1, \tag{5.67}$$

i.e.

$$G(x) := Ax^{m_1^+} + \hat{V}_1(x) - \hat{V}_2(x) + g_{12} \geq 0, \quad \forall\, 0 < x < x_1.$$

Since $A > 0$, $0 < p < 1 < m_1^+$, $K_2 - K_1 > 0$, a direct derivation shows that the second derivative of $G$ is positive, i.e. $G$ is strictly convex. By (5.63), we have $G'(x_1) = 0$ and so $G'$ is negative, i.e. $G$ is strictly decreasing on $(0, x_1)$. Now, by (5.62), we have $G(x_1) = 0$ and thus $G$ is positive on $(0, x_1)$, which proves (5.67).

By definition of $w_1$ on $(0, x_1)$, we have in the classical sense

$$\beta w_1 - \mathcal{L}_1 w_1 - f_1 = 0, \quad \text{on } (0, x_1). \tag{5.68}$$

We now check that

$$\beta w_{1} - \mathcal{L}_1 w_{1} - f_1 \geq 0, \quad \text{on } (x_{1}, \infty), \tag{5.69}$$

holds true in the classical sense, and so a fortiori in the viscosity sense. By definition of $w_1$ on $(x_1, \infty)$, and $K_1$, we have for all $x > x_1$,

$$\beta w_{1}(x) - \mathcal{L}_1 w_{1}(x) - f_1(x) = \frac{K_2 - K_1}{K_1} x^p - \beta g_{12}, \quad \forall x > x_1,$$

so that (5.69) is satisfied iff $\frac{K_2 - K_1}{K_1} x_1^p - \beta g_{12} \geq 0$ or equivalently by (5.64):

$$\frac{m_1^+}{m_1^+ - p} \geq \beta K_1 = \frac{\beta}{\beta - \rho_1 p + \frac{1}{2}\gamma_1^2 p(1 - p)} \tag{5.70}$$

Now, since $p < 1 < m_1^+$, and by definition of $m_1^+$, we have

$$\frac{1}{2}\gamma_1^2 m_1^+(p - 1) < \frac{1}{2}\gamma_1^2 m_1^+(m_1^+ - 1) = \beta - \rho_1 m_1^+,$$

which proves (5.70) and thus (5.69).

From relations (5.67), (5.68) and (5.69), we see that the viscosity solution property of $w_1$ to (5.66) is satisfied at any point $x \in (0, x_1) \cup (x_1, \infty)$. It remains to check the viscosity property of $w_1$ to (5.66) at $x_1$, and this will follow from the smooth-fit condition of $w_1$ at $x_1$. Indeed, since $w_1(x_1) - \hat{V}_2(x_1) - g_{12}$, the viscosity subsolution property is trivial. For the viscosity supersolution property, take some smooth $C^2$ function $\varphi$ s.t. $x_1$ is a local minimum of $w_1 - \varphi$. Since $w_1$ is $C^2$ on $(0, x_1) \cup (x_1, \infty)$, and $C^1$ on $x_1$, we have $w_1'(x_1) = \varphi'(x_1)$ and $\varphi''(x_1) \leq w_1''(x_1^-)$. By sending $x \nearrow x_1$ into (5.68), we obtain

$$\beta w_1(x_1) - \mathcal{L}\varphi(x_1) - f_1(x_1) \geq 0,$$

which is the required supersolution inequality. The the required assertion (5.66) is then proved.

$\star$ On the other hand, we check that

$$\hat{V}_2(x) > w_1(x) - g_{21}, \quad \forall x > 0, \tag{5.71}$$

which amounts to showing

$$H(x) := A x^{m_1^+} + \hat{V}_1(x) - \hat{V}_2(x) - g_{21} < 0, \quad \forall 0 < x < x_1.$$

Since $A > 0$, $0 < p < 1 < m_1^+$, $K_2 - K_1 > 0$, a direct derivation shows that the second derivative of $H$ is positive, i.e. $H$ is strictly convex. By (5.63), we have $H'(x_1) = 0$ and so $H'$ is negative, i.e. $H$ is strictly decreasing on $(0, x_1)$. Now, we have $H(0) = -g_{21} \leq 0$ and thus $H$ is negative on $(0, x_1)$, which proves (5.71). Recalling that $\hat{V}_2$ is a solution to $\beta \hat{V}_2 - \mathcal{L}_2 \hat{V}_2 - f_2 = 0$ on $(0, \infty)$, we deduce obviously from (5.71) that $\hat{V}_2$ is a classical, hence a viscosity solution to

$$\min\left\{\beta \hat{V}_2 - \mathcal{L}_2 \hat{V}_2 - f_2, \hat{V}_2 - (w_1 - g_{21})\right\} = 0, \quad \text{on } (0, \infty). \tag{5.72}$$

$\star$ Since $w_1(0^+) = 0 = (-g_{12})_+$, $\hat{V}_2(0^+) = 0 = (-g_{21})_+$, and $w_1$, $\hat{V}_2$ satisfy a linear growth condition, we deduce from (5.66), (5.72), and uniqueness to the PDE system (5.51)-(5.52), that

$$v_1 = w_1, \quad v_2 = \hat{V}_2, \quad \text{on } (0, \infty).$$

This proves $\underline{x}_1^* = \underline{x}_1$, $\mathcal{S}_1 = [\underline{x}_1, \infty)$ and $\mathcal{S}_2 = \emptyset$.

• *Case*: $g_{21} < 0$.

We shall prove that $\mathcal{S}_2 = (0, \bar{x}_2^*]$. To this end, let us consider the functions

$$w_1(x) = \begin{cases} Ax^{m_1^+} + \hat{V}_1(x), & x < \underline{x}_1 \\ w_2(x) - g_{12}, & x \geq \underline{x}_1 \end{cases}$$

$$w_2(x) = \begin{cases} w_1(x) - g_{21}, & x \leq \bar{x}_2 \\ Bx^{m_2^-} + \hat{V}_2(x), & x > \bar{x}_2, \end{cases}$$

where the positive constants $A$, $B$, $\underline{x}_1 > \bar{x}_2$, solutions to

$$A\underline{x}_1^{m_1^+} + \hat{V}_1(\underline{x}_1) = w_2(\underline{x}_1) - g_{12} = B\underline{x}_1^{m_2^-} + \hat{V}_2(\underline{x}_1) - g_{12} \qquad (5.73)$$

$$Am_1^+ \underline{x}_1^{m_1^+ - 1} + \hat{V}_1'(\underline{x}_1) = w_2'(\underline{x}_1) = Bm_2^- \underline{x}_1^{m_2^- - 1} + \hat{V}_2'(\underline{x}_1) \qquad (5.74)$$

$$A\bar{x}_2^{m_1^+} + \hat{V}_1(\bar{x}_2) - g_{21} = w_1(\bar{x}_2) - g_{21} = B\bar{x}_2^{m_2^-} + \hat{V}_2(\bar{x}_2) \qquad (5.75)$$

$$Am_1^+ \bar{x}_2^{m_1^+ - 1} + \hat{V}_1'(\bar{x}_2) = w_1'(\bar{x}_2) = Bm_2^- \bar{x}_2^{m_2^- - 1} + \hat{V}_2'(\bar{x}_2), \qquad (5.76)$$

exist and are explicitly determined after some calculations by

$$\bar{x}_2 = \left[ \frac{-m_2^- (g_{21} + g_{12}y^{m_1^+})}{(K_1 - K_2)(p - m_2^-)(1 - y^{m_1^+ - p})} \right]^{\frac{1}{p}} \qquad (5.77)$$

$$\underline{x}_1 = \frac{\bar{x}_2}{y} \qquad (5.78)$$

$$B = \frac{(K_1 - K_2)(m_1^+ - p)\underline{x}_1^{p - m_2^-} + m_1^+ g_{12}\underline{x}_1^{-m_2^-}}{m_1^+ - m_2^-} \qquad (5.79)$$

$$A = B\underline{x}_1^{m_2^- - m_1^+} - (K_1 - K_2)\underline{x}_1^{p - m_1^+} - g_{12}\underline{x}_1^{-m_1^+}, \qquad (5.80)$$

with $y$ solution in $\left(0, \left(-\frac{g_{21}}{g_{12}}\right)^{\frac{1}{m_1^+}}\right)$ to the equation

$$m_1^+ (p - m_2^-)(1 - y^{m_1^+ - p})(g_{12}y^{m_2^-} + g_{21})$$
$$+ m_2^- (m_1^+ - p)(1 - y^{m_2^- - p})(g_{12}y^{m_1^+} + g_{21}) = 0. \qquad (5.81)$$

Using (5.25), we have $y < \left(-\frac{g_{21}}{g_{12}}\right)^{\frac{1}{m_1^+}} < 1$. As such, $0 < \bar{x}_2 < \underline{x}_1$. Furthermore, by using (5.78) and the equation (5.81) satisfied by $y$, we may easily check that $A$ and $B$ are positive constants.

Notice that by construction, $w_1$ (resp. $w_2$) is $C^2$ on $(0, \underline{x}_1) \cup (\underline{x}_1, \infty)$ (resp. $(0, \bar{x}_2) \cup (\bar{x}_2, \infty)$) and $C^1$ at $\underline{x}_1$ (resp. $\bar{x}_2$).

$\star$ Let us now show that $w_i$, $i = 1, 2$, are viscosity solutions to the system:

$$\min\{\beta w_i - \mathcal{L}_i w_i - f_i, w_i - (w_j - g_{ij})\} = 0, \quad \text{on } (0,\infty), \; i,j = 1,2, \; j \neq i. \quad (5.82)$$

Since the proof is similar for both $w_i$, $i = 1, 2$, we only prove the result for $w_1$. We first check that

$$w_1 \geq w_2 - g_{12}, \quad \forall\, 0 < x < \underline{x}_1. \quad (5.83)$$

From the definition of $w_1$ and $w_2$ and using the fact that $g_{12} + g_{21} > 0$, it is straightforward to see that

$$w_1 \geq w_2 - g_{12}, \quad \forall\, 0 < x \leq \bar{x}_2. \quad (5.84)$$

Now, we need to prove that

$$G(x) := A x^{m_1^+} + \hat{V}_1(x) - B x^{m_2^-} - \hat{V}_2(x) + g_{12} \geq 0, \quad \forall\, \bar{x}_2 < x < \underline{x}_1. \quad (5.85)$$

We have $G(\bar{x}_2) = g_{12} + g_{21} > 0$ and $G(\underline{x}_1) = 0$. Suppose that there exists some $x_0 \in (\bar{x}_2, \underline{x}_1)$ such that $G(x_0) = 0$. We then deduce that there exists $x_3 \in (\bar{x}_0, \underline{x}_1)$ such that $G'(x_3) = 0$. As such, the equation $G'(x) = 0$ admits at least three solutions in $[\bar{x}_2, \underline{x}_1]$: $\{\bar{x}_2, x_3, \underline{x}_1\}$. However, a straightforward study of the function $G$ shows that $G'$ can take the value zero at most at two points in $(0, \infty)$. This leads to a contradiction, proving therefore (5.85).

By definition of $w_1$, we have in the classical sense

$$\beta w_1 - \mathcal{L}_1 w_1 - f = 0, \quad \text{on } (0, \underline{x}_1). \quad (5.86)$$

We now check that

$$\beta w_1 - \mathcal{L}_1 w_1 - f \geq 0, \quad \text{on } (\underline{x}_1, \infty) \quad (5.87)$$

holds true in the classical sense, and so a fortiori in the viscosity sense. By definition of $w_1$ on $(\underline{x}_1, \infty)$, and $K_1$, we have for all $x > \underline{x}_1$,

$$H(x) := \beta w_1(x) - \mathcal{L}_1 w_1(x) - f(x) = \frac{K_2 - K_1}{K_1} x^p + m_2^- L B x^{m_2^-} - \beta g_{12}, \quad (5.88)$$

where $L = \frac{1}{2}(\gamma_2^2 - \gamma_1^2)(m_2^- - 1) + \rho_2 - \rho_1$.
We distinguish two cases:
- First, if $L \geq 0$, the function $H$ would be non-decreasing on $(0, \infty)$ with $\lim_{x \to 0^+} H(x) = -\infty$ and $\lim_{x \to \infty} H(x) = \infty$. As such, it suffices to show that $H(\underline{x}_1) \geq 0$. From (5.73)-(5.74), we have

$$H(\underline{x}_1) = (K_2 - K_1)\left[\frac{m_1^+ - m_2^-}{K_1} - (m_1^+ - p)m_2^- L\right] - \beta g_{12} + m_1^+ m_2^- g_{12} L.$$

Using relations (5.70), (5.73), (5.74), (5.78) and the definition of $m_1^+$ and $m_2^-$, we then obtain

$$H(\underline{x}_1) = \frac{m_1^+(m_1^+ - m_2^-)}{K_1(m_1^+ - p)} - \beta \geq \frac{m_1^+}{K_1(m_1^+ - p)} - \beta \geq 0.$$

- Second, if $L < 0$, it suffices to show that

$$\frac{K_2 - K_1}{K_1} x^p - \beta g_{12} \geq 0, \quad \forall\, x > \underline{x}_1,$$

which is rather straightforward from (5.70) and (5.78).

Relations (5.83), (5.86), and (5.87) show the viscosity property of $w_1$ to (5.82) for $i = 1$, at any point $x \in (0, \underline{x}_1) \cup (\underline{x}_1, \infty)$. By the smooth-fit property of $w_1$ at $\underline{x}_1$, we also get the viscosity property of $w_1$ at this point, and so the required assertion (5.82) is proved.

⋆ Since $w_1(0^+) = 0 = (-g_{12})_+$, $w_2(0^+) = -g_{21} = (-g_{21})_+$, and $w_1$, $\hat{V}_2$ satisfy a linear growth condition, we deduce from (5.82) and uniqueness to the PDE system (5.51)-(5.52), that

$$v_1 = w_1, \quad v_2 = w_2, \quad \text{on } (0, \infty).$$

This proves $\underline{x}_1^* = \underline{x}_1$, $\mathcal{S}_1 = [\underline{x}_1, \infty)$ and $\bar{x}_2^* = \bar{x}_2$, $\mathcal{S}_2 = (0, \bar{x}_2]$.    □

**Different profit functions with identical diffusion regimes**

We suppose that the diffusion regimes are identical corresponding to a geometric Brownian motion

$$b_i(x) = \rho x, \quad \sigma_i(x) = \gamma x, \quad \text{for some constants } \rho,\, \gamma > 0,\, i = 1, 2,$$

and we denote by $\mathcal{L} = \mathcal{L}_1 = \mathcal{L}_2$ the associated generator. We also set $m^+ = m_1^+ = m_2^+$, $m^- = m_1^- = m_2^-$, and $\hat{X}^x = \hat{X}^{x,1} = \hat{X}^{x,2}$. Notice that in this case, the set $Q_{ij}$, $i, j = 1, 2$, $i \neq j$, introduced in Lemma 5.3.9, satisfies

$$Q_{ij} = \{x \in \mathcal{C}_j : (f_j - f_i)(x) - \beta g_{ij} \geq 0\}$$
$$\subset \hat{Q}_{ij} := \{x > 0 : (f_j - f_i)(x) - \beta g_{ij} \geq 0\}. \tag{5.89}$$

Once we are given the profit functions $f_i$, $f_j$, the set $\hat{Q}_{ij}$ can be explicitly computed. Moreover, we prove in the next key lemma that the structure of $\hat{Q}_{ij}$, when it is connected, determines the same structure for the switching region $\mathcal{S}_i$.

**Lemma 5.3.11** Let $i, j = 1, 2$, $i \neq j$ and assume that

$$\sup_{x>0}(\hat{V}_j - \hat{V}_i)(x) > g_{ij}. \tag{5.90}$$

(1) If there exists $0 < \underline{x}_{ij} < \infty$ such that

$$\hat{Q}_{ij} = [\underline{x}_{ij}, \infty), \tag{5.91}$$

then $0 < \underline{x}_i^* < \infty$ and

$$\mathcal{S}_i = [\underline{x}_i^*, \infty).$$

(2) If $g_{ij} \leq 0$ and there exists $0 < \bar{x}_{ij} < \infty$ such that

$$\hat{Q}_{ij} = (0, \bar{x}_{ij}], \tag{5.92}$$

*then $0 < \bar{x}_i^* < \infty$ and*

$$S_i = (0, \bar{x}_i^*].$$

*(3) If there exist $0 < \underline{x}_{ij} < \bar{x}_{ij} < \infty$ such that*

$$\hat{Q}_{ij} = [\underline{x}_{ij}, \bar{x}_{ij}], \tag{5.93}$$

*then $0 < \underline{x}_i^* < \bar{x}_i^* < \infty$ and*

$$S_i = [\underline{x}_i^*, \bar{x}_i^*].$$

*(4) If $g_{ij} \le 0$ and $\hat{Q}_{ij} = (0, \infty)$, then $S_i = (0, \infty)$ and $S_j = \emptyset$.*

**Proof.** First, we observe from Lemma 5.3.8 that under (5.90), the set $S_i$ is nonempty.
(1) Consider the case of condition (5.91). Since $S_i \ne \emptyset \subset \hat{Q}_{ij}$ by Lemma 5.3.9, this implies $\underline{x}_i^* = \inf S_i \in [\underline{x}_{ij}, \infty)$. By definition of $\underline{x}_i^*$, we already know that $(0, \underline{x}_i^*) \subset C_i$. We prove actually the equality, i.e. $S_i = [\underline{x}_i^*, \infty)$ or $v_i(x) = v_j(x) - g_{ij}$ for all $x \ge \underline{x}_i^*$. Consider the function $w_i(x) = v_j(x) - g_{ij}$ on $[\underline{x}_i^*, \infty)$, and let us check that $w_i$ is a viscosity solution to

$$\min\{\beta w_i - \mathcal{L}w_i - f_i , \ w_i - (v_j - g_{ij})\} = 0 \quad \text{on } (\underline{x}_i^*, \infty). \tag{5.94}$$

For this, take some point $\bar{x} > \underline{x}_i^*$ and some smooth test function $\varphi$ s.t. $\bar{x}$ is a local minimum of $w_i - \varphi$. Then, $\bar{x}$ is a local minimum of $v_j - (\varphi + g_{ij})$, and by the viscosity solution property of $v_j$ to its Bellman PDE, we have

$$\beta v_j(\bar{x}) - \mathcal{L}\varphi(\bar{x}) - f_j(\bar{x}) \ge 0.$$

Now, since $\underline{x}_i^* \ge \underline{x}_{ij}$, we have $\bar{x} > \underline{x}_{ij}$ and so by (5.91), $\bar{x} \in \hat{Q}_{ij}$. Hence,

$$(f_j - f_i)(\bar{x}) - \beta g_{ij} \ge 0.$$

By adding the two previous inequalities, we obtain

$$\beta w_i(\bar{x}) - \mathcal{L}\varphi(\bar{x}) - f_i(\bar{x}) \ge 0,$$

which proves the supersolution property of $w_i$ to

$$\beta w_i - \mathcal{L}w_i - f_i \ge 0, \quad \text{on } (\underline{x}_i^*, \infty),$$

and therefore the viscosity solution property of $w_i$ to (5.94). Since $w_i(\underline{x}_i^*) = v_i(\underline{x}_i^*)$ ($= v_j(\underline{x}_i^*) - g_{ij}$), and $w_i$ satisfies a linear growth condition, we deduce from uniqueness to the PDE (5.94) that $w_i$ is equal to $v_i$. In particular, we have $v_i(x) = v_j(x) - g_{ij}$ for $x \ge \underline{x}_i^*$, which shows that $S_i = [\underline{x}_i^*, \infty)$.

(2) The case of condition (5.92) is dealt with the same arguments as above: we first observe that $0 < \bar{x}_i^* = \sup S_i \le \bar{x}_{ij}$. Then, we show that the function $w_i(x) = v_j(x) - g_{ij}$ is a viscosity solution to

$$\min\{\beta w_i - \mathcal{L}w_i - f_i , \ w_i - (v_j - g_{ij})\} = 0 \quad \text{on } (0, \bar{x}_i^*).$$

Then, under the condition that $g_{ij} \leq 0$, we see that $g_{ji} > 0$ by (5.25), and so $v_i(0^+)$
$= -g_{ij} = (-g_{ji})_+ - g_{ij} = v_j(0^+) - g_{ij} = w_i(0^+)$. We also have $v_i(\bar{x}_i^*) = w_i(\bar{x}_i^*)$. By
uniqueness, we conclude that $v_i = w_i$ on $(0, \bar{x}_i^*]$, and so $\mathcal{S}_i = (0, \bar{x}_i^*]$.

(3) By Lemma 5.3.9 and (5.89), the condition (5.93) implies $0 < \underline{x}_{ij} \leq \underline{x}_i^* \leq \bar{x}_i^* \leq \bar{x}_{ij}$
$< \infty$. We claim that $\underline{x}_i^* < \bar{x}_i^*$. Otherwise, $\mathcal{S}_i = \{\bar{x}_i^*\}$ and $v_i$ would satisfy $\beta v_i - \mathcal{L}v_i - f_i$
$= 0$ on $(0, \bar{x}_i^*) \cup (\bar{x}_i^*, \infty)$. By continuity and the smooth-fit condition of $v_i$ at $\bar{x}_i^*$, this
implies that $v_i$ satisfies actually

$$\beta v_i - \mathcal{L}v_i - f_i = 0, \quad x \in (0, \infty),$$

and so by uniqueness, $v_i$ is equal to $\hat{V}_i$. Recalling that $v_i \leq v_j - g_{ij} \leq \hat{V}_j - g_{ij}$, this is in
contradiction with (5.90). By the same arguments as in Cases 1 or 2, we prove that $\mathcal{S}_i$
$= [\underline{x}_i^*, \bar{x}_i^*]$. It suffices to consider the function $w_i(x) = v_j(x) - g_{ij}$, and to check that it is
a viscosity solution to

$$\min \{\beta w_i - \mathcal{L}w_i - f_i \, , \, w_i - (v_j - g_{ij})\} = 0 \quad \text{on} \quad (\underline{x}_i^*, \bar{x}_i^*).$$

Since $w_i(\underline{x}_i^*) = v_i(\underline{x}_i^*)$, $w_i(\bar{x}_i^*) = v_i(\bar{x}_i^*)$, we conclude by uniqueness that $v_i = w_i$ on
$[\underline{x}_i^*, \bar{x}_i^*]$, and so $\mathcal{S}_i = [\underline{x}_i^*, \bar{x}_i^*]$.

(4) Suppose that $g_{ij} \leq 0$ and $\hat{Q}_{ij} = (0, \infty)$. We shall prove that $\mathcal{S}_i = (0, \infty)$ and $\mathcal{S}_j = \emptyset$.
To this end, we consider the smooth functions $w_i = \hat{V}_j - g_{ij}$ and $w_j = \hat{V}_j$. Then, recalling
the ODE satisfied by $\hat{V}_j$, and inequality (5.25), we get

$$\beta w_j - \mathcal{L}w_j - f_j = 0, \quad w_j - (w_i - g_{ji}) = g_{ij} + g_{ji} \geq 0.$$

Therefore $w_j$ is a smooth (and so a viscosity) solution to

$$\min \left[\beta w_j - \mathcal{L}w_j - f_j, w_j - (w_i - g_{ji})\right] = 0 \quad \text{on} \quad (0, \infty).$$

On the other hand, by definition of $\hat{Q}_{ij}$, which is supposed equal to $(0, \infty)$, we have

$$\beta w_i(x) - \mathcal{L}w_i(x) - f_i(x) = \beta \hat{V}_j(x) - \mathcal{L}\hat{V}_j(x) - f_j(x) + f_j(x) - f_i(x) - \beta g_{ij}$$
$$= f_j(x) - f_i(x) - \beta g_{ij} \geq 0, \quad \forall x > 0.$$

Moreover, by construction we have $w_i = w_j - g_{ij}$. Therefore $w_i$ is a smooth (and so a
viscosity) solution to

$$\min \left[\beta w_i - \mathcal{L}w_i - f_i, w_i - (w_j - g_{ij})\right] = 0 \quad \text{on} \quad (0, \infty).$$

Notice also that $g_{ji} > 0$ by (5.25) and since $g_{ij} \leq 0$. Hence, $w_i(0^+) = -g_{ij} = (-g_{ij})_+ =$
$v_i(0^+)$, $w_j(0^+) = 0 = (-g_{ji})_+ = v_j(0^+)$. By uniqueness, we conclude that $v_i = w_i$, $v_j =$
$w_j$, which proves that $\mathcal{S}_i = (0, \infty)$, $\mathcal{S}_j = \emptyset$.                                    $\square$

We shall now provide explicit solutions to the switching problem under general assumptions on the running profit functions, which include several interesting cases for
applications:

(**HF**)   There exists $\hat{x} \in \mathbb{R}_+$ s.t the function  $F := f_2 - f_1$
  is decreasing on $(0, \hat{x})$, increasing on $[\hat{x}, \infty)$,
  and $F(\infty) := \lim_{x \to \infty} F(x) > 0, \quad g_{12} > 0.$

Under (**HF**), there exists some $\bar{x} \in \mathbb{R}_+$ ($\bar{x} > \hat{x}$ if $\hat{x} > 0$ and $\bar{x} = 0$ if $\hat{x} = 0$) from which $F$ is positive: $F(x) > 0$ for $x > \bar{x}$. Economically speaking, condition (**HF**) means that the profit in regime 2 is "better" than profit in regime 1 from a certain level $\bar{x}$, and the improvement then becomes better and better. Moreover, since profit in regime 2 is better than the one in regime 1, it is natural to assume that the corresponding switching cost $g_{12}$ from regime 1 to 2 should be positive. However, we shall consider both cases where $g_{21}$ is positive and nonpositive. Notice that $F(\hat{x}) < 0$ if $\hat{x} > 0$, $F(\hat{x}) = 0$ if $\hat{x} = 0$, and we do not assume necessarily $F(\infty) = \infty$.

**Example 5.3.1** A typical example of different running profit functions satisfying (**HF**) is given by

$$f_i(x) = k_i x^{p_i}, \quad i = 1, 2, \quad \text{with } 0 < p_1 < p_2 < 1, \quad k_1 \in \mathbb{R}_+, \; k_2 > 0. \quad (5.95)$$

In this case, $\hat{x} = \left( \frac{k_1 p_1}{k_2 p_2} \right)^{\frac{1}{p_2 - p_1}}$, and $\lim_{x \to \infty} F(x) = \infty$.

Another example of profit functions of interest in applications is the case where the profit function in regime 1 is $f_1 = 0$, and the other $f_2$ is increasing. In this case, assumption (**HF**) is satisfied with $\hat{x} = 0$.

The next proposition states the form of the switching regions in regimes 1 and 2, depending on the parameter values.

**Proposition 5.3.3** *Assume that* (**HF**) *holds.*
*(1) Structure in regime 1:*

- (i) *If* $\beta g_{12} \geq F(\infty)$, *then* $\underline{x}_1^* = \infty$, *i.e.*

$$\mathcal{S}_1 = \emptyset.$$

- (ii) *If* $\beta g_{12} < F(\infty)$, *then* $\underline{x}_1^* \in (0, \infty)$ *and*

$$\mathcal{S}_1 = [\underline{x}_1^*, \infty).$$

*(2) Structure in regime 2:*

- *Positive switching cost:*
  (i) *If* $\beta g_{21} \geq -F(\hat{x})$, *then*

$$\mathcal{S}_2 = \emptyset.$$

  (ii) *If* $0 < \beta g_{21} < -F(\hat{x})$, *then* $0 < \underline{x}_2^* < \bar{x}_2^* < \underline{x}_1^*$, *and*

$$\mathcal{S}_2 = [\underline{x}_2^*, \bar{x}_2^*].$$

- *Nonpositive switching cost:*
  *(iii) If* $g_{21} \leq 0$ *and* $-F(\infty) < \beta g_{21} < -F(\hat{x})$, *then* $0 = \underline{x}_2^* < \bar{x}_2^* < \underline{x}_1^*$, *and*

  $$\mathcal{S}_2 = (0, \bar{x}_2^*].$$

  *(iv) If* $\beta g_{21} \leq -F(\infty)$, *then*

  $$\mathcal{S}_2 = (0, \infty).$$

**Proof.** (1) From Lemma 5.3.9, we have

$$\hat{Q}_{12} = \{x > 0 : F(x) \geq \beta g_{12}\}. \tag{5.96}$$

Since $g_{12} > 0$, and $f_i(0) = 0$, we have $F(0) = 0 < \beta g_{12}$. Under **(HF)**, we then distinguish the following two cases:
(i) If $\beta g_{12} \geq F(\infty)$, then $\hat{Q}_{12} = \emptyset$, and so by Lemma 5.3.9 and (5.89), $\mathcal{S}_1 = \emptyset$.
(ii) If $\beta g_{12} < F(\infty)$, then there exists $\hat{x}_{12} \in (0, \infty)$ such that

$$\hat{Q}_{12} = [\underline{x}_{12}, \infty). \tag{5.97}$$

Moreover, since

$$(\hat{V}_2 - \hat{V}_1)(x) = E\Big[\int_0^\infty e^{-\beta t} F(\hat{X}_t^x) dt\Big], \quad \forall x > 0,$$

we obtain by the dominated convergence theorem

$$\lim_{x \to \infty} (\hat{V}_2 - \hat{V}_1)(x) = E\Big[\int_0^\infty e^{-\beta t} F(\infty) dt\Big] = \frac{F(\infty)}{\beta} > g_{12}.$$

Hence, conditions (5.90)-(5.91) with $i = 1$, $j = 2$, are satisfied, and we obtain the first assertion by Lemma 5.3.11 1).
(2) From Lemma 5.3.9, we have

$$\hat{Q}_{21} = \{x > 0 : -F(x) \geq \beta g_{21}\}. \tag{5.98}$$

Under **(HF)**, we distinguish the following cases:

⋆ (i1) If $\beta g_{21} > -F(\hat{x})$, then $\hat{Q}_{21} = \emptyset$, and so $\mathcal{S}_2 = \emptyset$.

⋆ (i2) If $\beta g_{21} = -F(\hat{x})$, then either $\hat{x} = 0$ and so $\mathcal{S}_2 = \hat{Q}_{21} = \emptyset$, or $\hat{x} > 0$, and so $\hat{Q}_{21} = \{\hat{x}\}$, $\mathcal{S}_2 \subset \{\hat{x}\}$. In this last case, $v_2$ satisfies $\beta v_2 - \mathcal{L}v_2 - f_2 = 0$ on $(0, \hat{x}) \cup (\hat{x}, \infty)$. By continuity and smooth-fit condition of $v_2$ at $\hat{x}$, this implies that $v_2$ satisfies actually

$$\beta v_2 - \mathcal{L}v_2 - f_2 = 0, \quad x \in (0, \infty),$$

and so by uniqueness, $v_2 = \hat{V}_2$. Since $v_2 \geq v_1 - g_{21} \geq \hat{V}_1 - g_{21}$, this would imply $(\hat{V}_2 - \hat{V}_1)(x) \geq g_{21}$ for all $x > 0$, and so by sending $x$ to infinity, we get a contradiction: $F(\infty) \geq \beta g_{21} = -F(\hat{x})$.

⋆ If $\beta g_{21} < -F(\hat{x})$, we need to distinguish three subcases depending on $g_{21}$:

- If $g_{21} > 0$, then there exist $0 < \underline{x}_{21} < \hat{x} < \bar{x}_{21} < \infty$ such that

$$\hat{Q}_{21} = [\underline{x}_{21}, \bar{x}_{21}]. \tag{5.99}$$

We then conclude with Lemma 5.3.11 2) for $i = 2$, $j = 1$.
- If $g_{21} \leq 0$ with $\beta g_{21} > -F(\infty)$, then there exists $\bar{x}_{21} < \infty$ s.t.

$$\hat{Q}_{21} = (0, \bar{x}_{21}].$$

Moreover, we clearly have $\sup_{x>0}(\hat{V}_1 - \hat{V}_2)(x) > (\hat{V}_1 - \hat{V}_2)(0) = 0 \geq g_{21}$. Hence, conditions (5.90) and (5.92) with $i = 2$, $j = 1$ are satisfied, and we deduce from Lemma 5.3.11 (1) that $\mathcal{S}_2 = (0, \bar{x}_2^*]$ with $0 < \bar{x}_2^* < \infty$.
- If $\beta g_{21} \leq -F(\infty)$, then $\hat{Q}_{21} = (0, \infty)$, and we deduce from Lemma 5.3.11 3) for $i = 2$, $j = 1$, that $\mathcal{S}_2 = (0, \infty)$.

Finally, in the two above subcases when $\mathcal{S}_2 - [\underline{x}_2^*, \bar{x}_2^*]$ or $(0, \bar{x}_2^*]$, we notice that $\bar{x}_2^* < \underline{x}_1^*$ since $\mathcal{S}_2 \subset \mathcal{C}_1 = (0, \infty) \setminus \mathcal{S}_1$, which is equal, from 1), either to $(0, \infty)$ when $\underline{x}_1^* = \infty$ or to $(0, \underline{x}_1^*)$. □

**Economic interpretation.**
The previous proposition shows that, under **(HF)**, the switching region in regime 1 has two forms depending on the size of its corresponding positive switching cost: If $g_{12}$ is larger than the "maximum net" profit $F(\infty)$ that one can expect by changing regime (case (1) (i), which may occur only if $F(\infty) < \infty$), then one has no interest in switching regime, and one always stays in regime 1, i.e. $\mathcal{C}_1 - (0, \infty)$. However, if this switching cost is smaller than $F(\infty)$ (case (1) (ii), which always holds true when $F(\infty) = \infty$ ), then there is some positive threshold from which it is optimal to change regime.

The structure of the switching region in regime 2 exhibits several different forms depending on the sign and size of its corresponding switching cost $g_{21}$ with respect to the values $-F(\infty) < 0$ and $-F(\hat{x}) \geq 0$. If $g_{21}$ is nonnegative and larger than $-F(\hat{x})$ (case (2) (i)), then one has no interest in switching regime, and one always stays in regime 2, i.e. $\mathcal{C}_2 = (0, \infty)$. If $g_{21}$ is positive, but not too large (case (2) (ii)), then there exists some bounded closed interval, which is not a neighborhood of zero, where it is optimal to change regime. Finally, when the switching cost $g_{21}$ is negative, it is optimal to switch to regime 1 at least for small values of the state. Actually, if the negative cost $g_{21}$ is larger than $-F(\infty)$ (case (2) (iii)), which always holds true for negative cost when $F(\infty) = \infty$), then the switching region is a bounded neighborhood of 0. Moreover, if the cost is negative and large enough (case (2) (iv), which may occur only if $F(\infty) < \infty$), then it is optimal to change regime for every value of the state.

By combining the different cases for regimes 1 and 2, and observing that case 2) (iv) is not compatible with case 1) (ii) by (5.25), we then have a priori seven different forms for both switching regions. These forms reduce actually to three when $F(\infty) = \infty$. The various structures of the switching regions are depicted in Figure II.

Finally, we complete the results of Proposition 5.3.3 by providing the explicit solutions for the value functions and the corresponding boundaries of the switching regions in the seven different cases depending on the model parameter values.

**Theorem 5.3.5** *Assume that* **(HF)** *holds.*

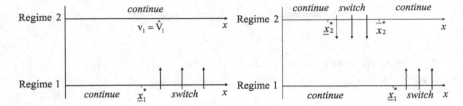

Fig. II.1: $\beta g_{12} < F(\infty), \beta g_{21} \geq -F(\hat{x})$    Fig. II.2: $\beta g_{12} < F(\infty), 0 < \beta g_{21} < -F(\hat{x})$

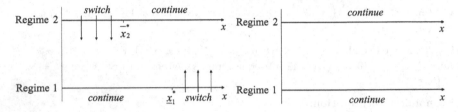

Fig. II.3: $\beta g_{12} < F(\infty), g_{21} \leq 0, -F(\infty) < \beta g_{21} < -F(\hat{x})$    Fig. II.4: $\beta g_{12} \geq F(\infty), \beta g_{21} > -F(\hat{x})$

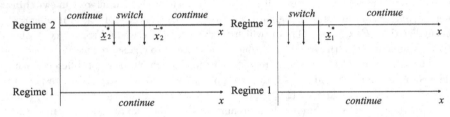

Fig. II.5: $\beta g_{12} \geq F(\infty), 0 < \beta g_{21} < -F(\hat{x})$    Fig. II.6: $\beta g_{12} \geq F(\infty), g_{21} \leq 0, F(\infty) < \beta g_{21} < -F(\hat{x})$

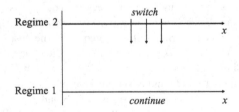

Fig. II.7: $\beta g_{12} \geq F(\infty), g_{21} \leq -F(\infty)$

Fig. II

*(1) If $\beta g_{12} < F(\infty)$ and $\beta g_{21} \geq -F(\hat{x})$, then*

$$v_1(x) = \begin{cases} Ax^{m^+} + \hat{V}_1(x), & x < \underline{x}_1^* \\ v_2(x) - g_{12}, & x \geq \underline{x}_1^* \end{cases}$$
$$v_2(x) = \hat{V}_2(x),$$

*where the constants $A$ and $\underline{x}_1^*$ are determined by the continuity and smooth-fit conditions of $v_1$ at $\underline{x}_1^*$:*

$$A(\underline{x}_1^*)^{m^+} + \hat{V}_1(\underline{x}_1^*) = \hat{V}_2(\underline{x}_1^*) - g_{12}$$
$$Am^+(\underline{x}_1^*)^{m^+-1} + \hat{V}_1'(\underline{x}_1^*) = \hat{V}_2'(\underline{x}_1^*).$$

*In regime 1, it is optimal to switch to regime 2 whenever the state process $X$ exceeds the threshold $\underline{x}_1^*$, while when we are in regime 2, it is optimal never to switch.*

*(2) If $\beta g_{12} < F(\infty)$ and $0 < \beta g_{21} < -F(\hat{x})$, then*

$$v_1(x) = \begin{cases} A_1x^{m^+} + \hat{V}_1(x), & x < \underline{x}_1^* \\ v_2(x) - g_{12}, & x \geq \underline{x}_1^* \end{cases} \tag{5.100}$$

$$v_2(x) = \begin{cases} A_2x^{m^+} + \hat{V}_2(x), & x < \underline{x}_2^* \\ v_1(x) - g_{21}, & \underline{x}_2^* \leq x \leq \bar{x}_2^* \\ B_2x^{m^-} + \hat{V}_2(x), & x > \bar{x}_2^*, \end{cases} \tag{5.101}$$

*where the constants $A_1$ and $\underline{x}_1^*$ are determined by the continuity and smooth-fit conditions of $v_1$ at $\underline{x}_1^*$, and the constants $A_2$, $B_2$, $\underline{x}_2^*$, $\bar{x}_2^*$ are determined by the continuity and smooth-fit conditions of $v_2$ at $\underline{x}_2^*$ and $\bar{x}_2^*$:*

$$A_1(\underline{x}_1^*)^{m^+} + \hat{V}_1(\underline{x}_1^*) = B_2(\underline{x}_1^*)^{m^-} + \hat{V}_2(\underline{x}_1^*) - g_{12} \tag{5.102}$$
$$A_1m^+(\underline{x}_1^*)^{m^+-1} + \hat{V}_1'(\underline{x}_1^*) = B_2m^-(\underline{x}_1^*)^{m^--1} + \hat{V}_2'(\underline{x}_1^*) \tag{5.103}$$
$$A_2(\underline{x}_2^*)^{m^+} + \hat{V}_2(\underline{x}_2^*) = A_1(\underline{x}_2^*)^{m^+} + \hat{V}_1(\underline{x}_2^*) - g_{21} \tag{5.104}$$
$$A_2m^+(\underline{x}_2^*)^{m^+-1} + \hat{V}_2'(\underline{x}_2^*) = A_1m^+(\underline{x}_2^*)^{m^+-1} + \hat{V}_1'(\underline{x}_2^*) \tag{5.105}$$
$$A_1(\bar{x}_2^*)^{m^+} + \hat{V}_1(\bar{x}_2^*) - g_{21} = B_2(\bar{x}_2^*)^{m^-} + \hat{V}_2(\bar{x}_2^*) \tag{5.106}$$
$$A_1m^+(\bar{x}_2^*)^{m^+-1} + \hat{V}_1'(\bar{x}_2^*) = B_2m^-(\bar{x}_2^*)^{m^--1} + \hat{V}_2'(\bar{x}_2^*). \tag{5.107}$$

*In regime 1, it is optimal to switch to regime 2 whenever the state process $X$ exceeds the threshold $\underline{x}_1^*$, while when we are in regime 2, it is optimal to switch to regime 1 whenever the state process lies between $\underline{x}_2^*$ and $\bar{x}_2^*$.*

*(3) If $\beta g_{12} < F(\infty)$ and $g_{21} \leq 0$ with $-F(\infty) < \beta g_{21} < -F(\hat{x})$, then*

$$v_1(x) = \begin{cases} Ax^{m^+} + \hat{V}_1(x), & x < \underline{x}_1^* \\ v_2(x) - g_{12}, & x \geq \underline{x}_1^* \end{cases}$$

$$v_2(x) = \begin{cases} v_1(x) - g_{21}, & 0 < x \leq \bar{x}_2^* \\ Bx^{m^-} + \hat{V}_2(x), & x > \bar{x}_2^*, \end{cases}$$

*where the constants $A$ and $\underline{x}_1^*$ are determined by the continuity and smooth-fit conditions of $v_1$ at $\underline{x}_1^*$, and the constants $B$ and $\bar{x}_2^*$ are determined by the continuity and smooth-fit conditions of $v_2$ at $\bar{x}_2^*$:*

$$A(\underline{x}_1^*)^{m^+} + \hat{V}_1(\underline{x}_1^*) = B(\underline{x}_1^*)^{m^-} + \hat{V}_2(\underline{x}_1^*) - g_{12}$$
$$Am^+(\underline{x}_1^*)^{m^+-1} + \hat{V}_1'(\underline{x}_1^*) = Bm^-(\underline{x}_1^*)^{m^--1} + \hat{V}_2'(\underline{x}_1^*)$$
$$A(\bar{x}_2^*)^{m^+} + \hat{V}_1(\bar{x}_2^*) - g_{21} = B(\bar{x}_2^*)^{m^-} + \hat{V}_2(\bar{x}_2^*)$$
$$Am^+(\bar{x}_2^*)^{m^+-1} + \hat{V}_1'(\bar{x}_2^*) = Bm^-(\bar{x}_2^*)^{m^--1} + \hat{V}_2'(\bar{x}_2^*).$$

*(4) If $\beta g_{12} \geq F(\infty)$ and $\beta g_{21} \geq -F(\hat{x})$, then $v_1 = \hat{V}_1$, $v_2 = \hat{V}_2$. It is optimal never to switch in both regimes 1 and 2.*

*(5) If $\beta g_{12} \geq F(\infty)$ and $0 < \beta g_{21} < -F(\hat{x})$, then*

$$v_1(x) = \hat{V}_1(x)$$
$$v_2(x) = \begin{cases} Ax^{m^+} + \hat{V}_2(x), & x < \underline{x}_2^* \\ v_1(x) - g_{21}, & \underline{x}_2^* \leq x \leq \bar{x}_2^* \\ Bx^{m^-} + \hat{V}_2(x), & x > \bar{x}_2^*, \end{cases}$$

*where the constants $A$, $B$, $\underline{x}_2^*$, $\bar{x}_2^*$ are determined by the continuity and smooth-fit conditions of $v_2$ at $\underline{x}_2^*$ and $\bar{x}_2^*$:*

$$A(\underline{x}_2^*)^{m^+} + \hat{V}_2(\underline{x}_2^*) = \hat{V}_1(\underline{x}_2^*) - g_{21}$$
$$Am^+(\underline{x}_2^*)^{m^+-1} + \hat{V}_2'(\underline{x}_2^*) = \hat{V}_1'(\underline{x}_2^*)$$
$$\hat{V}_1(\bar{x}_2^*) - g_{21} = B(\bar{x}_2^*)^{m^-} + \hat{V}_2(\bar{x}_2^*)$$
$$\hat{V}_1'(\bar{x}_2^*) = Bm^-(\bar{x}_2^*)^{m^--1} + \hat{V}_2'(\bar{x}_2^*).$$

*In regime 1, it is optimal never to switch, while when we are in regime 2, it is optimal to switch to regime 1 whenever the state process lies between $\underline{x}_2^*$ and $\bar{x}_2^*$.*

*(6) If $\beta g_{12} \geq F(\infty)$ and $g_{21} \leq 0$ with $-F(\infty) < \beta g_{21} < -F(\hat{x})$, then*

$$v_1(x) = \hat{V}_1(x)$$
$$v_2(x) = \begin{cases} v_1(x) - g_{21}, & 0 < x \leq \bar{x}_2^* \\ Bx^{m^-} + \hat{V}_2(x), & x > \bar{x}_2^*, \end{cases}$$

*where the constants $B$ and $\bar{x}_2^*$ are determined by the continuity and smooth-fit conditions of $v_2$ at $\bar{x}_2^*$:*

$$\hat{V}_1(\bar{x}_2^*) - g_{21} = B(\bar{x}_2^*)^{m^-} + \hat{V}_2(\bar{x}_2^*)$$
$$\hat{V}_1'(\bar{x}_2^*) = Bm^-(\bar{x}_2^*)^{m^--1} + \hat{V}_2'(\bar{x}_2^*).$$

*In regime 1, it is optimal never to switch, while when we are in regime 2, it is optimal to switch to regime 1 whenever the state process lies below $\bar{x}_2^*$.*

*(7) If $\beta g_{12} \geq F(\infty)$ and $\beta g_{21} \leq -F(\infty)$, then $v_1 = \hat{V}_1$ and $v_2 = v_1 - g_{12}$. In regime 1, it is optimal never to switch, while when we are in regime 2, it is always optimal to switch to regime 1.*

**Proof.** We prove the result only for case 2 since the other cases are dealt similarly and are even simpler. Case 2 corresponds to the combination of cases 1 (ii) and 2 (ii) in Proposition 5.3.3. We then have $\mathcal{S}_1 = [\underline{x}_1^*, \infty)$, which means that $v_1 = v_2 - g_{12}$ on $[\underline{x}_1^*, \infty)$ and $v_1$ is a solution to $\beta v_1 - \mathcal{L}v_1 - f_1 = 0$ on $(0, \underline{x}_1^*)$. Since $0 \le v_1(0^+) < \infty$, $v_1$ should have the form expressed in (5.100). Moreover, $\mathcal{S}_2 = [\underline{x}_2^*, \bar{x}_2^*]$, which means that $v_2 = v_1 - g_{21}$ on $[\underline{x}_2^*, \bar{x}_2^*]$, and $v_2$ satisfies on $\mathcal{C}_2 = (0, \underline{x}_2^*) \cup (\bar{x}_2^*, \infty)$: $\beta v_2 - \mathcal{L}v_2 - f_2 = 0$. Recalling again that $0 \le v_2(0^+) < \infty$ and $v_2$ satisfies a linear growth condition, we deduce that $v_2$ has the form expressed in (5.101). Finally, the constants $A_1$, $\underline{x}_1^*$, which characterize completely $v_1$, and the constants $A_2$, $B_2$, $\underline{x}_2^*$, $\bar{x}_2^*$, which characterize completely $v_2$, are determined by the six relations (5.102), (5.103), (5.104), (5.105), (5.106) and (5.107) resulting from the continuity and smooth-fit conditions of $v_1$ at $\underline{x}_1^*$ and $v_2$ at $\underline{x}_2^*$ and $\bar{x}_2^*$, and recalling that $\bar{x}_2^* < \underline{x}_1^*$. $\qquad\square$

## 5.4 Bibliographical remarks

The connection between optimal stopping problems and free boundary problems is classical (see e.g. the book by Bensoussan and Lions [BL82]), and is applied to American options in Jaillet, Lamberton, Lapeyre [JLL90]. For an updated presentation, we refer to the recent book by Peskir and Shiryaev [PeSh06]. The viscosity solutions approach for optimal stopping was developed by Pham [Pha98] and Øksendal and Reikvam [OR98]. The explicit solution for the perpetual American options is already presented in McKean [Mac65]. The example of the optimal selling of an asset is developed in the textbook by Øksendal [Oks00], and the application to the valuation of natural resources is due to Knudsen, Meister and Zervos [KMZ98]. We revisit their results by the viscosity solutions method. Other explicit examples of optimal stopping problems in finance can be found in Guo and Shepp [GS01].

Optimal switching problems and their connections with a system of variational inequalities were studied by Bensoussan and Lions [BL82], and [TY93]. The smooth-fit property is proved in Pham [Pha07]. Applications to real options and firm's investment under uncertainty were considered in Brekke and Oksendal [BO94], Duckworth and Zervos [DZ01], and the solutions are obtained by a verification theorem approach. The viscosity solutions approach for determining an explicit solution in the two-regime case is due to Ly Vath and Pham [LP07], and generalizes the previous results. Some extensions to the multi-regime case are studied in Pham, Ly Vath and Zhou [PLZ07].

# 6

# Backward stochastic differential equations and optimal control

## 6.1 Introduction

The theory of backward stochastic differential equations (BSDEs) was pioneered by Pardoux and Peng [PaPe90]. It became now very popular, and is an important field of research due to its connections with stochastic control, mathematical finance, and partial differential equations. BSDEs provide a probabilistic representation of nonlinear PDEs, which extends the famous Feynman-Kac formula for linear PDEs. As a consequence, BSDEs can be used for designing numerical algorithms to nonlinear PDEs.

This chapter is an introduction to the theory of BSDEs and its applications to mathematical finance and stochastic optimization. In Section 6.2, we state general results about existence and uniqueness of BSDEs, and useful comparison principles. Section 6.3 develops the connection between BSDEs and viscosity solutions to nonlinear PDEs. We show in Section 6.4 how BSDEs may be used for solving stochastic optimal control. Section 6.5 introduces the notion of reflected BSDEs, and shows how it is related to optimal stopping problems. Finally, Section 6.6 gives some illustrative examples of applications of BSDEs in finance.

## 6.2 General properties

### 6.2.1 Existence and uniqueness results

Let $W = (W_t)_{0 \le t \le T}$ be a standard $d$-dimensional Brownian motion on a filtered probability space $(\Omega, \mathcal{F}, \mathbb{F}, P)$ where $\mathbb{F} = (\mathcal{F}_t)_{0 \le t \le T}$ is the natural filtration of $W$, and $T$ is a fixed finite horizon.

We denote by $\mathbb{S}^2(0, T)$ the set of real-valued progressively measurable processes $Y$ such that

$$E\Big[ \sup_{0 \le t \le T} |Y_t|^2 \Big] < \infty,$$

and by $\mathbb{H}^2(0, T)^d$ the set of $\mathbb{R}^d$-valued progressively measurable processes $Z$ such that

$$E\Big[ \int_0^T |Z_t|^2 dt \Big] < \infty.$$

H. Pham, *Continuous-time Stochastic Control and Optimization with Financial Applications*, Stochastic Modelling and Applied Probability 61, DOI 10.1007/978-3-540-89500-8_6, © Springer-Verlag Berlin Heidelberg 2009

We are given a pair $(\xi, f)$, called the terminal condition and generator (or driver), satisfying:

- (A) $\xi \in L^2(\Omega, \mathcal{F}_T, P; \mathbb{R})$
- (B) $f : \Omega \times [0,T] \times \mathbb{R} \times \mathbb{R}^d \to \mathbb{R}$ s.t.:

  - $f(., t, y, z)$, written for simplicity $f(t, y, z)$, is progressively measurable for all $y, z$
  - $f(t, 0, 0) \in \mathbb{H}^2(0, T)$
  - $f$ satisfies a uniform Lipschitz condition in $(y, z)$, i.e. there exists a constant $C_f$

such that

$$|f(t, y_1, z_1) - f(t, y_2, z_2)| \le C_f\left(|y_1 - y_2| + |z_1 - z_2|\right), \quad \forall y_1, y_2, \ \forall z_1, z_2, \quad dt \otimes dP \text{ a.e.}$$

We consider the (unidimensional) backward stochastic differential equations (BSDE):

$$-dY_t = f(t, Y_t, Z_t)dt - Z_t.dW_t, \quad Y_T = \xi. \tag{6.1}$$

**Definition 6.2.1** *A solution to the BSDE (6.1) is a pair $(Y, Z) \in \mathbb{S}^2(0, T) \times \mathbb{H}^2(0, T)^d$ satisfying*

$$Y_t = \xi + \int_t^T f(s, Y_s, Z_s)ds - \int_t^T Z_s.dW_s, \quad 0 \le t \le T.$$

We prove an existence and uniqueness result for the above BSDE.

**Theorem 6.2.1** *Given a pair $(\xi, f)$ satisfying (A) and (B), there exists a unique solution $(Y, Z)$ to the BSDE (6.1).*

**Proof.** We give a proof based on a fixed point method. Let us consider the function $\Phi$ on $\mathbb{S}^2(0, T)^m \times \mathbb{H}^2(0, T)^d$, mapping $(U, V) \in \mathbb{S}^2(0, T) \times \mathbb{H}^2(0, T)^d$ to $(Y, Z) = \Phi(U, V)$ defined by

$$Y_t = \xi + \int_t^T f(s, U_s, V_s)ds - \int_t^T Z_s.dW_s. \tag{6.2}$$

More precisely, the pair $(Y, Z)$ is constructed as follows: we consider the martingale $M_t = E[\xi + \int_0^T f(s, U_s, V_s)ds|\mathcal{F}_t]$, which is square integrable under the assumptions on $(\xi, f)$. We may apply the Itô martingale representation theorem, which gives the existence and uniqueness of $Z \in \mathbb{H}^2(0, T)^d$ such that

$$M_t = M_0 + \int_0^t Z_s.dW_s. \tag{6.3}$$

We then define the process $Y$ by

$$Y_t = E\left[\xi + \int_t^T f(s, U_s, V_s)ds \Big| \mathcal{F}_t\right] = M_t - \int_0^t f(s, U_s, V_s)ds, \quad 0 \le t \le T.$$

By using the representation (6.3) of $M$ in the previous relation, and noting that $Y_T = \xi$, we see that $Y$ satisfies (6.2). Observe by Doob's inequality that

$$E\Big[\sup_{0\le t\le T}\Big|\int_t^T Z_s.dW_s\Big|^2\Big]\le 4E\Big[\int_0^T |Z_s|^2 ds\Big] < \infty.$$

Under the conditions on $(\xi, f)$, we deduce that $Y$ lies in $\mathbb{S}^2(0,T)$. Hence, $\Phi$ is a well-defined function from $\mathbb{S}^2(0,T)\times\mathbb{H}^2(0,T)^d$ into itself. We then see that $(Y,Z)$ is a solution to the BSDE (6.1) if and only if it is a fixed point of $\Phi$.

Let $(U,V), (U',V')\in\mathbb{S}^2(0,T)\times\mathbb{H}^2(0,T)^d$ and $(Y,Z)=\Phi(U,V)$, $(Y',Z')=\Phi(U',V')$. We set $(\bar{U},\bar{V})=(U-U',V-V')$, $(\bar{Y},\bar{Z})=(Y-Y',Z-Z')$ and $\bar{f}_t = f(t,U_t,V_t) - f(t,U'_t,V'_t)$. Take some $\beta > 0$ to be chosen later, and apply Itô's formula to $e^{\beta s}|\bar{Y}_s|^2$ between $s=0$ and $s=T$:

$$\begin{aligned}
|\bar{Y}_0|^2 = &-\int_0^T e^{\beta s}\left(\beta|\bar{Y}_s|^2 - 2\bar{Y}_s.\bar{f}_s\right)ds\\
&-\int_0^T e^{\beta s}|\bar{Z}_s|^2 ds - 2\int_0^T e^{\beta s}\bar{Y}'_s\bar{Z}_s.dW_s.
\end{aligned}\tag{6.4}$$

Observe that

$$E\Big[\Big(\int_0^T e^{2\beta t}|Y_t|^2|Z_t|^2 dt\Big)^{\frac12}\Big]\le \frac{e^{\beta T}}{2}E\Big[\sup_{0<t<T}|Y_t|^2 + \int_0^T |Z_t|^2 dt\Big] < \infty,$$

which shows that the local martingale $\int_0^t e^{\beta s}\bar{Y}'_s\bar{Z}_s.dW_s$ is actually a uniformly integrable martingale from the Burkholder-Davis-Gundy inequality. By taking the expectation in (6.4), we get

$$\begin{aligned}
E|\bar{Y}_0|^2 + E\Big[\int_0^T e^{\beta s}\Big(\beta|\bar{Y}_s|^2 + |\bar{Z}_s|^2\Big)ds\Big] &= 2E\Big[\int_0^T e^{\beta s}\bar{Y}_s.\bar{f}_s ds\Big]\\
&\le 2C_f E\Big[\int_0^T e^{\beta s}|\bar{Y}_s|(|\bar{U}_s|+|\bar{V}_s|)ds\Big]\\
&\le 4C_f^2 E\Big[\int_0^T e^{\beta s}|\bar{Y}_s|^2 ds\Big] + \frac12 E\Big[\int_0^T e^{\beta s}(|\bar{U}_s|^2 + |\bar{V}_s|^2)ds\Big]
\end{aligned}$$

Now, we choose $\beta = 1 + 4C_f^2$, and obtain

$$E\Big[\int_0^T e^{\beta s}\Big(|\bar{Y}_s|^2 + |\bar{Z}_s|^2\Big)ds\Big]\le \frac12 E\Big[\int_0^T e^{\beta s}(|\bar{U}_s|^2 + |\bar{V}_s|^2)ds\Big].$$

This shows that $\Phi$ is a strict contraction on the Banach space $\mathbb{S}^2(0,T)\times\mathbb{H}^2(0,T)^d$ endowed with the norm

$$\|(Y,Z)\|_\beta = \Big(E\Big[\int_0^T e^{\beta s}(|Y_s|^2 + |Z_s|^2)ds\Big]\Big)^{\frac12}.$$

We conclude that $\Phi$ admits a unique fixed point, which is the solution to the BSDE (6.1).
□

## 6.2.2 Linear BSDE

We consider the particular case where the generator $f$ is linear in $y$ and $z$. The linear BSDE is written in the form

$$- dY_t = (A_t Y_t + Z_t.B_t + C_t) \, dt - Z_t.dW_t, \quad Y_T = \xi, \tag{6.5}$$

where $A$, $B$ are bounded progressively measurable processes valued in $\mathbb{R}$ and $\mathbb{R}^d$, and $C$ is a process in $\mathbb{H}^2(0, T)$. We can solve this BSDE explicitly.

**Proposition 6.2.1** *The unique solution* $(Y, Z)$ *to the linear BSDE* (6.5) *is given by*

$$\Gamma_t Y_t = E\left[ \Gamma_T \xi + \int_t^T \Gamma_s C_s ds \Big| \mathcal{F}_t \right], \tag{6.6}$$

*where* $\Gamma$ *is the adjoint (or dual) process, solution to the linear SDE*

$$d\Gamma_t = \Gamma_t \left( A_t dt + B_t.dW_t \right), \quad \Gamma_0 = 1.$$

**Proof.** By Itô's formula to $\Gamma_t Y_t$, we get

$$d(\Gamma_t Y_t) = -\Gamma_t C_t dt + \Gamma_t (Z_t + Y_t B_t).dW_t,$$

and so

$$\Gamma_t Y_t + \int_0^t \Gamma_s C_s ds = Y_0 + \int_0^t \Gamma_s (Z_s + Y_s B_s).dW_t. \tag{6.7}$$

Since $A$ and $B$ are bounded, we see that $E[\sup_t |\Gamma_t|^2] < \infty$, and by denoting by $b_\infty$ the upper-bound of $B$, we have

$$E\left[ \left( \int_0^T \Gamma_s^2 |Z_s + Y_s B_s|^2 ds \right)^{\frac{1}{2}} \right] \leq \frac{1}{2} E\left[ \sup_t |\Gamma_t|^2 + 2 \int_0^T |Z_t|^2 dt + 2b_\infty^2 \int_0^T |Y_t|^2 dt \right]$$
$$< \infty.$$

From the Burkholder-Davis-Gundy inequality, this shows that the local martingale in (6.7) is a uniformly integrable martingale. By taking the expectation, we obtain

$$\Gamma_t Y_t + \int_0^t \Gamma_s C_s ds = E\left[ \Gamma_T Y_T + \int_0^T \Gamma_s C_s ds \Big| \mathcal{F}_t \right]$$
$$= E\left[ \Gamma_T \xi + \int_0^T \Gamma_s C_s ds \Big| \mathcal{F}_t \right], \tag{6.8}$$

which gives the expression (6.6) for $Y$. Finally, $Z$ is given via the Itô martingale representation (6.7) of the martingale in (6.8). $\quad\square$

### 6.2.3 Comparison principles

We state a very useful comparison principle for BSDEs.

**Theorem 6.2.2** *Let* $(\xi^1, f^1)$ *and* $(\xi^2, f^2)$ *be two pairs of terminal conditions and generators satisfying conditions (A) and (B), and let* $(Y^1, Z^1)$, $(Y^2, Z^2)$ *be the solutions to their corresponding BSDEs. Suppose that:*

- $\xi^1 \leq \xi^2$ *a.s.*
- $f^1(t, Y_t^1, Z_t^1) \leq f^2(t, Y_t^1, Z_t^1) \, dt \otimes dP$ *a.e.*

- $f^2(t, Y_t^1, Z_t^1) \in \mathbb{H}^2(0, T)$.

Then $Y_t^1 \leq Y_t^2$ for all $0 \leq t \leq T$, a.s.

Furthermore, if $Y_0^2 \leq Y_0^1$, then $Y_t^1 = Y_t^2$, $0 \leq t \leq T$. In particular, if $P(\xi^1 < \xi^2) > 0$ or $f^1(t,.,.) < f^2(t,.,.)$ on a set of strictlly positive measure $dt \otimes dP$, then $Y_0^1 < Y_0^2$.

**Proof.** To simplify the notation, we assume $d = 1$. We define $\bar{Y} = Y^2 - Y^1$, $\bar{Z} = Z^2 - Z^1$. Then $(\bar{Y}, \bar{Z})$ satisfies the linear BSDE

$$- d\bar{Y}_t = \left( \Delta_t^y \bar{Y}_t + \Delta_t^z \bar{Z}_t + \bar{f}_t \right) dt - \bar{Z}_t dW_t, \quad \bar{Y}_T = \xi^2 - \xi^1. \tag{6.9}$$

where

$$\Delta_t^y = \frac{f^2(t, Y_t^2, Z_t^2) - f^2(t, Y_t^1, Z_t^2)}{Y_t^2 - Y_t^1} 1_{Y_t^2 - Y_t^1 \neq 0}$$

$$\Delta_t^z = \frac{f^2(t, Y_t^1, Z_t^2) - f^2(t, Y_t^1, Z_t^1)}{Z_t^2 - Z_t^1} 1_{Z_t^2 - Z_t^1 \neq 0}$$

$$\bar{f}_t = f^2(t, Y_t^1, Z_t^1) - f^1(t, Y_t^1, Z_t^1).$$

Since the generator $f^2$ is uniformly Lipschitz in $y$ and $z$, the processes $\Delta^y$ and $\Delta^z$ are bounded. Moreover, $\bar{f}_t$ is a process in $\mathbb{H}^2(0, T)$. From 6.2.1, $\bar{Y}$ is then given by

$$\Gamma_t \bar{Y}_t = E\left[ \Gamma_T(\xi^2 - \xi^1) + \int_t^T \Gamma_s \bar{f}_s ds \Big| \mathcal{F}_t \right],$$

where the adjoint process $\Gamma$ is strictly positive. We conclude from this expectation formula for $Y$, and the positivity of $\xi^2 - \xi^1$ and $\bar{f}$.   □

**Remark 6.2.1** Notice that in the proof of Theorem 6.2.2, it is not necessary to suppose regularity conditions on the generator $f_1$. The uniform Lipschitz condition is only required for $f_2$.

**Corollary 6.2.1** If the pair $(\xi, f)$ satisfies $\xi \geq 0$ a.s. and $f(t, 0, 0) \geq 0$ $dt \otimes dP$ a.e., then $Y_t \geq 0$, $0 \leq t \leq T$ a.s. Moreover, if $P[\xi > 0] > 0$ or $f(t, 0, 0) > 0$ $dt \otimes dP$ a.e., then $Y_0 > 0$.

**Proof.** This is an immediate consequence of the comparison theorem 6.2.2 with $(\xi^1, f^1) = (0, 0)$, whose solution is obviously $(Y^1, Z^1) = (0, 0)$.   □

## 6.3 BSDE, PDE and nonlinear Feynman-Kac formulae

We recall the well-known result (see Section 1.3.3) that the solution to the linear parabolic PDE

$$-\frac{\partial v}{\partial t} - \mathcal{L}v - f(t, x) = 0, \quad (t, x) \in [0, T) \times \mathbb{R}^n,$$

$$v(T, x) = g(x), \quad x \in \mathbb{R}^n,$$

has the probabilistic Feynman-Kac representation

$$v(t,x) = E\Big[ \int_t^T f(s, X_s^{t,x})ds + g(X_T^{t,x}) \Big], \qquad (6.10)$$

where $\{X_s^{t,x}, t \leq s \leq T\}$ is the solution to the SDE

$$dX_s = b(X_s)ds + \sigma(X_s)dW_s, \quad t \leq s \leq T, \ X_t = x,$$

and $\mathcal{L}$ is the second-order operator

$$\mathcal{L}v = b(x).D_x v + \frac{1}{2}\mathrm{tr}(\sigma(x)\sigma'(x)D_{xx}^2 v).$$

In the previous chapter, we derived a generalization of this linear Feynman-Kac formula for nonlinear PDEs in the form

$$-\frac{\partial v}{\partial t} - \sup_{a \in A}\big[\mathcal{L}^a v + f(t,x,a)\big] = 0, \quad (t,x) \in [0,T) \times \mathbb{R}^n, \qquad (6.11)$$

$$v(T,x) = g(x), \quad x \in \mathbb{R}^n, \qquad (6.12)$$

where, for any $a \in A$, subset of $\mathbb{R}^m$,

$$\mathcal{L}^a v = b(x,a).D_x v + \frac{1}{2}\mathrm{tr}(\sigma(x,a)\sigma'(x,a)D_{xx}^2 v).$$

The solution (in the viscosity sense) to (6.11)-(6.12) may be represented by means of a stochastic control problem as

$$v(t,x) = \sup_{\alpha \in \mathcal{A}} E\Big[ \int_t^T f(s, X_s^{t,x}, \alpha_s)ds + g(X_T^{t,x}) \Big],$$

where $\mathcal{A}$ is the set of progressively measurable processes $\alpha$ valued in $A$, and for $\alpha \in \mathcal{A}$, $\{X_s^{t,x}, t \leq s \leq T\}$ is the controlled diffusion

$$dX_s = b(X_s, \alpha_s)ds + \sigma(X_s, \alpha_s)dW_s, \quad t \leq s \leq T, \ X_t = x.$$

In this chapter, we study another extension of Feynman-Kac formula for semilinear PDE in the form

$$-\frac{\partial v}{\partial t} - \mathcal{L}v - f(t,x,v,\sigma' D_x v) = 0, \quad (t,x) \in [0,T) \times \mathbb{R}^n, \qquad (6.13)$$

$$v(T,x) = g(x), \quad x \in \mathbb{R}^n. \qquad (6.14)$$

We shall represent the solution to this PDE by means of the BSDE

$$-dY_s = f(s, X_s, Y_s, Z_s)ds - Z_s.dW_s, \quad t \leq s \leq T, \ Y_T = g(X_T), \qquad (6.15)$$

and the forward SDE valued in $\mathbb{R}^n$:

$$dX_s = b(X_s)ds + \sigma(X_s)dW_s. \qquad (6.16)$$

The functions $b$ and $\sigma$ satisfy a Lipschitz condition on $\mathbb{R}^n$; $f$ is a continuous function on $[0,T] \times \mathbb{R}^n \times \mathbb{R} \times \mathbb{R}^d$ satisfying a linear growth condition in $(x,y,z)$, and a Lipschitz condition in $(y,z)$, uniformly in $(t,x)$. The continuous function $g$ satisfies a linear growth condition. Hence, by a standard estimate on the second moment of $X$, we see that the terminal condition and the generator of the BSDE (6.15) satisfy the conditions (A) and (B) stated in Section 6.2. By the Markov property of the diffusion $X$, and uniqueness of a solution $(Y, Z, K)$ to the BSDE (6.15), we notice that $Y_t = v(t, X_t)$, $0 \leq t \leq T$, where

$$v(t, x) := Y_t^{t,x} \tag{6.17}$$

is a deterministic function of $(t, x)$ in $[0, T] \times \mathbb{R}^n$, $\{X_s^{t,x}, t \leq s \leq T\}$ is the solution to (6.16) starting from $x$ at $t$, and $\{(Y_s^{t,x}, Z_s^{t,x}), t \leq s \leq T\}$ is the solution to the BSDE (6.15) with $X_s = X_s^{t,x}$, $t \leq s \leq T$. We call this framework a Markovian case for the BSDE.

The next verification result for the PDE (6.13) is analogous to the verification theorem for Hamilton-Jacobi-Bellman equations (6.11), and shows that a classical solution to the semilinear PDE provides a solution to the BSDE.

**Proposition 6.3.2** *Let $v \in C^{1,2}([0, T) \times \mathbb{R}^n) \cap C^0([0, T] \times \mathbb{R}^n)$ be a classical solution to (6.13)-(6.14), satisfying a linear growth condition and such that for some positive constants $C, q$, $|D_x v(t, x)| \leq C(1 + |x|^q)$ for all $x \in \mathbb{R}^n$. Then, the pair $(Y, Z)$ defined by*

$$Y_t = v(t, X_t), \quad Z_t = \sigma'(X_t) D_x v(t, X_t), \quad 0 \leq t \leq T,$$

*is the solution to the BSDE (6.15).*

**Proof.** This is an immediate consequence of Itô's formula applied to $v(t, X_t)$, and noting from the growth conditions on $v$, $D_x v$ that $(Y, Z)$ lie in $\mathbb{S}^2(0, T) \times \mathbb{H}^2(0, T)^d$.  $\square$

We now study the converse property by proving that the solution to the BSDE (6.15) provides a solution to the PDE (6.13)-(6.14).

**Theorem 6.3.3** *The function $v(t, x) = Y_t^{t,x}$ in (6.17) is a continuous function on $[0, T] \times \mathbb{R}^n$, and is a viscosity solution to (6.13)-(6.14).*

**Proof.** 1) For $(t_1, x_1)$, $(t_2, x_2) \in [0, T] \times \mathbb{R}^n$, with $t_1 \leq t_2$, we write $X_s^i = X_s^{t_i, x_i}$, $i = 1, 2$, with the convention that $X_s^2 = x_2$ if $t_1 \leq s \leq t_2$, and $(Y_s^i, Z_s^i) = (Y_s^{t_i, x_i}, Z_s^{t_i, x_i})$, $i = 1, 2$, which is then well-defined for $t_1 \leq s \leq T$. By applying Itô's formula to $|Y_s^1 - Y_s^2|^2$ between $s = t \in [t_1, T]$ and $s = T$, we get

$$|Y_t^1 - Y_t^2|^2 = |g(X_T^1) - g(X_T^2)|^2 - \int_t^T |Z_s^1 - Z_s^2|^2 ds$$

$$+ 2 \int_t^T (Y_s^1 - Y_s^2) \cdot (f(s, X_s^1, Y_s^1, Z_s^1) - f(s, X_s^2, Y_s^2, Z_s^2)) ds$$

$$- 2 \int_t^T (Y_s^1 - Y_s^2)'(Z_s^1 - Z_s^2) dW_s.$$

As in the proof of Theorem 6.2.1, the local martingale $\int_t^s (Y_u^1 - Y_u^2)'(Z_u^1 - Z_u^2) dW_u$, $t \leq s \leq T$, is actually uniformly integrable, and so by taking the expectation in the above relation, we derive

$$E\left[|Y_t^1 - Y_t^2|^2\right] + E\left[\int_t^T |Z_s^1 - Z_s^2|^2 ds\right]$$

$$= E\left[|g(X_T^1) - g(X_T^2)|^2\right]$$

$$+ 2 E\left[\int_t^T (Y_s^1 - Y_s^2).(f(s, X_s^1, Y_s^1, Z_s^1) - f(s, X_s^2, Y_s^2, Z_s^2))ds\right]$$

$$\le E\left[|g(X_T^1) - g(X_T^2)|^2\right]$$

$$+ 2 E\left[\int_t^T |Y_s^1 - Y_s^2| \, |f(s, X_s^1, Y_s^1, Z_s^1) - f(s, X_s^2, Y_s^1, Z_s^1)|ds\right]$$

$$+ 2C_f E\left[\int_t^T |Y_s^1 - Y_s^2| \, (|Y_s^1 - Y_s^2| + |Z_s^1 - Z_s^2|) \, ds\right]$$

$$\le E\left[|g(X_T^1) - g(X_T^2)|^2\right]$$

$$+ E\left[\int_t^T |f(s, X_s^1, Y_s^1, Z_s^1) - f(s, X_s^2, Y_s^1, Z_s^1)|^2 ds\right]$$

$$+ (1 + 4C_f^2)E\left[\int_t^T |Y_s^1 - Y_s^2|^2 ds + \frac{1}{2} E\int_t^T |Z_s^1 - Z_s^2|^2 ds\right],$$

where $C_f$ is the uniform Lipschitz constant of $f$ with respect to $y$ and $z$. This yields

$$E\left[|Y_t^1 - Y_t^2|^2\right] \le E\left[|g(X_T^1) - g(X_T^2)|^2\right] + E\left[\int_t^T |f(s, X_s^1, Y_s^1, Z_s^1) - f(s, X_s^2, Y_s^1, Z_s^1)|^2 ds\right]$$

$$+ (1 + 4C_f^2)E\left[\int_t^T |Y_s^1 - Y_s^2|^2 ds\right]$$

and so, by Gronwall's lemma

$$E\left[|Y_t^1 - Y_t^2|^2\right] \le C\left\{E\left[|g(X_T^1) - g(X_T^2)|^2\right]\right.$$

$$\left. + E\left[\int_t^T |f(s, X_s^1, Y_s^1, Z_s^1) - f(s, X_s^2, Y_s^1, Z_s^1)|^2 ds\right]\right\}.$$

This last inequality, combined with continuity of $f$ and $g$ in $x$, continuity of $X^{t,x}$ in $(t, x)$, shows the mean-square continuity of $\{Y_s^{t,x}, x \in \mathbb{R}^n, 0 \le t \le s \le T\}$, and so the continuity of $(t, x) \to v(t, x) = Y_t^{t,x}$. The terminal condition (6.14) is trivially satisfied.

2) We next show that $v(t, x) = Y_t^{t,x}$ is a viscosity solution to (6.13). We check the viscosity subsolution property, the viscosity supersolution property is then proved similarly. Let $\varphi$ be a smooth test function and $(t, x) \in [0, T) \times \mathbb{R}^n$ such that $(t, x)$ is a local maximum of $v - \varphi$ with $u(t, x) = \varphi(t, x)$. We argue by contradiction by assuming that

$$-\frac{\partial \varphi}{\partial t}(t, x) - \mathcal{L}\varphi(t, x) - f(t, x, v(t, x), (D_x\varphi)'(t, x)\sigma(x)) > 0.$$

By continuity of $f$, $\varphi$ and its derivatives, there exists $h, \varepsilon > 0$ such that for all $t \le s \le t + h$, $|x - y| \le \varepsilon$,

$$v(s, y) \le \varphi(s, y) \tag{6.18}$$

$$-\frac{\partial \varphi}{\partial t}(s, y) - \mathcal{L}\varphi(s, y) - f(s, y, v(s, y), (D_x\varphi)'(s, y)\sigma(y)) > 0. \tag{6.19}$$

Let $\tau = \inf\{s \geq t : |X_s^{t,x} - x| \geq \varepsilon\} \wedge (t+h)$, and consider the pair

$$(Y_s^1, Z_s^1) = (Y_{s\wedge\tau}^{t,x}, 1_{[0,\tau]}(s)Z_s^{t,x}), \quad t \leq s \leq t+h.$$

By construction, $(Y_s^1, Z_s^1)$ solves the BSDE

$$-dY_s^1 = 1_{[0,\tau]}(s)f(s, X_s^{t,x}, u(s, X_s^{t,x}), Z_s^1)ds - Z_s^1 dW_s, \quad t \leq s \leq t+h,$$
$$Y_{t+h}^1 = u(\tau, X_\tau^{t,x}).$$

On the other hand, the pair

$$(Y_s^2, Z_s^2) = (\varphi(s, X_{s\wedge\tau}^{t,x}), 1_{[0,\tau]}(s)D_x\varphi(s, X_s^{t,x})'\sigma(X_s^{t,x})), \quad t \leq s \leq t+h.$$

satisfies, by Itô's formula, the BSDE

$$-dY_s^2 = -1_{[0,\tau]}(s)(\frac{\partial\varphi}{\partial t} + \mathcal{L}\varphi)(s, X_s^{t,x}) - Z_s^2 dW_s, \quad t \leq s \leq t+h,$$
$$Y_{t+h}^1 = \varphi(\tau, X_\tau^{t,x}).$$

From the inequalities (6.18)-(6.19), and the strict comparison principle in Theorem 6.2.2, we deduce $Y_0^1 < Y_0^2$, i.e. $u(t,x) < \varphi(t,x)$, a contradiction.     □

## 6.4 Control and BSDE

In this section, we show how BSDEs may be used for dealing with stochastic control problems.

### 6.4.1 Optimization of a family of BSDEs

**Theorem 6.4.4** *Let $(\xi, f)$ and $(\xi^\alpha, f^\alpha)$, $\alpha \in \mathcal{A}$ a subset of progressively measurable processes, be a family of the pair terminal condition-generator, and $(Y, Z)$, $(Y^\alpha, Z^\alpha)$ the solutions to their associated BSDEs. Suppose that there exists $\hat{\alpha} \in \mathcal{A}$ such that*

$$f(t, Y_t, Z_t) = \operatorname*{ess\,inf}_\alpha f^\alpha(t, Y_t, Z_t) = f^{\hat{\alpha}}(t, Y_t, Z_t), \quad dt \otimes dP \ a.e.$$
$$\xi = \operatorname*{ess\,inf}_\alpha \xi^\alpha = \xi^{\hat{\alpha}}.$$

*Then,*

$$Y_t = \operatorname*{ess\,inf}_\alpha Y_t^\alpha = Y_t^{\hat{\alpha}}, \quad 0 \leq t \leq T, \ a.s.$$

**Proof.** From the comparison theorem 6.2.2, since $\xi \leq \xi^\alpha$ and $f(t, Y_t, Z_t) \leq f^\alpha(t, Y_t, Z_t)$, we have $Y_t \leq Y_t^\alpha$ for all $\alpha$, and so

$$Y_t \leq \operatorname*{ess\,inf}_\alpha Y_t^\alpha.$$

Moreover, if there exists $\hat{\alpha}$ such that $\xi = \xi^{\hat{\alpha}}$ and $f(t, Y_t, Z_t) = f^{\hat{\alpha}}(t, Y_t, Z_t)$, then $(Y, Z)$ and $(Y^{\hat{\alpha}}, Z^{\hat{\alpha}})$ are both solutions to the same BSDE with terminal condition-generator: $(\xi^{\hat{\alpha}}, f^{\hat{\alpha}})$. By uniqueness, we deduce that these solutions coincide, and so

$$\operatorname*{ess\,inf}_{\alpha} Y_t^{\alpha} \leq Y_t^{\hat{\alpha}} = Y_t \leq \operatorname*{ess\,inf}_{\alpha} Y_t^{\alpha},$$

which ends the proof.    □

By means of the above result, we show how the solution to a BSDE with concave generator may be represented as the value function of a control problem.

Let $f(t, y, z)$ be a generator, concave in $(y, z)$ and $(Y, Z)$ the solution to the BSDE associated to the pair $(\xi, f)$. We consider the Fenchel-Legendre transform of $f$

$$F(t, b, c) = \sup_{(y,z) \in \mathbb{R} \times \mathbb{R}^d} [f(t, y, z) - yb - z.c], \quad (b, c) \in \mathbb{R} \times \mathbb{R}^d. \tag{6.20}$$

Since $f$ is concave, we have the duality relation

$$f(t, y, z) = \inf_{(b,c) \in \mathbb{R} \times \mathbb{R}^d} [F(t, b, c) + yb + z.c], \quad (y, z) \in \mathbb{R} \times \mathbb{R}^d. \tag{6.21}$$

We denote by $\mathcal{A}$ the set of bounded progressively measurable processes $(\beta, \gamma)$, valued in $\mathbb{R} \times \mathbb{R}^d$ such that

$$E\left[\int_0^T |F(t, \beta_t, \gamma_t)|^2 dt\right] < \infty.$$

The boundedness condition on $\mathcal{A}$ means that for any $(\beta, \gamma) \in \mathcal{A}$, there exists a constant (dependent of $(\beta, \gamma)$) such that $|\beta_t| + |\gamma_t| \leq C$, $dt \otimes dP$ a.e. Let us consider the family of linear generators

$$f^{\beta,\gamma}(t, y, z) = F(t, \beta_t, \gamma_t) + y\beta_t + z.\gamma_t, \quad (\beta, \gamma) \in \mathcal{A}.$$

Given $(\beta, \gamma) \in \mathcal{A}$, we denote by $(Y^{\beta,\gamma}, Z^{\beta,\gamma})$ the solution to the linear BSDE associated to the pair $(\xi, f^{\beta,\gamma})$.

**Theorem 6.4.5** *Y is equal to the value function of the control problem*

$$Y_t = \operatorname*{ess\,inf}_{\beta,\gamma \in \mathcal{A}} Y_t^{\beta,\gamma}, \quad 0 \leq t \leq T, \; a.s. \tag{6.22}$$

$$Y_t^{\beta,\gamma} = E^{Q^\gamma}\left[\int_t^T e^{\int_t^s \beta_u du} F(s, \beta_s, \gamma_s) ds + e^{\int_t^T \beta_u du} \xi \Big| \mathcal{F}_t\right],$$

*where $Q^\gamma$ is the probability measure with density process*

$$dL_t = L_t \gamma_t . dW_t, \quad L_0 = 1.$$

**Proof.** (1) Observe from relation (6.21) that $f(t, Y_t, Z_t) \leq f^{\beta,\gamma}(t, Y_t, Z_t)$ for all $(\beta, \gamma) \in \mathcal{A}$. Moreover, since $F$ is convex with a linear growth condition, for each $(t, \omega, y, z)$, the infimum in the relation (6.21) is attained at $(\hat{b}(t, y, z), \hat{c}(t, y, z))$ belonging to the subdifferential of $-f$, and so is bounded by the Lipschitz constant of $f$. By a measurable selection theorem (see e.g. Appendix in Chapter III of Dellacherie and Meyer [DM75]), since $Y, Z$ are progressively measurable, we may find a pair of bounded progressively measurable processes $(\hat{\beta}, \hat{\gamma})$ such that

$$f(t, Y_t, Z_t) = f^{\hat{\beta}, \hat{\gamma}}(t, Y_t, Z_t) = F(t, \hat{\beta}_t, \hat{\gamma}_t) + Y_t \hat{\beta}_t + Z_t . \hat{\gamma}_t, \quad 0 \le t \le T, \ a.s.$$

We then obtain the relation (6.22) by Theorem 6.4.4.

(2) Moreover, by Proposition 6.2.1, the solution $Y^{\beta,\gamma}$ to the linear BSDE associated to the pair $(\xi, f^{\beta,\gamma})$ is explicitly written as

$$\Gamma_t Y_t^{\beta,\gamma} = E\Big[ \int_t^T \Gamma_s F(s, \beta_s, \gamma_s) ds + \Gamma_T \xi \Big| \mathcal{F}_t \Big],$$

where $\Gamma$ is the adjoint (dual) process given by the SDE:

$$d\Gamma_t = \Gamma_t \left( \beta_t dt + \gamma_t . dW_t \right), \quad \Gamma_0 = 1.$$

We conclude by observing that $\Gamma_t = e^{\int_0^t \beta_u du} L_t$, and using the Bayes formula. $\qquad \square$

## 6.4.2 Stochastic maximum principle

In the previous chapter, we studied how to solve a stochastic control problem by the dynamic programming method. We present here an alternative approach, called Pontryagin maximum principle, and based on optimality conditions for controls.

We consider the framework of a stochastic control problem on a finite horizon as defined in Chapter 3: let $X$ be a controlled diffusion on $\mathbb{R}^n$ governed by

$$dX_s = b(X_s, \alpha_s)ds + \sigma(X_s, \alpha_s)dW_s, \tag{6.23}$$

where $W$ is a $d$-dimensional standard Brownian motion, and $\alpha \in \mathcal{A}$, the control process, is a progressively measurable valued in $A$. The gain functional to maximize is

$$J(\alpha) = E\Big[ \int_0^T f(t, X_t, \alpha_t)dt + g(X_T) \Big],$$

where $f : [0, T] \times \mathbb{R}^n \times A \to \mathbb{R}$ is continuous in $(t, x)$ for all $a$ in $A$, $g : \mathbb{R}^n \to \mathbb{R}$ is a concave $C^1$ function, and $f$, $g$ satisfy a quadratic growth condition in $x$.

We define the generalized Hamiltonian $\mathcal{H} : [0, T] \times \mathbb{R}^n \times A \times \mathbb{R}^n \times \mathbb{R}^{n \times d} \to \mathbb{R}$ by

$$\mathcal{H}(t, x, a, y, z) = b(x, a).y + \mathrm{tr}(\sigma'(x, a)z) + f(t, x, a), \tag{6.24}$$

and we assume that $\mathcal{H}$ is differentiable in $x$ with derivative denoted by $D_x \mathcal{H}$. We consider for each $\alpha \in \mathcal{A}$, the BSDE, called the adjoint equation:

$$- dY_t = D_x \mathcal{H}(t, X_t, \alpha_t, Y_t, Z_t)dt - Z_t dW_t, \quad Y_T = D_x g(X_T). \tag{6.25}$$

**Theorem 6.4.6** *Let $\hat{\alpha} \in \mathcal{A}$ and $\hat{X}$ the associated controlled diffusion. Suppose that there exists a solution $(\hat{Y}, \hat{Z})$ to the associated BSDE (6.25) such that*

$$\mathcal{H}(t, \hat{X}_t, \hat{\alpha}_t, \hat{Y}_t, \hat{Z}_t) = \max_{a \in A} \mathcal{H}(t, \hat{X}_t, a, \hat{Y}_t, \hat{Z}_t), \quad 0 \le t \le T, \ a.s. \tag{6.26}$$

*and*

$$(x, a) \to \mathcal{H}(t, x, a, \hat{Y}_t, \hat{Z}_t) \quad \text{is a concave function,} \tag{6.27}$$

*for all $t \in [0, T]$. Then $\hat{\alpha}$ is an optimal control, i.e.*

$$J(\hat{\alpha}) = \sup_{\alpha \in \mathcal{A}} J(\alpha).$$

**Proof.** For any $\alpha \in \mathcal{A}$, we write

$$J(\hat{\alpha}) - J(\alpha) = E\Big[\int_0^T f(t, \hat{X}_t, \hat{\alpha}_t) - f(t, X_t, \alpha_t)dt + g(\hat{X}_T) - g(X_T)\Big]. \quad (6.28)$$

By concavity of $g$ and Itô's formula, we have

$$E\Big[g(\hat{X}_T) - g(X_T)\Big] \geq E\Big[(\hat{X}_T - X_T).D_x g(\hat{X}_T)\Big] = E\Big[(\hat{X}_T - X_T).\hat{Y}_T\Big]$$

$$= E\Big[\int_0^T (\hat{X}_t - X_t).d\hat{Y}_t + \int_0^T \hat{Y}_t.(d\hat{X}_t - dX_t) + \int_0^T \mathrm{tr}\big[(\sigma(\hat{X}_t, \hat{\alpha}_t) - \sigma(X_t, \alpha_t))'\hat{Z}_t\big]dt\Big]$$

$$= E\Big[\int_0^T (\hat{X}_t - X_t).(-D_x \mathcal{H}(t, \hat{X}_t, \hat{\alpha}_t, \hat{Y}_t, \hat{Z}_t))dt + \int_0^T \hat{Y}_t.(b(\hat{X}_t, \hat{\alpha}_t) - b(X_t, \alpha_t))dt$$

$$+ \int_0^T \mathrm{tr}\big[(\sigma(\hat{X}_t, \hat{\alpha}_t) - \sigma(X_t, \alpha_t))'\hat{Z}_t\big]dt\Big]. \quad (6.29)$$

Moreover, by definition of $\mathcal{H}$, we have

$$E\Big[\int_0^T f(t, \hat{X}_t, \hat{\alpha}_t) - f(t, X_t, \alpha_t)dt\Big] = E\Big[\int_0^T \mathcal{H}(t, \hat{X}_t, \hat{\alpha}_t, \hat{Y}_t, \hat{Z}_t) - \mathcal{H}(t, X_t, \alpha_t, \hat{Y}_t, \hat{Z}_t)dt$$

$$- \int_0^T (b(\hat{X}_t, \hat{\alpha}_t) - b(X_t, \alpha_t)).\hat{Y}_t$$

$$- \int_0^T \mathrm{tr}\Big[(\sigma(\hat{X}_t, \hat{\alpha}_t) - \sigma(X_t, \alpha_t))'\hat{Z}_t\Big]dt\Big]. \quad (6.30)$$

By adding (6.29) and (6.30) into (6.28), we obtain

$$J(\hat{\alpha}) - J(\alpha) \geq E\Big[\int_0^T \mathcal{H}(t, \hat{X}_t, \hat{\alpha}_t, \hat{Y}_t, \hat{Z}_t) - \mathcal{H}(t, X_t, \alpha_t, \hat{Y}_t, \hat{Z}_t)dt$$

$$- \int_0^T (\hat{X}_t - X_t).D_x \mathcal{H}(t, \hat{X}_t, \hat{\alpha}_t, \hat{Y}_t, \hat{Z}_t)dt\Big].$$

Under the conditions (6.26) and (6.27), the term between the bracket in the above relation is nonpositive, which ends the proof. $\square$

We shall illustrate in Section 6.6.2 how to use Theorem 6.4.6 for solving a control problem in finance arising in mean-variance hedging.

We conclude this section by providing the connection between maximum principle and dynamic programming. The value function of the stochastic control problem considered above is defined by

$$v(t, x) = \sup_{\alpha \in \mathcal{A}} E\Big[\int_t^T f(s, X_s^{t,x}, \alpha_s)ds + g(X_T^{t,x})\Big], \quad (6.31)$$

where $\{X_s^{t,x}, t \leq s \leq T\}$ is the solution to (6.23) starting from $x$ at $t$. Recall that the associated Hamilton-Jacobi-Bellman equation is

$$-\frac{\partial v}{\partial t} - \sup_{a \in A}\big[\mathcal{G}(t, x, a, D_x v, D_x^2 v)\big] = 0, \quad (6.32)$$

where for $(t, x, a, p, M) \in [0, T] \times \mathbb{R}^n \times A \times \mathbb{R}^n \times \mathcal{S}_n$,

$$G(t, x, a, p, M) = b(x, a).p + \frac{1}{2}\mathrm{tr}(\sigma\sigma'(x, a)M) + f(t, x, a). \tag{6.33}$$

**Theorem 6.4.7** *Suppose that* $v \in C^{1,3}([0, T) \times \mathbb{R}^n) \cap C^0([0, T] \times \mathbb{R}^n)$, *and there exists an optimal control* $\hat{\alpha} \in \mathcal{A}$ *to* (6.31) *with associated controlled diffusion* $\hat{X}$. *Then*

$$G(t, \hat{X}_t, \hat{\alpha}_t, D_x v(t, \hat{X}_t), D_x^2 v(t, \hat{X}_t)) = \max_{a \in A} G(t, \hat{X}_t, a, D_x v(t, \hat{X}_t), D_x^2 v(t, \hat{X}_t)), \tag{6.34}$$

*and the pair*

$$(\hat{Y}_t, \hat{Z}_t) = (D_x v(t, \hat{X}_t) \;,\; D_x^2 v(t, \hat{X}_t)\, \sigma(\hat{X}_t, \hat{\alpha}_t)), \tag{6.35}$$

*is solution to the adjoint BSDE* (6.25).

**Proof.** Since $\hat{\alpha}$ is an optimal control, we have

$$\begin{aligned} v(t, \hat{X}_t) &= E\Big[ \int_t^T f(s, \hat{X}_s, \hat{\alpha}_s)ds + g(\hat{X}_T)\Big|\mathcal{F}_t\Big] \\ &= -\int_0^t f(s, \hat{X}_s, \hat{\alpha}_s)ds + M_t, \quad 0 \le t \le T, \text{ a.s.} \end{aligned} \tag{6.36}$$

where $M$ is the martingale $M_t = E\Big[ \int_0^T f(s, \hat{X}_s, \hat{\alpha}_s)ds + g(\hat{X}_T)\Big|\mathcal{F}_t\Big]$. By applying Itô's formula to $v(t, \hat{X}_t)$, and identifying the terms in $dt$ in relation (6.36), we get

$$-\frac{\partial v}{\partial t}(t, \hat{X}_t) - G(t, \hat{X}_t, \hat{\alpha}_t, D_x v(t, \hat{X}_t), D_x^2 v(t, \hat{X}_t)) = 0. \tag{6.37}$$

Since $v$ is smooth, it satisfies the HJB equation (6.32), which yields (6.34).

From (6.32) and (6.37), we have

$$\begin{aligned} 0 &= \frac{\partial v}{\partial t}(t, \hat{X}_t) + G(t, \hat{X}_t, \hat{\alpha}_t, D_x v(t, \hat{X}_t), D_x^2 v(t, \hat{X}_t)) \\ &\ge \frac{\partial v}{\partial t}(t, x) + G(t, x, \hat{\alpha}_t, D_x v(t, x), D_x^2 v(t, x)), \quad \forall x \in \mathbb{R}^n. \end{aligned}$$

Since $v$ is $C^{1,3}$, the optimality condition for the above relation implies

$$\frac{\partial}{\partial x}\left( \frac{\partial v}{\partial t}(t, x) + G(t, x, \hat{\alpha}_t, D_x v(t, x), D_x^2 v(t, x)) \right)\Big|_{x = \hat{X}_t} = 0.$$

By recalling the expressions (6.33) and (6.24) of $G$ and $\mathcal{H}$, the previous equality is written as

$$\begin{aligned} \frac{\partial^2 v}{\partial t \partial x}(t, \hat{X}_t) &+ D_x^2 v(t, \hat{X}_t) b(\hat{X}_t, \hat{\alpha}_t) + \frac{1}{2}\mathrm{tr}(\sigma\sigma'(\hat{X}_t, \hat{\alpha}_t) D_x^3 v(t, \hat{X}_t)) \\ &+ D_x^2 \mathcal{H}(t, \hat{X}_t, \hat{\alpha}_t, D_x v(t, \hat{X}_t), D_x^2 v(t, \hat{X}_t)\, \sigma(\hat{X}_t, \hat{\alpha}_t)) = 0. \end{aligned} \tag{6.38}$$

By applying Itô's formula to $D_x v(t, \hat{X}_t)$, and using (6.38), we then get

$$-dD_x v(t, \hat{X}_t) = -\left[\frac{\partial^2 v}{\partial t \partial x}(t, \hat{X}_t) + D_x^2 v(t, \hat{X}_t) b(\hat{X}_t, \hat{\alpha}_t) + \frac{1}{2}\mathrm{tr}(\sigma\sigma'(\hat{X}_t, \hat{\alpha}_t) D_x^3 v(t, \hat{X}_t))\right] dt$$
$$- D_x^2 v(t, \hat{X}_t)\ \sigma(\hat{X}_t, \hat{\alpha}_t) dW_t$$
$$= D_x \mathcal{H}(t, \hat{X}_t, \hat{\alpha}_t, D_x v(t, \hat{X}_t), D_x^2 v(t, \hat{X}_t)\ \sigma(\hat{X}_t, \hat{\alpha}_t))\ dt$$
$$- D_x^2 v(t, \hat{X}_t)\ \sigma(\hat{X}_t, \hat{\alpha}_t) dW_t.$$

Moreover, since $v(T, .) = g(.)$, we have

$$D_x v(T, \hat{X}_T) = D_x g(\hat{X}_T),$$

and this proves the result (6.35).    $\square$

## 6.5 Reflected BSDEs and optimal stopping problems

We consider a class of BSDEs where the solution $Y$ is constrained to stay above a given process, called obstacle. An increasing process is introduced for pushing the solution upwards, above the obstacle. This leads to the notion of reflected BSDE, which is formalized as follows.

Let $W = (W_t)_{0 \leq t \leq T}$ be a standard $d$-dimensional Brownian motion on a filtered probability space $(\Omega, \mathcal{F}, \mathbb{F}, P)$ where $\mathbb{F} = (\mathcal{F}_t)_{0 \leq t \leq T}$ is the natural filtration of $W$, and $T$ is a fixed finite horizon. We are given a pair $(\xi, f)$ satisfying conditions (A) and (B), and in addition a continuous process $(L_t)_{0 \leq t \leq T}$, satisfying $\xi \geq L_T$ and

(C)    $\qquad\qquad L \in \mathbb{S}^2(0, T)$,    i.e.    $E[\sup_{0 \leq t \leq T} |L_t|^2] < \infty$.

A solution to the reflected BSDE with terminal condition-generator $(\xi, f)$ and obstacle $L$ is a triple $(Y, Z, K)$ of progressively measurable processes valued in $\mathbb{R} \times \mathbb{R}^d \times \mathbb{R}_+$ such that $Y \in \mathbb{S}^2(0, T)$, $Z \in \mathbb{H}^2(0, T)^d$, $K$ is a continuous increasing process, $K_0 = 0$, and satisfying

$$Y_t = \xi + \int_t^T f(s, Y_s, Z_s) ds + K_T - K_t - \int_t^T Z_s . dW_s, \quad 0 \leq t \leq T \qquad (6.39)$$

$$Y_t \geq L_t, \quad 0 \leq t \leq T, \qquad\qquad\qquad (6.40)$$

$$\int_0^T (Y_t - L_t) dK_t = 0. \qquad\qquad\qquad (6.41)$$

**Remark 6.5.2** The condition (6.41) means that the push of the increasing process $K$ is minimal in the sense that it is active only when the constraint is saturated, i.e. when $Y_t = L_t$. There is another formulation of this minimality condition for defining a solution to a reflected BSDE: we say that $(Y, Z, K)$ is a minimal solution to the reflected BSDE, if it satisfies (6.39)-(6.40), and for any other solution $(\tilde{Y}, \tilde{Z}, \tilde{K})$ satisfying (6.39)-(6.40), we have $Y_t \leq \tilde{Y}_t$, $0 \leq t \leq T$ a.s. We shall discuss the equivalence of this formulation in Remark 6.5.3.

In the special case where the generator $f$ does not depend on $y, z$, the notion of a reflected BSDE is directly related to optimal stopping problems, as stated in the following proposition.

**Proposition 6.5.3** *Suppose that $f$ does not depend on $y, z$, and $f \in \mathbb{H}^2(0, T)$. Then, there exists a unique solution $(Y, Z, K)$ to the reflected BSDE (6.39), (6.40) and (6.41), and $Y$ has the explicit optimal stopping time representation*

$$Y_t = \operatorname{ess\,sup}_{\tau \in \mathcal{T}_{t,T}} E\left[\int_t^T f(s)ds + L_\tau 1_{\tau < T} + \xi 1_{\tau = T} \Big| \mathcal{F}_t\right], \quad 0 \le t \le T. \tag{6.42}$$

**Proof.** Let us consider the process $Y$ defined by (6.42), and observe that $Y_t + \int_0^t f(s)ds$ is the Snell envelope of the process

$$H_t = \int_0^t f(s)ds + L_t 1_{t < T} + \xi 1_{t = T}, \quad 0 < t \le T.$$

From the conditions (A), (B), and (C) on $f, \xi$ and $L$, and since $\xi \ge L_T$, we notice that the process $H$ is continuous on $[0, T)$, with a positive jump at $T$, and satisfies $E[\sup_{0 \le t \le T} |H_t|^2] < \infty$, in particular of class (DL). Hence, by Proposition 1.1.8, the process $Y_t + \int_0^t f(s)ds$ is a continuous supermartingale dominating $H$, i.e. $Y_t \ge L_t$, $0 \le t \le T$, and for any $t \in [0, T]$, the stopping time

$$\tau_t = \inf\{s \ge t : Y_s = L_s\} \wedge T,$$

is optimal, in the sense that

$$Y_t + \int_0^t f(s)ds = E\left[\int_0^{\tau_t} f(s)ds + L_{\tau_t} 1_{\tau_t < T} + \xi 1_{\tau_t = T} \Big| \mathcal{F}_t\right]. \tag{6.43}$$

On the other hand, by applying the Doob-Meyer decomposition to the continuous supermartingale $S_t = Y_t + \int_0^t f(s)ds$ of class (DL), we get the existence of a continuous martingale $M$ and an adapted continuous nondecreasing process $K$, $K_0 = 0$ such that

$$Y_t + \int_0^t f(s)ds = M_t - K_t, \quad 0 \le t \le T. \tag{6.44}$$

By observing that $Y_{\tau_t} = L_{\tau_t} 1_{\tau_t < T} + \xi 1_{\tau_t = T}$, and from the optional sampling theorem for the martingale $M$: $M_t = E[M_{\tau_t} | \mathcal{F}_t]$, we deduce

$$Y_t + \int_0^t f(s)ds = E\left[\int_0^{\tau_t} f(s)ds + L_{\tau_t} 1_{\tau_t < T} + \xi 1_{\tau_t = T} + K_{\tau_t} - K_t \Big| \mathcal{F}_t\right].$$

By comparing with (6.43), it follows that $E[K_{\tau_t} - K_t | \mathcal{F}_t] = 0$, and so $K_{\tau_t} = K_t$, or equivalently by definition of $\tau_t$

$$\int_0^T (Y_t - L_t)dK_t = 0.$$

Moreover, since $H$ lies in $\mathbb{S}^2(0, T)$, we easily see that $Y$ also lies in $\mathbb{S}^2(0, T)$. Then, in the decomposition (6.44), the martingale $M$ is square-integrable, and $K_T$ is also square-integrable. By the Itô representation theorem, there exists $Z \in \mathbb{H}^2(0, T)$ such that

$$M_t = M_0 + \int_0^t Z_s.dW_s, \quad 0 \le t \le T.$$

Plugging into (6.44), and recalling that $Y_T = \xi$, we deduce that $(Y, Z, K)$ solves (6.39), (6.40) and (6.41).

It remains to check uniqueness. Let $(Y, Z, K)$ and $(\bar Y, \bar Z, \bar K)$ be two solutions of (6.39), (6.40) and (6.41), and define $\Delta Y = Y - \bar Y$, $\Delta Z = Z - \bar Z$, $\Delta K = K - \bar K$. Then, $(\Delta Y, \Delta Z, \Delta K)$ satisfies

$$\Delta Y_t = -\int_t^T \Delta Z_s.dW_s + \Delta K_T - \Delta K_t, \quad 0 \le t \le T.$$

Moreover, by (6.40)-(6.41), we have for all $t \in [0, T]$

$$\int_t^T \Delta Y_s d\Delta K_s = \int_t^T (Y_s - L_s + L_s - \bar Y_s)(dK_s - d\bar K_s)$$

$$= -\int_t^T (Y_s - L_s)d\bar K_s - \int_t^T (\bar Y_s - L_s)dK_s \le 0. \qquad (6.45)$$

By Itô's formula to $|\Delta Y_t|^2$, we then obtain

$$|\Delta Y_t|^2 + \int_t^T |\Delta Z_s|^2 ds = 2\int_t^T \Delta Y_s d\Delta K_s - 2\int_t^T \Delta Y_s \Delta Z_s.dW_s$$

$$\le -2\int_t^T \Delta Y_s \Delta Z_s.dW_s. \qquad (6.46)$$

From the integrability conditions $\Delta Y \in \mathbb{S}^2(0, T)$, $\Delta Z \in \mathbb{H}^2(0, T)^d$, and the Burkholder-Davis-Gundy inequality, we observe that the local martingale $\int_0^t \Delta Y_s \Delta Z_s.dW_s$ is uniformly integrable, thus a martingale. By taking the expectation in (6.46), we conclude that

$$E\left[|\Delta Y_t|^2 + \int_t^T |\Delta Z_s|^2 ds\right] \le 0, \quad 0 \le t \le T,$$

which proves that $Y = \bar Y$, $Z = \bar Z$, and $K = \bar K$. $\qquad\qquad\square$

In the sequel, we consider the general case where $f$ may depend on $y, z$. We shall prove the existence of a solution to the reflected BSDE, and in the Markovian case, we relate this solution to a variational inequality extending the free boundary problem for optimal stopping problems.

### 6.5.1 Existence and approximation via penalization

In this section, we prove the existence and uniqueness of a solution to the reflected BSDE (6.39), (6.40) and (6.41), based on approximation via penalization. For each $n \in \mathbb{N}$, we consider the BSDE

$$Y_t^n = \xi + \int_t^T f(s, Y_s^n, Z_s^n)ds + n\int_t^T (Y_n^s - L_s)^- ds - \int_t^T Z_s^n.dW_s. \qquad (6.47)$$

Notice that the generator $f_n(t, y, z) = f(t, y, z) + n(y - L_t)^-$ satisfies condition (B). From Theorem 6.2.1, there exists for each $n$, a unique solution $(Y^n, Z^n)$ to the BSDE (6.47). We define

$$K_t^n = n \int_0^t (Y_n^s - L_s)^- ds, \quad 0 \le t \le T,$$

which is a continuous nondecreasing process, and is square integrable. Formally, the solution $Y^n$ is penalized (by the factor $n$) once it falls below the obstacle $L$. The rest of this section is devoted to the convergence of the sequence $(Y^n, Z^n, K^n)_n$ to the solution to the reflected BSDE.

We first state a priori uniform estimates on the sequence $(Y^n, Z^n, K^n)_n$.

**Lemma 6.5.1** *There exists a constant $C$ such that*

$$E\Big[\sup_{0 \le t \le T} |Y_t^n|^2 + \int_0^T |Z_t^n|^2 dt + |K_T^n|^2\Big] \le C, \quad \forall n \in \mathbb{N}.$$

**Proof.** By applying Itô's formula to $|Y_t^n|^2$, we get

$$E\Big[|Y_t^n|^2 + \int_t^T |Z_s^n|^2 ds\Big] = E[|\xi|^2] + 2E\Big[\int_t^T f(s, Y_s^n, Z_s^n) Y_s^n ds\Big] + 2E\Big[\int_t^T Y_s^n dK_s^n\Big].$$

Now, by definition of $K^n$, we have: $\int_t^T Y_s^n dK_s^n \le \int_t^T L_s dK_s^n \le \sup_{0 \le t \le T} |L_t|(K_T^n - K_t^n)$. From the Lipschitz property of $f$ in condition (B), and using the inequality $2ab \le \frac{1}{\alpha} a^2 + \alpha b^2$ for any constant $\alpha > 0$, we then obtain

$$E\Big[|Y_t^n|^2 + \int_t^T |Z_s^n|^2 ds\Big]$$

$$\le E[|\xi|^2] + 2E\Big[\int_t^T \big(f(s, 0, 0) + C_f(|Y_s^n| + |Z_s^n|) Y_s^n ds\Big] + 2E\Big[\sup_{0 \le t \le T} |L_t|(K_T^n - K_t^n)\Big]$$

$$\le C\Big(1 + E\Big[\int_t^T |Y_s^n|^2 ds\Big]\Big) + \frac{1}{2} E\Big[\int_t^T |Z_s^n|^2 ds\Big] + \frac{1}{\alpha} E\Big[\sup_{0 \le t \le T} |L_t|^2\Big] + \alpha E|K_T^n - K_t^n|^2,$$

and so

$$E\Big[|Y_t^n|^2 + \frac{1}{2} \int_t^T |Z_s^n|^2 ds\Big] \le C\Big(1 + E\Big[\int_t^T |Y_s^n|^2 ds\Big]\Big) + \alpha E|K_T^n - K_t^n|^2. \quad (6.48)$$

Moreover, from the relation

$$K_T^n - K_t^n = Y_t^n - \xi - \int_t^T f(s, Y_s^n, Z_s^n) ds + \int_t^T Z_s^n . dW_s,$$

and conditions (A) and (B) on $\xi$ and $f$, there exists a constant $C_1$ such that

$$E|K_T^n - K_t^n|^2 \le C_1\Big(1 + E|Y_t^n|^2 + E\Big[\int_t^T |Y_s^n|^2 + |Z_s^n|^2 ds\Big]\Big).$$

By choosing $\alpha = 1/4C_1$, and plugging into (6.48), we get

$$\frac{3}{4}E\Big[|Y_t^n|^2 + \frac{1}{4}\int_t^T |Z_s^n|^2 ds\Big] \leq C\Big(1 + E\Big[\int_t^T |Y_s^n|^2 ds\Big]\Big).$$

By Gronwall's lemma, this implies

$$\sup_{0 \leq t \leq T} E|Y_t^n|^2 + E\Big[\int_t^T |Z_s^n|^2 ds\Big] + E|K_T^n|^2 \leq C. \tag{6.49}$$

Finally, by writing from (6.47) that

$$\sup_{0 \leq t \leq T} |Y_t^n|^2 \leq C\Big(|\xi|^2 + \int_0^T |f(s, Y_s^n, Z_s^n)|^2 ds + |K_T^n|^2 + \sup_{0 \leq t \leq T}\Big|\int_0^T Z_s.dW_s\Big|^2\Big),$$

we obtain the required result from the Burkholder-Davis-Gundy inequality, conditions (A) and (B) on $(\xi, f)$, and estimate (6.49).     □

We next focus on the convergence of the sequence $(Y^n)_n$.

**Lemma 6.5.2** *The sequence $(Y^n)_n$ converges increasingly to a process $Y \in \mathbb{S}^2(0, T)$, and the convergence also holds in $\mathbb{H}^2(0, T)$, i.e.*

$$\lim_{n \to \infty} E\Big[\int_0^T |Y_t^n - Y_t|^2 dt\Big] = 0. \tag{6.50}$$

*Furthermore, $Y_t \geq L_t$, $0 \leq t \leq T$, a.s., and*

$$\lim_{n \to \infty} E\Big[\sup_{0 \leq t \leq T} (Y_t^n - L_t)^-\Big] = 0. \tag{6.51}$$

**Proof.** Since the generator $f_n$ of the BSDE for $Y_n$ is nondecreasing in $n$: $f_n(t, y, z) \leq f_{n+1}(t, y, z)$, we deduce from the comparison principle (Theorem 6.2.2) that $Y_t^n \leq Y_t^{n+1}$, $0 \leq t \leq T$ a.s. Together with the uniform estimate for $(Y_n)_n$ in $\mathbb{S}^2(0, T)$ in Lemma 6.5.1, this shows that the nondecreasing limit

$$Y_t := \lim_{n \to \infty} Y_t^n, \quad 0 \leq t \leq T,$$

exists a.s., and this defines an adapted process $Y \in \mathcal{S}^2$. From the dominated convergence theorem, we also get the convergence (6.50).

Notice that since the sequence $K_T^n = n \int_0^T (Y_t^n - L_t)^- dt$ is bounded in $L^2(\Omega, \mathcal{F}, P)$ by Lemma 6.5.1, then $E\Big[\int_0^T (Y_t - L_t)^- dt\Big] = 0$, which implies that $Y_t \geq L_t$ $dt \otimes dP$ a.e. We want to prove the stronger result $Y_t \geq L_t$, $0 \leq t \leq T$, a.s., and so use another argument. Let us consider the solution $(\tilde{Y}^n, \tilde{Z}^n)$ to the linear BSDE

$$\tilde{Y}_t^n = \xi + \int_t^T f(s, Y_s^n, Z_s^n) ds + n \int_t^T (L_s - \tilde{Y}_n^s) ds - \int_t^T \tilde{Z}_s^n.dW_s.$$

The generator $\tilde{f}_n(t, \tilde{y}, \tilde{z}) = f(t, Y_t^n, Z_t^n) + n(L_t - \tilde{y})$ of this BSDE satisfies $\tilde{f}_n(t, \tilde{Y}_t^n, \tilde{Z}_t^n) \leq f_n(t, \tilde{Y}_t^n, \tilde{Z}_t^n)$, so that by the comparison principle (Theorem 6.2.2) $\tilde{Y}_t^n \leq Y_t^n$, $0 \leq t \leq T$. Moreover, by Proposition 6.2.1, the solution to this linear BSDE is explicitly given by

$$\tilde{Y}_\tau^n = E\Big[e^{-n(T-\tau)}\xi + \int_\tau^T e^{-n(s-\tau)}(f(Y_s^n, Z_s^n) + nL_s) ds \Big| \mathcal{F}_\tau\Big],$$

for any stopping time $\tau$ valued in $[0,T]$. It is not difficult (left to the reader) to check that as $n$ goes to infinity

$$\tilde{Y}^n_\tau \to \xi 1_{\tau=T} + L_\tau 1_{\tau<T} \geq L_\tau \quad \text{in } L^2(\Omega, \mathcal{F}, P).$$

Therefore $Y_\tau \geq L_\tau$ a.s. From that and section theorem (see Theorem 1.1.1), we deduce that $Y_t \geq L_t$, $0 \leq t \leq T$, a.s. This implies $(Y^n_t - L_t)^- \downarrow 0$, $0 \leq t \leq T$ a.s., and this convergence is also uniform in $t$ by Dini's theorem: $\sup_{t\in[0,T]}(Y^n_t - L_t)^- \downarrow 0$ a.s. Finally, we obtain the result (6.51) by the monotone convergence theorem.    $\square$

We can finally state the main result of this section.

**Theorem 6.5.8** *There exists a unique $(Y, Z, K)$ solution to the reflected BSDE* (6.39), (6.40) *and* (6.41), *and this triple $(Y, Z, K)$ is the limit of the sequence $(Y^n, Z^n, Z^n)_n$ in* $\mathbb{S}^2(0,T) \times \mathbb{H}^2(0,T)^d \times \mathbb{S}^2(0,T)$, *i.e.*

$$\lim_{n\to\infty} E\Big[ \sup_{0\leq t\leq T} |Y^n_t - Y_t|^2 + \int_0^T |Z^n_t - Z_t|^2 dt + \sup_{0\leq t\leq T} |K^n_t - K_t|^2 \Big] = 0. \quad (6.52)$$

**Proof.** For any $n, p \in \mathbb{N}$, we apply Itô's formula to $|Y^n_t - Y^p_t|^2$:

$$|Y^n_t - Y^p_t|^2 + \int_t^T |Z^n_s - Z^p_s|^2 ds$$

$$= 2\int_t^T \big(f(s, Y^n_s, Z^n_s) - f(s, Y^p_s, Z^p_s)\big)(Y^n_s - Y^p_s) ds - 2\int_t^T (Y^n_s - Y^p_s)(Z^n_s - Z^p_s).dW_s$$

$$+ 2\int_t^T (Y^n_s - Y^p_s)d(K^n_s - K^p_s)$$

$$\leq C\int_t^T |Y^n_s - Y^p_s|^2 ds + \frac{1}{2}\int_t^T |Z^n_s - Z^p_s|^2 ds - 2\int_t^T (Y^n_s - Y^p_s)(Z^n_s - Z^p_s).dW_s$$

$$+ 2\int_t^T (Y^n_s - L_s)^- dK^p_s + 2\int_t^T (Y^p_s - L_s)^- dK^n_s, \quad (6.53)$$

where we used the Lipschitz condition on $f$, the inequality $2ab \leq \alpha a^2 + \frac{1}{\alpha}b^2$ (for suitable choice of $\alpha > 0$), and the definitions of $K^n$, $K^p$. By taking the expectation, this yields

$$E\Big[\int_t^T |Z^n_s - Z^p_s|^2 ds\Big] \leq CE\Big[\int_t^T |Y^n_s - Y^p_s|^2 ds\Big]$$

$$+ 4E\Big[\int_t^T (Y^n_s - L_s)^- dK^p_s + \int_t^T (Y^p_s - L_s)^- dK^n_s\Big].$$

From Lemma 6.5.1 and (6.51), we have

$$E\Big[\int_t^T (Y^n_s - L_s)^- dK^p_s + \int_t^T (Y^p_s - L_s)^- dK^n_s\Big] \to 0, \quad \text{as } n, p \to \infty. \quad (6.54)$$

We deduce with (6.50) that

$$E\Big[\int_t^T |Y^n_s - Y^p_s|^2 ds + \int_t^T |Z^n_s - Z^p_s|^2 ds\Big] \to 0, \quad \text{as } n, p \to \infty. \quad (6.55)$$

Now, from (6.53) and the Burholder-Davis-Gundy inequality, we get

$$E\Big[\sup_{0\le t\le T}|Y_t^n - Y_t^p|^2\Big] \le CE\Big[\int_t^T |Y_s^n - Y_s^p|^2 ds + \int_t^T |Z_s^n - Z_s^p|^2 ds\Big]$$

$$+ 2E\Big[\int_t^T (Y_s^n - L_s)^- dK_s^p + \int_t^T (Y_s^p - L_s)^- dK_s^n\Big]$$

$$+ CE\Big[\sup_{0\le t\le T}|Y_s^n - Y_s^p|\Big(\int_t^T |Z_s^n - Z_s^p|^2\Big)^{\frac{1}{2}}\Big]$$

$$\le CE\Big[\int_t^T |Y_s^n - Y_s^p|^2 ds + \int_t^T |Z_s^n - Z_s^p|^2 ds\Big]$$

$$+ 2E\Big[\int_t^T (Y_s^n - L_s)^- dK_s^p + \int_t^T (Y_s^p - L_s)^- dK_s^n\Big]$$

$$+ \frac{1}{2}E\Big[\sup_{0\le t\le T}|Y_t^n - Y_t^p|^2\Big] + CE\Big[\int_t^T |Z_s^n - Z_s^p|^2 ds\Big],$$

where we used again the inequality $2ab \le \alpha a^2 + \frac{1}{\alpha}b^2$. Together with (6.54) and (6.55), this proves that

$$E\Big[\sup_{0\le t\le T}|Y_t^n - Y_t^p|^2\Big] \to 0, \quad \text{as } n, p \to \infty. \tag{6.56}$$

By writing from (6.47) that

$$K_t^n - K_t^p = Y_0^n - Y_0^p - (Y_t^n - Y_t^p) - \int_0^t (f(s, Y_s^n, Z_s^n) - f(s, Y_s^p, Z_s^p)) ds$$

$$+ \int_0^t (Z_s^n - Z_s^p).dW_s,$$

we then obtain by the Lipschitz condition on $f$, (6.55) and (6.56) that

$$E\Big[\sup_{0\le t\le T}|K_t^n - K_t^p|^2\Big] \to 0, \quad \text{as } n, p \to \infty.$$

Consequently, $(Z^n, K^n)_n$ is a Cauchy sequence in the Banach space $\mathbb{H}^2(0,T)^d \times \mathbb{S}^2(0,T)$, and this gives the existence of a $(Z, K) \in \mathbb{H}^2(0,T)^d \times \mathbb{S}^2(0,T)$ such that

$$\lim_{n\to\infty} E\Big[\int_0^T |Z_t^n - Z_t|^2 dt + \sup_{0\le t\le T}|K_t^n - K_t|^2\Big] = 0.$$

By (6.56), we also know that the convergence of the sequence $(Y^n)$ to the limit $Y$ in Lemma 6.5.2, holds in $\mathbb{S}^2(0,T)$: $E\big[\sup_{0\le t\le T}|Y_t^n - Y_t|^2\big] \to 0$. Notice that the limit $K$ of $K^n$ in $\mathbb{S}^2(0,T)$, inherits from $K^n$ the nondecreasing and continuity path properties. We can then pass to the (strong) limit in (6.47), and deduce that $(Y, Z, K)$ solves (6.39)-(6.40). Let us now check the condition (6.41). The convergence of $(Y^n, K^n)$ to $(Y, K)$ in $\mathbb{S}^2(0,T) \times \mathbb{S}^2(0,T)$ implies by the Tchebyshev inequality that the convergence also holds uniformly in $t$ in probability. Then the measure $dK^n$ tends to $dK$ weakly in probability, and so $\int_0^T (Y_t^n - L_t)dK_t^n \to \int_0^T (Y_t - L_t)dK_t$ in probability as $n$ goes to infinity. Moreover, since $Y$ satisfies (6.40), we have $\int_0^T (Y_t - L_t)dK_t \ge 0$ a.s. On the other hand, by definition of $K^n$, we have $\int_0^T (Y_t - L_t)dK_t^n \leq 0$. We conclude that $\int_0^T (Y_t - L_t)dK_t = 0$ a.s., and this shows that $(Y, Z, K)$ is a solution to the reflected BSDE (6.39)-(6.40)-(6.41).

We finally turn to uniqueness. Let $(Y, Z, K)$ and $(\bar{Y}, \bar{Z}, \bar{K})$ be two solutions of (6.39), (6.40) and (6.41), and define $\Delta Y = Y - \bar{Y}$, $\Delta Z = Z - \bar{Z}$, $\Delta K = K - \bar{K}$. Then, $(\Delta Y, \Delta Z, \Delta K)$ satisfies

$$\Delta Y_t = \int_t^T (f(s, Y_s, Z_s) - f(s, \bar{Y}_s, \bar{Z}_s))ds - \int_t^T \Delta Z_s . dW_s + \Delta K_T - \Delta K_t, \quad 0 \leq t \leq T.$$

By applying Itô's formula to $|\Delta Y_t|^2$, and using similar computations as for $|Y_t^n - Y_t^p|^2$ and recalling $\int_t^T \Delta Y_s d\Delta K_s \leq 0$ (see (6.45)), we have

$$E\left[|\Delta Y_t|^2 + \frac{1}{2}\int_t^T |\Delta Z_s|^2 ds\right] \leq CE\left[\int_t^T |\Delta Y_s|^2 ds\right].$$

By Gronwall's lemma, we conclude that $\Delta Y = 0$, $\Delta Z = 0$ and so $\Delta K = 0$.   □

**Remark 6.5.3** For any triple $(\tilde{Y}, \tilde{Z}, \tilde{K})$ satisfying (6.39)-(6.40), and for $(Y^n, Z^n)$ solution to the penalized BSDE (6.47) one can prove by a comparison principle that $Y_t^n \leq \tilde{Y}_t$, $0 \leq t \leq T$ a.s. By passing to the limit, this shows that $Y_t \leq \tilde{Y}_t$, $0 \leq t \leq T$ a.s. Therefore, the solution to the reflected BSDE (6.39), (6.40) and (6.41) is also a minimal solution in the sense defined in Remark 6.5.2.

### 6.5.2 Connection with variational inequalities

We put our reflected BSDE in a Markovian framework in the sense that the terminal condition, the generator and the obstacle are functions of a forward SDE. More precisely, we are given a diffusion on $\mathbb{R}^n$

$$dX_s = b(X_s)ds + \sigma(X_s)dW_s, \tag{6.57}$$

with Lipschitz coefficents $b$ and $\sigma$ on $\mathbb{R}^n$, and we consider the reflected BSDE

$$Y_t = g(X_T) + \int_t^T f(s, X_s, Y_s, Z_s)ds + K_T - K_t - \int_t^T Z_s . dW_s, \quad 0 \leq t \leq T \tag{6.58}$$

$$Y_t \geq h(X_t), \quad 0 \leq t \leq T, \tag{6.59}$$

$$\int_0^T (Y_t - h(X_t))dK_t = 0, \tag{6.60}$$

where $f$ is a continuous function on $[0, T] \times \mathbb{R}^n \times \mathbb{R} \times \mathbb{R}^d$, satisfying a linear growth condition in $(x, y, z)$, a Lipschitz condition in $(y, z)$ uniformly in $(t, x)$, $g$ is a measurable function on $\mathbb{R}^n$ with a linear growth condition, and $h$ is a continuous function on $\mathbb{R}^n$ with a linear growth condition, and $g \geq h$.

By the Markov property of the diffusion $X$, and uniqueness of a solution to the reflected BSDE, we see that $Y_t = v(t, X_t)$, $0 \leq t \leq T$, where

$$v(t,x) := Y_t^{t,x} \qquad (6.61)$$

is a deterministic function of $(t,x) \in [0,T] \times \mathbb{R}^n$, $\{X_s^{t,x}, t \le s \le T\}$ denotes the solution to (6.57) starting from $x$ at $t$, and $\{(Y_s^{t,x}, Z_s^{t,x}, K_s^{t,x}), t \le s \le T\}$ is the solution to the reflected BSDE (6.58), (6.59) and (6.60) with $X_s = X_s^{t,x}$, $t \le s \le T$. We shall relate this reflected BSDE to the variational inequality

$$\min\Big[ -\frac{\partial v}{\partial t} - \mathcal{L}v - f(.,v,\sigma' D_x v) \,,\, v - h\Big] = 0, \quad \text{on } [0,T) \times \mathbb{R}^n \qquad (6.62)$$

$$v(T,.) = g \quad \text{on } \mathbb{R}^n, \qquad (6.63)$$

where $\mathcal{L}$ is the second-order operator associated to the diffusion $X$:

$$\mathcal{L}v = b(x).D_x v + \frac{1}{2}\mathrm{tr}(\sigma\sigma'(x)D_x^2 v).$$

The following result is the analog of Proposition 6.3.2 for BSDEs, and shows that a classical solution to the variational inequality provides a solution to the reflected BSDE.

**Proposition 6.5.4** *Suppose that $v \in C^{1,2}([0,T) \times \mathbb{R}^n) \cap C^0([0,T] \times \mathbb{R}^n)$ is a classical solution to (6.62)-(6.63), satisfying a linear growth condition and such that for some positive constants $C, q > 0$: $|D_x v(t,x)| \le C(1+|x|^q)$, for all $x \in \mathbb{R}^n$. Then, the triple $(Y,Z,K)$ defined by*

$$Y_t = v(t,X_t), \quad Z_t = \sigma'(X_t)D_x v(t,X_t), \quad 0 \le t \le T,$$

$$K_t = \int_0^t \Big( -\frac{\partial v}{\partial t}(s,X_s) - \mathcal{L}v(s,X_s) - f(t,X_s,Y_s,Z_s)\Big)ds,$$

*is the solution to the reflected BSDE (6.58), (6.59) and (6.60).*

**Proof.** By Itô's formula applied to $v(t,X_t)$ and from the terminal condition (6.63), we immediately see that $(Y,Z,K)$ satisfies the relation (6.58). Since $v$ satisfies (6.62), the term in the bracket of $K$ is nonnegative, and so $K$ is nondecreasing. The obstacle constraint (6.59) is also clearly satisfied. Moreover, the minimality condition (6.60) follows from the equality in (6.62). Finally, the integrability conditions on $(Y,Z) \in \mathbb{S}^2(0,T) \times \mathbb{H}^2(0,T)^d$ are direct consequences of the growth conditions on $v$ and $D_x v$. □

We now focus on the converse property, and prove that a solution to the reflected BSDE provides a solution to the variational inequality.

**Theorem 6.5.9** *The function $v(t,x) = Y_t^{t,x}$ in (6.61) is continuous on $[0,T] \times \mathbb{R}^n$, and is a viscosity solution to (6.62)-(6.63).*

**Proof.** The continuity of $v$ is proved similarly as in Theorem 6.3.3, and the terminal condition (6.63) is obviously satisfied from the terminal condition on the BSDE. In order to prove the viscosity property to the variational inequality, we use the approximation by the penalized BSDE. For any $(t,x) \in [0,T] \times \mathbb{R}^n$, $m \in \mathbb{N}$, we denote by $\{(Y^{m,t,x}, Z^{m,t,x}), t \le s \le T\}$ the solution to the penalized BSDE

$$Y_t^m = g(X_T) + \int_t^T f(s,X_s,Y_s^m,Z_s^m)ds + m\int_t^T (Y_s^m - h(X_s))^- ds - \int_t^T Z_s^m.dW_s,$$

with $X_s = X_s^{t,x}$. From Theorem 6.3.3, we know that the function

$$v_m(t,x) := Y_t^{m,t,x}$$

is a continuous viscosity solution to the semilinear PDE

$$-\frac{\partial v_m}{\partial t} - \mathcal{L}v_m - f(., v_m, \sigma'D_x v_m) - m(v_m - h)^- = 0, \quad \text{on } [0,T) \times \mathbb{R}^n \quad (6.64)$$

$$v_m(T,.) = g, \quad \text{on } \mathbb{R}^n. \quad (6.65)$$

From the convergence result of the penalized BSDEs proved in the previous section, we know that for any $(t,x) \in [0,T] \times \mathbb{R}^n$, $v_m(t,x)$ converges increasingly to $v(t,x)$ as $m$ goes to infinity. Since $v_m$ and $v$ are continuous, this convergence is uniform on compacts of $[0,T] \times \mathbb{R}^n$ by Dini's theorem.

We prove the viscosity solution property of $v$ by using the definition-characterization by super(sub)-jets (see Lemma 4.4.5). We first show the viscosity supersolution property. Let $(t,x) \in [0,T) \times \mathbb{R}^n$, and $(q,p,M) \in \bar{\mathcal{P}}^{2,-}v(t,x)$. From Lemma 6.1 in [CIL92], there exist sequences

$$m_j \to \infty, \quad (t_j, x_j) \to (t,x), \quad (q_j, p_j, M_j) \in \mathcal{P}^{2,-}v_{m_j}(t_j, x_j),$$

such that

$$(v_{m_j}(t_j, x_j), q_j, p_j, M_j) \to (v(t,x), q, p, M).$$

From the viscosity supersolution property of $v_{m_j}$ to (6.64), we have

$$-q_j - b(x_j).p_j - \frac{1}{2}\mathrm{tr}(\sigma\sigma'(x_j)M_j) - f(t_j, x_j, v_{m_j}(t_j, x_j), \sigma'(x_j)p_j)$$
$$-m_j(v_{m_j}(t_j, x_j) - h(x_j))^- \geq 0,$$

and so

$$-q_j - b(x_j).p_j - \frac{1}{2}\mathrm{tr}(\sigma\sigma'(x_j)M_j) - f(t_j, x_j, v_{m_j}(t_j, x_j), \sigma'(x_j)p_j) \geq 0.$$

By sending $j$ to infinity, we then obtain

$$-q - b(x).p - \frac{1}{2}\mathrm{tr}(\sigma\sigma'(x)M) - f(t,x,v(t,x), \sigma'(x)p) \geq 0.$$

Since we already know that $v(t,x) \geq h(x)$ by the obstacle condition on the reflected BSDE, this proves that $v$ is a viscosity supersolution to (6.62).

We conclude by showing the viscosity subsolution property. Let $(t,x) \in [0,T) \times \mathbb{R}^n$, and $(q,p,M) \in \bar{\mathcal{P}}^{2,+}v(t,x)$ such that $v(t,x) > h(x)$. As above, there exist sequences

$$m_j \to \infty, \quad (t_j, x_j) \to (t,x), \quad (q_j, p_j, M_j) \in \mathcal{P}^{2,-}v_{m_j}(t_j, x_j),$$

such that

$$(v_{m_j}(t_j, x_j), q_j, p_j, M_j) \to (v(t,x), q, p, M).$$

From the viscosity subsolution property of $v_{m_j}$ to (6.64), we have

$$-q_j - b(x_j).p_j - \frac{1}{2}\mathrm{tr}(\sigma\sigma'(x_j)M_j) - f(t_j, x_j, v_{m_j}(t_j, x_j), \sigma'(x_j)p_j) \quad (6.66)$$

$$-m_j(v_{m_j}(t_j, x_j) - h(x_j))^- \leq 0. \quad (6.67)$$

Since $v(t,x) > h(x)$, then for $j$ large enough, $v_{m_j}(t_j, x_j) > h(x_j)$ and so $(v_{m_j}(t_j, x_j) - h(x_j))^- = 0$. By sending $j$ to infinity into (6.66), this yields

$$-q - b(x).p - \frac{1}{2}\mathrm{tr}(\sigma\sigma'(x)M) - f(t, x, v(t,x), \sigma'(x)p) \leq 0,$$

which proves the required result.                                    □


## 6.6 Applications

### 6.6.1 Exponential utility maximization with option payoff

We consider a financial market with one riskless asset of price $S^0 = 1$ and one risky asset of price process

$$dS_t = S_t(b_t dt + \sigma_t dW_t),$$

where $W$ is a standard Brownian motion on $(\Omega, \mathcal{F}, \mathbb{F} = (\mathcal{F}_t)_t, P)$ equipped with the natural filtration $\mathbb{F}$ of $W$, $b$ and $\sigma$ are two bounded progressively measurable processes, $\sigma_t \geq \varepsilon$, for all $t$, a.s. with $\varepsilon > 0$. An agent, starting from a capital $x$, invests an amount $\alpha_t$ at any time $t$ in the risky asset. His wealth process, controlled by $\alpha$, is given by

$$X_t^{x,\alpha} = x + \int_0^t \alpha_u \frac{dS_u}{S_u} = x + \int_0^t \alpha_u(b_u du + \sigma_u dW_u), \quad 0 \leq t \leq T. \quad (6.68)$$

We denote by $\mathcal{A}$ the set of progressively measurable processes $\alpha$ valued in $\mathbb{R}$, such that $\int_0^T |\alpha_t|^2 dt < \infty$ a.s. and $X^{x,\alpha}$ is lower-bounded. The agent must provide at maturity $T$ an option payoff represented by a bounded random variable $\xi$ $\mathcal{F}_T$-measurable. Given his risk aversion characterized by an exponential utility

$$U(x) = -\exp(-\eta x), \quad x \in \mathbb{R}, \quad \eta > 0, \quad (6.69)$$

the objective of the agent is to solve the maximization problem:

$$v(x) = \sup_{\alpha \in \mathcal{A}} E[U(X_T^{x,\alpha} - \xi)]. \quad (6.70)$$

The approach adopted here for determining the value function $v$ and the optimal control $\hat{\alpha}$ is quite general, and is based on the following argument. We construct a family of processes $(J_t^\alpha)_{0 \leq t \leq T}$, $\alpha \in \mathcal{A}$, satisfying the properties:

(i) $J_T^\alpha = U(X_T^{x,\alpha} - \xi)$ for all $\alpha \in \mathcal{A}$

(ii) $J_0^\alpha$ is a constant independent of $\alpha \in \mathcal{A}$

*(iii)* $J^\alpha$ is a supermartingale for all $\alpha \in \mathcal{A}$, and there exists $\hat\alpha \in \mathcal{A}$ such that $J^{\hat\alpha}$ is a martingale.

Indeed, in this case, for such $\hat\alpha$, we have for any $\alpha \in \mathcal{A}$,

$$E[U(X_T^{x,\alpha} - \xi)] = E[J_T^\alpha] \le J_0^\alpha = J_0^{\hat\alpha} = E[J_T^{\hat\alpha}] = E[U(X_T^{x,\hat\alpha} - \xi)] = v(x),$$

which proves that $\hat\alpha$ is an optimal control, and $v(x) = J_0^{\hat\alpha}$.

We construct such a family $(J_t^\alpha)$ in the form

$$J_t^\alpha = U(X_t^{x,\alpha} - Y_t), \quad 0 \le t \le T, \ \alpha \in \mathcal{A}, \tag{6.71}$$

with $(Y, Z)$ solution to the BSDE

$$Y_t = \xi + \int_t^T f(s, Z_s)ds - \int_t^T Z_s dW_s, \quad 0 \le t \le T, \tag{6.72}$$

where $f$ is a generator to be determined. The conditions *(i)* and *(ii)* are clearly satisfied, and the value function is then given by

$$v(x) = J_0^\alpha = U(x - Y_0).$$

In order to satisfy the condition *(iii)*, we shall exploit the particular structure of the exponential utility function $U$. Indeed, by substituting (6.68), (6.72) into (6.71) with $U$ as in (6.69), we obtain

$$J_t^\alpha = M_t^\alpha C_t^\alpha,$$

where $M^\alpha$ is the (local) martingale given by

$$M_t^\alpha = \exp(-\eta(x - Y_0)) \exp\left(-\int_0^t \eta(\alpha_u\sigma_u - Z_u)dW_u - \frac{1}{2}\int_0^t |\eta(\alpha_u\sigma_u - Z_u)|^2 du\right),$$

and

$$C_t^\alpha = -\exp\left(\int_0^t \rho(u, \alpha_u, Z_u)du\right),$$

with

$$\rho(t, a, z) = \eta\left(\frac{\eta}{2}|a\sigma_t - z|^2 - ab_t - f(t,z)\right).$$

We are then looking for a generator $f$ such that the process $(C_t^\alpha)$ is nonincreasing for all $\alpha \in \mathcal{A}$, and constant for some $\hat\alpha \in \mathcal{A}$. In other words, the problem is reduced to findng $f$ such that

$$\rho(t, \alpha_t, Z_t) \ge 0, \quad 0 \le t \le T, \ \forall \alpha \in \mathcal{A} \tag{6.73}$$

and

$$\rho(t, \hat\alpha_t, Z_t) = 0, \quad 0 \le t \le T. \tag{6.74}$$

By rewriting $\rho$ in the form

$$\frac{1}{\eta}\rho(t,a,z) = \frac{\eta}{2}\left|a\sigma_t - z - \frac{1}{\eta}\frac{b_t}{\sigma_t}\right|^2 - z\frac{b_t}{\sigma_t} - \frac{1}{2\eta}\left|\frac{b_t}{\sigma_t}\right|^2 - f(t,z),$$

we clearly see that conditions (6.73) and (6.74) will be satisfied with

$$f(t,z) = -z\frac{b_t}{\sigma_t} - \frac{1}{2\eta}\left|\frac{b_t}{\sigma_t}\right|^2, \tag{6.75}$$

and

$$\hat{\alpha}_t = \frac{1}{\sigma_t}\left(Z_t + \frac{1}{\eta}\frac{b_t}{\sigma_t}\right), \quad 0 \le t \le T. \tag{6.76}$$

**Theorem 6.6.10** *The value function to problem (6.70) is equal to*

$$v(x) = U(x - Y_0) = -\exp(-\eta(x - Y_0)),$$

*where $(Y,Z)$ is the solution to the BSDE*

$$-dY_t = f(t,Z_t)dt - Z_tdW_t, \quad Y_T = \xi, \tag{6.77}$$

*with a generator $f$ given by (6.75). Moreover, the optimal control $\hat{\alpha}$ is given by (6.76).*

**Proof.** In view of the above arguments, it remains to check rigorously the condition (*iii*) on $J^\alpha$. Since $b/\sigma$ and $\xi$ are bounded, we first observe from (6.2.1) that the solution $(Y,Z)$ to the linear BSDE (6.77) is such that $Y$ is bounded. Moreover, for all $\alpha \in \mathcal{A}$, the process $M^\alpha$ is a local martingale, and there exists a sequence of stopping times $(\tau_n)$, $\tau_n \to \infty$ a.s., such that $(M^\alpha_{t\wedge\tau_n})$ is a (positive) martingale. With the choice of $f$ in (6.75), the process $C^\alpha$ is nonincreasing, and thus $(J^\alpha_{t\wedge\tau_n}) = (M^\alpha_{t\wedge\tau_n}C^\alpha_{t\wedge\tau_n})$ is a supermartingale. Since $X^{x,\alpha}$ is lower-bounded and $Y$ is bounded, the process $J^\alpha$, given by (6.71), is also lower-bounded. By Fatou's lemma, we deduce that $J^\alpha$ is a supermartingale.

Finally, with the choice of $\hat{\alpha}$ in (6.76), we have

$$J^{\hat{\alpha}}_t = M^{\hat{\alpha}}_t = \exp(-\eta(x - Y_0))\exp\left(-\int_0^t \frac{b_u}{\sigma_u}dW_u - \frac{1}{2}\int_0^t\left|\frac{b_u}{\sigma_u}\right|^2 du\right).$$

Since $b/\sigma$ is bounded, we conclude that $J^{\hat{\alpha}}$ is a martingale. ☐

**Remark 6.6.4** The financial model described in this example is a complete market model: any contingent claim $\xi$, $\mathcal{F}_T$-measurable and bounded, is perfectly replicable by means of a self-financed wealth process. In other words, there exists $\pi \in \mathcal{A}$ such that $\xi = X^{x_\xi,\pi}_T$ where $x_\xi$ is the arbitrage price of $\xi$ given by $x_\xi = E^Q[\xi]$, and $Q$ is the unique probability measure equivalent to $P$, which makes the price process $S$ a (local) martingale under $Q$, and called risk-neutral probability. The problem (6.70) may be then formulated as

$$v(x) = \sup_{\alpha \in \mathcal{A}} E[U(X^{x-x_\xi,\alpha-\pi}_T)].$$

We are thus reduced to an exponential utility maximization problem without option payoff. Hence, the optimal strategy (6.76) of the initial problem is decomposed into the sum $\alpha_t = \pi_t + \alpha_t^0$ of the hedging strategy $\pi_t = Z_t/\sigma_t$ for the contingent claim $\xi$ and the optimal strategy $\alpha_t^0 = \frac{1}{\eta} b_t/\sigma_t^2$ for the exponential utility maximization without option.

In a more general context of incomplete market, i.e. when the option $\xi$ is not perfectly replicable, the same approach (i), (ii), (iii), can be applied, but leads to a more complex generator $f$ involving a quadratic term in $z$, see El Karoui and Rouge [ElkR00].

### 6.6.2 Mean-variance criterion for portfolio selection

We consider a Black-Scholes financial model. There is one riskless asset of price process

$$dS_t^0 = rS_t^0 dt,$$

and one stock of price process

$$dS_t = S_t(bdt + \sigma dW_t),$$

with constants $b > r$ and $\sigma > 0$. An agent invests at any time $t$ an amount $\alpha_t$ in the stock, and his wealth process is governed by

$$
\begin{aligned}
dX_t &= \alpha_t \frac{dS_t}{S_t} + (X_t - \alpha_t) \frac{dS_t^0}{S_t^0} \\
&= [rX_t + \alpha_t(b - r)]\,dt + \sigma\alpha_t dW_t, \quad X_0 = x.
\end{aligned}
\tag{6.78}
$$

We denote by $\mathcal{A}$ the set of progressively measurable processes $\alpha$ valued in $\mathbb{R}$, such that $E[\int_0^T |\alpha_t|^2 dt] < \infty$.

The mean-variance criterion for portfolio selection consists in minimizing the variance of the wealth under the constraint that its expectation is equal to a given constant:

$$V(m) = \inf_{\alpha \in \mathcal{A}} \left\{ \text{Var}(X_T) : \ E(X_T) = m \right\}, \quad m \in \mathbb{R}. \tag{6.79}$$

We shall see in Proposition 6.6.5, by the Lagrangian method, that this problem is reduced to the resolution of an auxiliary control problem

$$\tilde{V}(\lambda) = \inf_{\alpha \in \mathcal{A}} E[X_T - \lambda]^2, \quad \lambda \in \mathbb{R}. \tag{6.80}$$

We shall solve problem (6.80) by the stochastic maximum principle described in Section 6.4.2. In this case, the Hamiltonian in (6.24) takes the form

$$\mathcal{H}(x, a, y, z) = [rx + a(b - r)]\,y + \sigma az.$$

The adjoint BSDE (6.25) is written for any $\alpha \in \mathcal{A}$ as

$$- dY_t = rY_t dt - Z_t dW_t, \quad Y_T = 2(X_T - \lambda). \tag{6.81}$$

Let $\hat{\alpha} \in \mathcal{A}$ a candidate for the optimal control, and $\hat{X}$, $(\hat{Y}, \hat{Z})$ the corresponding processes. Then,

$$\mathcal{H}(x, a, \hat{Y}_t, \hat{Z}_t) = rx\hat{Y}_t + a\left[(b - r)\hat{Y}_t + \sigma\hat{Z}_t\right].$$

Since this expression is linear in $a$, we see that conditions (6.26) and (6.27) will be satisfied iff

$$(b - r)\hat{Y}_t + \sigma\hat{Z}_t = 0, \quad 0 \le t \le T, \ a.s. \tag{6.82}$$

We are looking for the $(\hat{Y}, \hat{Z})$ solution to (6.81) in the form

$$\hat{Y}_t = \varphi(t)\hat{X}_t + \psi(t), \tag{6.83}$$

for some deterministic $C^1$ functions $\varphi$ and $\psi$. By substituting in (6.81), and using expression (6.78), we see that $\varphi$, $\psi$ and $\hat{\alpha}$ should satisfy

$$\varphi'(t)\hat{X}_t + \varphi(t)(r\hat{X}_t + \hat{\alpha}_t(b - r)) + \psi'(t) = -r(\varphi(t)\hat{X}_t + \psi(t)), \tag{6.84}$$

$$\varphi(t)\sigma\hat{\alpha}_t = \hat{Z}_t, \tag{6.85}$$

together with the terminal conditions

$$\varphi(T) = 2, \ \psi(T) = -2\lambda. \tag{6.86}$$

By using relations (6.82), (6.83) and (6.85), we obtain the expression of $\hat{\alpha}$:

$$\hat{\alpha}_t = \frac{(r - b)\hat{Y}_t}{\sigma^2\varphi(t)} = \frac{(r - b)(\varphi(t)\hat{X}_t + \psi(t))}{\sigma^2\varphi(t)}. \tag{6.87}$$

On the other hand, from (6.84), we have

$$\hat{\alpha}_t = \frac{(\varphi'(t) + 2r\varphi(t))\hat{X}_t + \psi'(t) + r\psi(t)}{(r - b)\varphi(t)}. \tag{6.88}$$

By comparing with (6.87), we get the ordinary differential equations satisfied by $\varphi$ and $\psi$:

$$\varphi'(t) + \left(2r - \frac{(b - r)^2}{\sigma^2}\right)\varphi(t) = 0, \quad \varphi(T) = 2 \tag{6.89}$$

$$\psi'(t) + \left(r - \frac{(b - r)^2}{\sigma^2}\right)\psi(t) = 0, \quad \varphi(T) = -2\lambda, \tag{6.90}$$

whose explicit solutions are (only $\psi = \psi_\lambda$ depends on $\lambda$)

$$\varphi(t) = 2\exp\left[\left(2r - \frac{(b - r)^2}{\sigma^2}\right)(T - t)\right], \tag{6.91}$$

$$\psi_\lambda(t) = \lambda\psi_1(t) = -2\lambda\exp\left[\left(r - \frac{(b - r)^2}{\sigma^2}\right)(T - t)\right]. \tag{6.92}$$

With this choice of $\varphi$, $\psi_\lambda$, the processes $(\hat{Y}, \hat{Z})$ solve the adjoint BSDE (6.81), and the conditions for the maximum principle in Theorem 6.4.6 are satisfied: the optimal control is given by (6.87), which is written in the Markovian form as

$$\hat{\alpha}_\lambda(t, x) = \frac{(r - b)(\varphi(t)x + \psi_\lambda(t))}{\sigma^2\varphi(t)}. \tag{6.93}$$

To compute the value function $\tilde{V}(\lambda)$, we proceed as follows. For any $\alpha \in \mathcal{A}$, we apply Itô's formula to $\frac{1}{2}\varphi(t)X_t^2 + \psi_\lambda(t)X_t$ between 0 and $T$, by using the dynamics (6.78) of $X$ and the ODE (6.89)-(6.90) satisfied by $\varphi$ and $\psi_\lambda$. By taking the expectation, we then obtain

$$E\left[X_T - \lambda\right]^2 = \frac{1}{2}\varphi(0)x^2 + \psi_\lambda(0)x + \lambda^2$$
$$+ E\left[\int_0^T \frac{\varphi(t)\sigma^2}{2}\left(\alpha_t - \frac{(r-b)(\varphi(t)X_t + \psi_\lambda(t))}{\sigma^2\varphi(t)}\right)^2 dt\right]$$
$$- \frac{1}{2}\int_0^T \left(\frac{b-r}{\sigma}\right)^2 \frac{\psi_\lambda(t)^2}{\varphi(t)}dt.$$

This shows again that the optimal control is given by (6.87), and the value function is equal to

$$\tilde{V}(\lambda) = \frac{1}{2}\varphi(0)x^2 + \psi_\lambda(0)x + \lambda^2 - \frac{1}{2}\int_0^T \left(\frac{b-r}{\sigma}\right)^2 \frac{\psi_\lambda(t)^2}{\varphi(t)}dt,$$

and so with the explicit expressions (6.91)-(6.92) of $\varphi$ and $\psi_\lambda$

$$\tilde{V}(\lambda) = e^{-\frac{(b-r)^2}{\sigma^2}T}(\lambda - e^{rT}x)^2, \quad \lambda \in \mathbb{R}. \tag{6.94}$$

We finally show how problems (6.79) and (6.80) are related.

**Proposition 6.6.5** *We have the conjugate relations*

$$\tilde{V}(\lambda) = \inf_{m\in\mathbb{R}}\left[V(m) + (m-\lambda)^2\right], \quad \lambda \in \mathbb{R}, \tag{6.95}$$
$$V(m) = \sup_{\lambda\in\mathbb{R}}\left[\tilde{V}(\lambda) - (m-\lambda)^2\right], \quad m \in \mathbb{R}. \tag{6.96}$$

*For any $m$ in $\mathbb{R}$, the optimal control of $V(m)$ is equal to $\hat{\alpha}_{\lambda_m}$ given by (6.93) where $\lambda_m$ attains the maximum in (6.96), i.e.*

$$\lambda_m = \frac{m - \exp\left[\left(r - \frac{(b-r)^2}{\sigma^2}\right)T\right]x}{1 - \exp\left[-\frac{(b-r)^2}{\sigma^2}T\right]}. \tag{6.97}$$

**Proof.** Notice first that for all $\alpha \in \mathcal{A}$, $\lambda \in \mathbb{R}$, we have

$$E[X_T - \lambda]^2 = \text{Var}(X_T) + (E(X_T) - \lambda)^2. \tag{6.98}$$

Fix an arbitrary $m \in \mathbb{R}$. By definition of $V(m)$, for all $\varepsilon > 0$, one can find $\alpha^\varepsilon \in \mathcal{A}$ with controlled diffusion $X^\varepsilon$, such that $E(X_T^\varepsilon) = m$ and $\text{Var}(X_T^\varepsilon) \leq V(m) + \varepsilon$. We deduce with (6.98) that

$$E[X_T^\varepsilon - \lambda]^2 \leq V(m) + (m-\lambda)^2 + \varepsilon,$$

and so

$$\tilde{V}(\lambda) \leq V(m) + (m-\lambda)^2, \quad \forall m, \lambda \in \mathbb{R}. \tag{6.99}$$

On the other hand, for $\lambda \in \mathbb{R}$, let $\hat{\alpha}_\lambda \in \mathcal{A}$ with controlled diffusion $\hat{X}^\lambda$, an optimal control for $\tilde{V}(\lambda)$. We set $m_\lambda = E(\hat{X}_T^\lambda)$. From (6.98), we then get

$$\tilde{V}(\lambda) = \mathrm{Var}(\hat{X}_T^\lambda) + (m_\lambda - \lambda)^2$$
$$\geq V(m_\lambda) + (m_\lambda - \lambda)^2.$$

This last inequality, combined with (6.99), proves (6.95):

$$\tilde{V}(\lambda) = \inf_{m \in \mathbb{R}} \left[ V(m) + (m - \lambda)^2 \right]$$
$$= V(m_\lambda) + (m_\lambda - \lambda)^2,$$

and also that $\hat{\alpha}_\lambda$ is solution to $V(m_\lambda)$.

We easily check that the function $V$ is convex in $m$. By writing the relation (6.95) under the form $(\lambda^2 - \tilde{V}(\lambda))/2 = \sup_m [m\lambda - (V(m) + m^2)/2]$, we see that the function $\lambda \to (\lambda^2 - \tilde{V}(\lambda))/2$ is the Fenchel-Legendre transform of the convex function $m \to (V(m) + m^2)/2$. We then have the duality relation $(V(m) + m^2)/2 = \sup_\lambda [m\lambda - (\lambda^2 - \tilde{V}(\lambda))/2]$, which gives (6.96).

Finally, for any $m \in \mathbb{R}$, let $\lambda_m \in \mathbb{R}$ be the argument maximum of $V(m)$ in (6.96), which is explicitly given by (6.97) from the expression (6.94) of $\tilde{V}$. Then, $m$ is an argument minimum of $\tilde{V}(\lambda_m)$ in (6.95). Since the function $m \to V(m) + (m - \lambda)^2$ is strictly convex, this argument minimum is unique, and so $m = m_{\lambda_m} = E(\hat{X}_T^{\lambda_m})$. We thus obtain

$$V(m) = \tilde{V}(\lambda_m) + (m - \lambda_m)^2$$
$$= E[\hat{X}_T^{\lambda_m} - \lambda_m]^2 + \left[ E(\hat{X}_T^{\lambda_m}) - \lambda_m \right]^2 = \mathrm{Var}(\hat{X}_T^{\lambda_m}),$$

which proves that $\hat{\alpha}_{\lambda_m}$ is a solution to $V(m)$. $\qquad\square$

**Remark 6.6.5** There is a financial interpretation of the optimal portfolio strategy (6.93) to problem (6.80). Indeed, observe that it is written also as

$$\hat{\alpha}_t^{(\lambda)} := \hat{\alpha}_\lambda(t, X_t) = -\frac{b-r}{\sigma^2}(X_t - R_\lambda(t)), \quad 0 \leq t \leq T,$$

where the (deterministic) process $R_\lambda(t) = -\psi_\lambda(t)/\varphi(t)$ is explicitly determined by

$$dR_\lambda(t) = rR_\lambda(t)dt, \quad R_\lambda(T) = \lambda.$$

$R_\lambda$ is the wealth process with zero investment in the stock, and replicates perfectly the constant option payoff $\lambda$. On the other hand, consider the problem of an investor with self-financed wealth process $\bar{X}_t$, who wants to minimize $E[(\bar{X}_T)^2]$ in this complete market model. His optimal strategy is the Merton portfolio allocation for a quadratic utility function $U(x) = -x^2$, and given by

$$\bar{\alpha}_t = -\frac{b-r}{\sigma^2}\bar{X}_t, \quad 0 \leq t \leq T. \tag{6.100}$$

The optimal strategy for the problem (6.80) is then equal to the stragegy according to (6.100) with wealth process $X_t - R_\lambda(t)$, and could be directly derived with this remark. We illustrated here in this simple example how one may apply the maximum principle for solving the mean-variance criterion. Actually, this approach succeeds for dealing with more complex cases of random coefficients on the price process in incomplete markets, and leads to BSDE for $\varphi(t)$ and $\psi_\lambda(t)$, see e.g. Kohlmann and Zhou [KZ00].

## 6.7 Bibliographical remarks

BSDEs were introduced in the linear case by Bismut [Bis76] as the adjoint equation associated to the stochastic version of the Pontryagin maximum principle in control theory. The general nonlinear case was studied in the seminal paper by Pardoux and Peng [PaPe90]. Since then, BSDEs generate a very active research area due to their various connections with mathematical finance, stochastic control and partial differential equations. Motivated by applications in mathematical finance, there were several extensions for relaxing the Lipschitz condition on the generator of BSDE. We cite in particular the paper by Kobylanski [Ko00], who shows the existence of a bounded solution when the generator satisfies a quadratic growth condition in $z$. We also refer to the book edited by El Karoui and Mazliak [ElkM97] or the monograph by Ma and Yong [MY00] for other extensions. The connection between BSDE and PDE is developed in more detail in the survey paper by Pardoux [Pa98].

Applications of BSDEs to control and mathematical finance were first developed in the work by El Karoui, Peng and Quenez [ElkPQ97]. The presentation of Section 6.4.1 is largely inspired by their paper. Other applications of BSDE for control are studied in Hamadène and Lepeltier [HL95]. The sufficient verification theorem for the maximum principle stated in Section 6.4.2 is also dealt with in the book by Yong and Zhou [YZ00].

Reflected BSDEs and their connection with optimal stopping problems were introduced in the reference article by El Karoui et al. [EKPPQ97]. Several extensions of BSDEs with reflections on one or two barriers were studied in the literature motivated especially by applications to real options in finance. We cite among others the papers by Lepeltier, Matoussi and Xu [LMX05], and Hamadène and Jeanblanc [HJ07]. Inspired by applications to hedging problems under portfolio constraints, BSDEs with constraints on the $Z$ component were considered in Cvitanic, Karatzas and Soner [CKS98] and Buckdahn and Hu [BuH98]. We also cite the important paper by Peng [Pe99], who developed nonlinear decomposition theorem of Doob Meyer type, and studied the general case of constraints on $(Y, Z)$.

The use of BSDE for the resolution of the exponential utility maximization problem with option payoff was studied in El Karoui and Rouge [ElkR00], see also the papers by Sekine [Se06] and Hu, Imkeller, Müller [HIM05] for power utility functions. The applications of BSDEs to mean-variance hedging problems, and more generally to control problems with linear state and quadratic costs, were initiated by Bismut [Bis78] and extended in the papers by Kohlmann and Zhou [KZ00], Zhou and Li [ZL00], Kohlmann and Tang [KT02], or Mania [Ma03].

# 7

## Martingale and convex duality methods

### 7.1 Introduction

In the optimization methods by dynamic programming or BSDEs studied in the previous chapters, the optimization carried essentially on the control process $\alpha$ influencing the state process. The basic idea of martingale methods is to reduce the initial problem to an optimization problem on the state variable by means of a linear representation under an expectation formula weighted by a variable, called a dual variable. Let us illustrate this idea in a simple example. Consider a state process $X$, controlled by a progressively measurable process $\alpha$, with dynamics

$$dX_t = \alpha_t(dt + dW_t), \quad 0 \le t \le T,$$

where $W$ is a standard Brownian motion on $(\Omega, \mathcal{F}, \mathbb{F}, P)$. We assume that $\mathbb{F} = (\mathcal{F}_t)_{0 \le t \le T}$ is the natural filtration of $W$. For $x \in \mathbb{R}_+$ and $\alpha$ control, we denote by $X^x$ the solution to the above SDE starting from $x$ at $t = 0$ and $\mathcal{A}(x)$ the set of control processes $\alpha$ such that $X_t^x \ge 0$, $0 \le t \le T$. Given a gain function, increasing and concave on $\mathbb{R}_+$, the optimization problem is

$$v(x) = \sup_{\alpha \in \mathcal{A}(x)} E[g(X_T^x)], \quad x \ge 0. \tag{7.1}$$

Let us introduce the probability measure $Q \sim P$, which makes the process $B_t = W_t + t$ a Brownian motion, by Girsanov's theorem. From the Itô representation theorem under $Q$, for any nonnegative random variable $X_T$, $\mathcal{F}_T$-measurable, denoted by $X_T \in L_+^0(\Omega, \mathcal{F}_T, P)$, and satisfying the constraint $E^Q[X_T] \le x$, there exists $\alpha \in \mathcal{A}(x)$ such that

$$X_T = E^Q[X_T] + \int_0^T \alpha_t dB_t \le X_T^x = x + \int_0^T \alpha_t dB_t.$$

Conversely, for any $\alpha \in \mathcal{A}(x)$, the process $X^x = x + \int \alpha dB$ is a nonnegative local martingale under $Q$, thus a $Q$ supermartingale, and so $E^Q[X_T^x] \le x$. We deduce that the optimization problem (7.1) can be formulated equivalently in

$$v(x) = \sup_{X_T \in L_+^0(\Omega, \mathcal{F}_T, P)} E[g(X_T)] \quad \text{under the constraint} \quad E\left[\frac{dQ}{dP} X_T\right] \le x. \tag{7.2}$$

H. Pham, *Continuous-time Stochastic Control and Optimization with Financial Applications*, Stochastic Modelling and Applied Probability 61, DOI 10.1007/978-3-540-89500-8_7, © Springer-Verlag Berlin Heidelberg 2009

We are then reduced to a concave optimization problem in $L^0_+(\Omega, \mathcal{F}_T, P)$ subject to a linear constraint represented by the dual variable $dQ/dP$. Thus, we may apply methods in convex analysis for solving (7.2).

The key tool in the above dual resolution approach is the Itô martingale representation theorem, which is also the central argument in the perfect replication of contingent claims in complete markets. The extension of this method to more general optimization problems is based on a powerful theorem in stochastic analysis, called optional decomposition for supermartingales. This theorem was initially motivated by the superreplication problem in incomplete markets, and was originally stated in the context of Itô processes by El Karoui and Quenez [ElkQ95]. It was then extended to the general framework of semimartingale processes. This result is stated in Section 7.2.

When the initial problem is transformed into a convex optimization (primal) problem under linear constraints, we can then use convex analysis methods. This leads to the formulation and resolution of a dual problem arising from the Lagrangian method on the constrained primal problem. In Section 7.3, we detail this dual resolution approach for the utility maximization problem from terminal wealth. We mention that this martingale duality approach allows us to obtain existence and characterization results in a general semimartingale model for asset prices, while the dynamic programming and Hamilton-Jacobi-Bellman approach requires us to consider a Markovian framework.

We study in Section 7.4 the mean-variance hedging problem in a general continuous semimartingale model. This is formulated as a projection in $L^2$ of a random variable into a space of stochastic integrals. We solve this problem by combining the Kunita-Watanabe projection theorem, duality methods and change of numéraire.

## 7.2 Dual representation for the superreplication cost

### 7.2.1 Formulation of the superreplication problem

Let $S$ be a continuous $\mathbb{R}^n$-valued semimartingale on a filtered probability space $(\Omega, \mathcal{F}, \mathbb{F} = (\mathcal{F}_t)_{0 \le t \le T}, P)$ satisfying the usual conditions. For simplicity, we assume that $\mathcal{F} = \mathcal{F}_T$ and $\mathcal{F}_0$ is trivial, i.e. $\mathcal{F}_0 = \{\emptyset, \Omega\}$. We fix a finite horizon $T < \infty$. $S$ represents the discounted price process of $n$ risky assets. We denote by $L(S)$ the set of progressively measurable processes, integrable with respect to $S$. An element $\alpha \in L(S)$ represents a portfolio strategy for an investor: $\alpha_t$ is the number of shares invested in the assets at time $t$. Thus, starting from some initial capital $x \in \mathbb{R}$, the wealth process of the investor following the portfolio strategy $\alpha$ is

$$x + \int_0^t \alpha_s dS_s, \quad 0 \le t \le T.$$

We say that a control $\alpha \in L(S)$ is admissible if $\int \alpha dS$ is lower-bounded, and we denote by $\mathcal{A}(S)$ the set of such controls. This admissibility condition prevents doubling strategies (see Harrison and Pliska [HP81]): indeed, otherwise, one could construct a sequence of portfolio strategies $(\alpha^n)_{n \ge 1} \in L(S)$ such that $\int_0^T \alpha^n_t dS_t \to \infty$ a.s., which represents a way to earn money as much as desired at time $T$ from a zero capital!

We are given a contingent claim of maturity $T$ characterized by an $\mathcal{F}_T$-measurable nonnegative random variable $X_T$. The superreplication problem of $X_T$ consists in finding the minimal initial capital that allows us to dominate (superhedge) the contingent claim at maturity. Mathematically, this problem is formulated as

$$v_0 = \inf\left\{x \in \mathbb{R} : \exists \alpha \in \mathcal{A}(S), \; x + \int_0^T \alpha_t dS_t \geq X_T \; a.s.\right\}. \qquad (7.3)$$

$v_0$ is called superreplication cost of $X_T$, and if $v_0$ attains the infimum in the above relation, the control $\alpha \in \mathcal{A}(S)$ such that $v_0 + \int_0^T \alpha_t dS_t \geq X_T$ a.s., is called a superreplication portfolio strategy.

We denote by $L^0_+(\Omega, \mathcal{F}_T, P)$ the space of $\mathcal{F}_T$-measurable nonnegative random variables. For any $x \in \mathbb{R}_+$, we define the set

$$\mathcal{C}(x) = \left\{X_T \in L^0_+(\Omega, \mathcal{F}_T, P) : \exists \alpha \in \mathcal{A}(S), \; x + \int_0^T \alpha_t dS_t \geq X_T \; a.s.\right\}. \qquad (7.4)$$

$\mathcal{C}(x)$ represents the set of contingent claims, which can be dominated from an initial capital $x$ and an admissible portfolio strategy.

The aim of this section is to provide a representation and characterization of $v_0$ and $\mathcal{C}(x)$ in terms of some dual space of probability measures.

### 7.2.2 Martingale probability measures and no arbitrage

We define

$$\mathcal{M}_e(S) = \left\{Q \sim P \text{ on } (\Omega, \mathcal{F}_T) : \; S \text{ is a } Q - \text{local martingale}\right\}.$$

$\mathcal{M}_e(S)$ is called set of martingale or risk-neutral probability measures.

In the rest of this chapter, we make the crucial standing assumption

$$\mathcal{M}_e(S) \neq \emptyset. \qquad (7.5)$$

This assumption is equivalent to the no free lunch condition, which is a refinement of the no arbitrage condition, and we refer to the seminal paper by Delbaen and Schachermayer [DS94] for this result, known as the first fundamental theorem of asset pricing. Let us simply mention here that for any $Q \in \mathcal{M}_e(S)$ and $\alpha \in \mathcal{A}(S)$, the lower-bounded stochastic integral $\int \alpha dS$ is a local martingale under $Q$, thus a $Q$-supermartingale by Fatou's lemma. We then get $E^Q[\int_0^T \alpha_t dS_t] \leq 0$. In consequence, the condition (7.5) implies

$$\not\exists \alpha \in \mathcal{A}(S), \quad \int_0^T \alpha_t dS_t \geq 0, \; a.s. \quad \text{and} \quad P\left[\int_0^T \alpha_t dS_t > 0\right] > 0.$$

In other words, we cannot find an admissible portfolio strategy, which allows us, starting from a null capital, to reach almost surely at $T$ a nonnegative wealth, with a nonzero probability of being strictly positive. This is the economical condition of no arbitrage.

### 7.2.3 Optional decomposition theorem and dual representation for the superreplication cost

The superreplication problem inspired a very nice result on stochastic analysis, which we state in the general continuous semimartingale case. We shall give below a detailed proof of this result, called the optional decomposition theorem for supermartingales, in the context of Itô processes.

**Theorem 7.2.1** *Let $X$ be a nonnegative càd-làg process, which is a supermartingale under any probability measure $Q \in \mathcal{M}_e(S) \neq \emptyset$. Then, there exists $\alpha \in L(S)$ and $C$ an adapted process, nondecreasing, $C_0 = 0$, such that*

$$X = X_0 + \int \alpha dS - C. \qquad (7.6)$$

**Remark 7.2.1** We recall that in the Doob-Meyer decomposition theorem of a super-martingale $X$ as the difference of a local $M$ and a nondecreasing process $C$: $X = M - C$, the process $C$ can be chosen predictable, and in this case the decomposition is unique. The decomposition (7.6) is universal in the sense that the process $M = X_0 + \int \alpha dS$ is a local martingale under any $Q \in \mathcal{M}_e(S)$. Moreover, the process $C$ is in general not predictable, but only optional, and it is not unique.

Let us now investigate how this theorem provides a dual representation for the superreplication cost of a contingent claim $X_T \in L_+^0(\Omega, \mathcal{F}_T, P)$. For this, we consider a càd-làg modification of the process

$$X_t = \text{ess} \sup_{Q \in \mathcal{M}_e(S)} E^Q[X_T | \mathcal{F}_t], \quad 0 \leq t \leq T, \qquad (7.7)$$

(there is no ambiguity of notation at time $T$ in the previous relation). We check that it is a supermartingale under any $Q \in \mathcal{M}_e(S)$, and we then apply the optional decomposition theorem.

**Theorem 7.2.2** *Let $X_T \in L_+^0(\Omega, \mathcal{F}_T, P)$. Then, its superreplication cost is equal to*

$$v_0 = \sup_{Q \in \mathcal{M}_e(S)} E^Q[X_T], \qquad (7.8)$$

*Furthermore, if $\sup_{Q \in \mathcal{M}_e(S)} E^Q[X_T] < \infty$, i.e. $v_0$ is finite, then $v_0$ attains its infimum in (7.3) with a superreplication portfolio strategy $\alpha$ given by the optional decomposition (7.6) of the process $X$ defined in (7.7).*

**Proof.** Notice that for all $\alpha \in \mathcal{A}(S)$ and $Q \in \mathcal{M}_e(S)$, the lower-bounded stochastic integral $\int \alpha dS$ is a local martingale under $Q$, and so a $Q$-supermartingale. It follows that for all $x \in \mathbb{R}_+$ such that $x + \int_0^T \alpha_t dS_t \geq X_T$ a.s. with $\alpha \in \mathcal{A}(S)$, $E^Q[X_T] \leq x$ for all $Q \in \mathcal{M}_e(S)$. This implies by definition of $v_0$ that

$$\sup_{Q \in \mathcal{M}_e(S)} E^Q[X_T] \leq v_0. \qquad (7.9)$$

If $\sup_{Q \in \mathcal{M}_e(S)} E^Q[X_T] = \infty$, the equality (7.8) is then obvious. We now suppose that

$$\sup_{Q \in \mathcal{M}_e(S)} E^Q[X_T] < \infty. \tag{7.10}$$

**1.** Let us first show that the process $(X_t)_{0 \le t \le T}$ defined in (7.7) is a supermartingale under any $Q \in \mathcal{M}_e(S)$, and that it admits a càd-làg modification. We consider the family of adapted processes $\{\Gamma_t^Q : 0 \le t \le T, \; Q \in \mathcal{M}_e(S)\}$ where

$$\Gamma_t^Q = E^Q[X_T | \mathcal{F}_t], \quad 0 \le t \le T, \quad Q \in \mathcal{M}_e(S),$$

is well-defined by (7.10).

(i) We check that for all $t \in [0, T]$, the set $\{\Gamma_t^Q : Q \in \mathcal{M}_e(S)\}$ is stable by supremum, i.e. for all $Q_1, Q_2 \in \mathcal{M}_e(S)$, there exists $Q \in \mathcal{M}_e(S)$ such that $\max(\Gamma_t^{Q_1}, \Gamma_t^{Q_2}) = \Gamma_t^Q$. For this, let us fix some element $Q^0 \in \mathcal{M}_e(S)$ with martingale density process $Z^0$, and define the process

$$Z_s = \begin{cases} Z_s^0, & s \le t \\ Z_t^0 \left( \dfrac{Z_s^1}{Z_t^1} 1_A + \dfrac{Z_s^2}{Z_t^2} 1_{\Omega \backslash A} \right), & t < s \le T, \end{cases}$$

where $Z^1$ (resp. $Z^2$) is the martingale density process of $Q^1$ (resp. $Q^2$), $A = \{\omega : \Gamma_t^{Q_1}(\omega) \ge \Gamma_t^{Q_2}(\omega)\} \in \mathcal{F}_t$. By using the law of iterated conditional expectations, it is easy to see that $Z$ inherits from $Z^0$, $Z^1$ and $Z^2$ the martingale property under $P$. Moreover, since $Z$ is strictly positive with $Z_0 = 1$, one can associate a probability measure $Q \sim P$ such that $Z$ is the martingale density process of $Q$. By definition of $\mathcal{M}_e(S)$, and from the Bayes formula, the processes $Z^1 S$ and $Z^2 S$ are local martingales under $P$, and so $Z S$ inherits this local martingale property. Thus, $Q \in \mathcal{M}_e(S)$. Moreover, we have

$$\begin{aligned}
\Gamma_t^Q = E^Q[X_T | \mathcal{F}_t] &= E\left[ \frac{Z_T}{Z_t} X_T \,\middle|\, \mathcal{F}_t \right] \\
&= E\left[ \frac{Z_T^1}{Z_t^1} X_T 1_A + \frac{Z_T^2}{Z_t^2} X_T 1_{\Omega \backslash A} \,\middle|\, \mathcal{F}_t \right] \\
&= 1_A E^{Q^1}[X_T | \mathcal{F}_t] + 1_{\Omega \backslash A} E^{Q^2}[X_T | \mathcal{F}_t] \\
&= 1_A \Gamma_t^{Q_1} + 1_{\Omega \backslash A} \Gamma_t^{Q_2} = \max(\Gamma_t^{Q_1}, \Gamma_t^{Q_2}),
\end{aligned}$$

which is the stability property by supremum. It follows that for all $t \in [0, T]$, there exists a sequence $(Q_k^t)_{k \ge 1}$ in $\mathcal{M}_e(S)$ such that

$$X_t := \operatorname*{ess\,sup}_{Q \in \mathcal{M}_e(S)} \Gamma_t^Q = \lim_{k \to \infty} \uparrow \Gamma_t^{Q_k^t}, \tag{7.11}$$

(the symbol $\lim_{k \to \infty} \uparrow$ means that the limit is increasing, i.e. $\Gamma_t^{Q_k^t} \le \Gamma_t^{Q_{k+1}^t}$.)

(ii) Let us now prove the universal supermartingale property. Let $Q_0$ be arbitrary in $\mathcal{M}_e(S)$ with martingale density process $Z^0$, and fix $0 \le u < t \le T$. Denote by $(Q_k^t)_{k \ge 1}$ the sequence given in (7.11) and $(Z^{k,t})_{k \ge 1}$ the associated sequence of martingale density processes. Observe that for all $k \ge 1$, the process defined by

$$\tilde{Z}_s^{k,t} = \begin{cases} Z_s^0, & s \le t \\ Z_t^0 \dfrac{Z_s^{k,t}}{Z_t^{k,t}}, & t < s \le T, \end{cases}$$

is a martingale (under $P$), strictly positive with initial value $\tilde{Z}_0^{k,t} = 1$, and is thus associated to a probability measure $\tilde{Q}_k^t \sim P$. Moreover, $\tilde{Z}^{k,t}S$ is a local martingale under $P$, and so $\tilde{Q}_k^t \in \mathcal{M}_e(S)$. We then have for all $k \geq 1$,

$$
\begin{aligned}
E^{Q_0}[\Gamma_t^{Q_k^t}|\mathcal{F}_u] = E\Big[\frac{Z_t^0}{Z_u^0}\Gamma_t^{Q_k^t}\Big|\mathcal{F}_u\Big] &= E\Big[\frac{Z_t^0}{Z_u^0}E\Big[\frac{Z_T^{k,t}}{Z_t^{k,t}}X_T\Big|\mathcal{F}_t\Big]\Big|\mathcal{F}_u\Big] \\
&= E\Big[\frac{Z_t^0}{Z_u^0}\frac{Z_T^{k,t}}{Z_t^{k,t}}X_T\Big|\mathcal{F}_u\Big] = E\Big[\frac{\tilde{Z}_T^{k,t}}{\tilde{Z}_u^{k,t}}X_T\Big|\mathcal{F}_u\Big] \\
&= E^{\tilde{Q}_k^t}[X_T|\mathcal{F}_u] = \Gamma_u^{\tilde{Q}_k^t}.
\end{aligned}
$$

From (7.11), we deduce by the monotone convergence theorem

$$
\begin{aligned}
E^{Q_0}[X_t|\mathcal{F}_u] = \lim_{k\to\infty} \uparrow E^{Q_0}[\Gamma_t^{Q_k}|\mathcal{F}_u] &= \lim_{k\to\infty} \uparrow \Gamma_u^{\tilde{Q}_k^t} \qquad (7.12) \\
&\leq \underset{Q\in\mathcal{M}_e(S)}{\mathrm{ess\ sup}}\ \Gamma_u^Q = X_u,
\end{aligned}
$$

which proves that $X$ is a $Q_0$-supermartingale.

(iii) It remains to check that $X$ admits a càd-làg modification. We know from Theorem 1.1.8 that this is indeed the case when the function $t \to E^{Q^0}[X_t]$ is right continuous. From (7.12) with $u = 0$, we have:

$$
E^{Q^0}[X_t] = \lim_{k\to\infty} \uparrow E^{\tilde{Q}_k^t}[X_T], \quad \forall t \in [0,T]. \qquad (7.13)
$$

Fix $t$ in $[0,T]$ and let $(t_n)_{n\geq 1}$ be a sequence in $[0,T]$ converging decreasingly to $t$. Since $X$ is a $Q^0$-supermatingale, we have

$$
\lim_{n\to\infty} E^{Q^0}[X_{t_n}] \leq E^{Q^0}[X_t].
$$

On the other hand, for all $\varepsilon > 0$, there exists, by (7.13), $\hat{k} = \hat{k}(\varepsilon) \geq 1$ such that

$$
E^{Q^0}[X_t] \leq E^{\tilde{Q}_{\hat{k}}^t}[X_T] + \varepsilon. \qquad (7.14)
$$

Notice that $\tilde{Z}_T^{\hat{k},t_n}$, the Radon-Nikodym density of $\tilde{Q}_{\hat{k}}^{t_n}$, converges a.s. to $\tilde{Z}_T^{\hat{k},t}$, the Radon-Nikodym density of $\tilde{Q}_{\hat{k}}^t$, as $n$ goes to infinity. By Fatou's lemma, we deduce with (7.14)

$$
\begin{aligned}
E^{Q^0}[X_t] &\leq \lim_{n\to\infty} E^{\tilde{Q}_{\hat{k}}^{t_n}}[X_T] + \varepsilon \\
&\leq \lim_{n\to\infty} E^{Q^0}[X_{t_n}] + \varepsilon
\end{aligned}
$$

where the second inequality follows by (7.13). Since $\varepsilon$ is arbitrary, this proves that $\lim_{n\to\infty} E^{Q^0}[X_{t_n}] = E^{Q^0}[X_t]$, i.e. the right continuity of $(E^{Q^0}[X_t])_{t\in[0,T]}$.

**2.** We can then apply the optional decomposition theorem to the càd-làg modification, still denoted by $X$, and obtain the existence of a process $\hat{\alpha} \in L(S)$, and an adapted nondecreasing process $C$, $C_0 = 0$ such that

$$
X_t = X_0 + \int_0^t \hat{\alpha}_s dS_s - C_t, \quad 0 \leq t \leq T, \ a.s. \qquad (7.15)
$$

Since $X$ are $C$ are nonnegative, this last relation shows that $\int \hat{\alpha} dS$ is lower-bounded (by $-X_0$), and so $\hat{\alpha} \in \mathcal{A}(S)$. Moreover, the relation (7.15) for $t = T$ yields

$$X_T \leq X_0 + \int_0^T \hat{\alpha}_s dS_s, \quad a.s.$$

This proves by definition of $v_0$ that

$$v_0 \leq X_0 = \sup_{Q \in \mathcal{M}_e(S)} E^Q[X_T].$$

We conclude the proof by recalling (7.9).                                      □

By means of this dual representation of the superreplication cost, we get immediately the following characterization of the sets $\mathcal{C}(x)$ introduced in (7.4).

**Corollary 7.2.1** *For all $x \in \mathbb{R}_+$, we have*

$$\mathcal{C}(x) = \left\{ X_T \in L_+^0(\Omega, \mathcal{F}_T, P) : \sup_{Q \in \mathcal{M}_e(S)} E^Q[X_T] \leq x \right\}. \tag{7.16}$$

*In particular, $\mathcal{C}(x)$ is closed for the topology of the convergence in measure, i.e. if $(X^n)_{n \geq 1}$ is a sequence in $\mathcal{C}(x)$ converging a.s. to $\hat{X}_T$, then $\hat{X}_T \in \mathcal{C}(x)$.*

We have a useful characterization of $\mathcal{C}(x)$: to know if a contingent claim can be dominated from an initial capital $x$, it is necessary and sufficient to test if its expectation under any martingale probability measure is less or equal to $x$. Mathematically, this characterization is the starting point for the resolution by duality methods of the utility maximization problem from terminal wealth. Furthermore, from this characterization, we get the closure property of the set $\mathcal{C}(x)$ in $L_+^0(\Omega, \mathcal{F}_T, P)$, which was not obvious from its original (primal) definition (7.4).

### 7.2.4 Itô processes and Brownian filtration framework

We consider the following model for the asset price process $S = (S^1, \ldots, S^n)$:

$$dS_t = \mu_t dt + \sigma_t dW_t, \quad 0 \leq t \leq T, \tag{7.17}$$

where $W$ is a $d$-dimensional Brownian motion on $(\Omega, \mathcal{F}, \mathbb{F}, P)$ with $\mathbb{F} = (\mathcal{F}_t)_{0 \leq t \leq T}$, the natural filtration of $W$, $d \geq n$, $\mu$, $\sigma$ are progressively measurable processes valued respectively in $\mathbb{R}^n$ and $\mathbb{R}^{n \times d}$, and such that $\int_0^T |\mu_t| dt + \int_0^T |\sigma_t|^2 dt < \infty$ a.s. We assume that for all $t \in [0, T]$, the matrix $\sigma_t$ is of full rank equal to $n$. The square $n \times n$ matrix, $\sigma_t \sigma_t'$, is thus invertible, and we define the progressively measurable process valued in $\mathbb{R}^d$:

$$\lambda_t = \sigma_t'(\sigma_t \sigma_t')^{-1} \mu_t, \quad 0 \leq t \leq T.$$

For simplicity, we assume (see Remark 7.2.4) that $\lambda$ is bounded.

**Remark 7.2.2** In the literature, in order to get a positive price process, we often consider an Itô dynamics in the form

$$dS_t = \text{diag}(S_t)\,(\tilde{\mu}_t dt + \tilde{\sigma}_t dW_t), \quad 0 \leq t \leq T, \tag{7.18}$$

where $\text{diag}(S_t)$ denotes the diagonal $n \times n$ matrix with diagonal elements $S_t^i$. The Black-Scholes model and stochastic volatility models considered in the previous chapters are particular examples of (7.18). Observe that the model (7.17) includes (7.18) with

$$\mu_t = \text{diag}(S_t)\tilde{\mu}_t, \quad \sigma_t = \text{diag}(S_t)\tilde{\sigma}_t.$$

In a first step, we give in this framework an explicit description of the set of martingale probability measures $\mathcal{M}_e(S)$. Let us consider the set

$$K(\sigma) = \left\{ \nu \in L^2_{loc}(W) : \sigma\nu = 0, \quad \text{on } [0,T] \times \Omega, \ dt \otimes dP \ a.e. \right\}.$$

For any $\nu \in K(\sigma)$, we define the exponential local martingale

$$Z_t^\nu = \exp\left( -\int_0^t (\lambda_u + \nu_u).dW_u - \frac{1}{2}\int_0^t |\lambda_u + \nu_u|^2 du \right), \quad 0 \leq t \leq T.$$

We also define the set

$$K_m(\sigma) = \left\{ \nu \in K(\sigma) : Z^\nu \ \text{is a martingale} \right\}.$$

**Remark 7.2.3** We recall (see Chapter 1, Section 1.2.5) that a sufficient condition ensuring that $Z^\nu$ is a martingale, i.e. $E[Z_T^\nu] = 1$, is the Novikov criterion:

$$E\left[ \exp\left( \frac{1}{2}\int_0^T |\lambda_u|^2 + |\nu_u|^2 du \right) \right] < \infty. \tag{7.19}$$

(Notice that since $\lambda$ and $\nu$ are orthogonal, i.e. $\lambda'\nu = 0$, then $|\lambda + \nu|^2 = |\lambda|^2 + |\nu|^2$.)

For any $\nu \in K_m(\sigma)$, one can define a probability measure $P^\nu \sim P$ with martingale density process $Z^\nu$. Moreover, by Girsanov's theorem, the process

$$W^\nu = W + \int \lambda + \nu \, dt$$

is a Brownian motion under $P^\nu$.

We then obtain the following explicit characterization of $\mathcal{M}_e(S)$.

**Proposition 7.2.1** *We have*

$$\mathcal{M}_e(S) = \{ P^\nu : \nu \in K_m(\sigma) \}.$$

**Proof.** (i) Since by definition, $\sigma\lambda = \mu$, and for all $\nu \in K_m(\sigma)$, $\sigma\nu = 0$, it follows that the dynamics of $S$ under $P^\nu$ is written as

$$dS_t = \sigma_t dW_t^\nu. \tag{7.20}$$

This shows that $S$ is a local martingale under $P^\nu$, i.e. $P^\nu \in \mathcal{M}_e(S)$.

(ii) Conversely, let $Q \in \mathcal{M}_e(S)$ and $Z$ its martingale density process (which is strictly positive). By the martingale Itô representation theorem, there exists $\rho \in L^2_{loc}(W)$ such that

$$Z_t = \exp\left(-\int_0^t \rho_u.dW_u - \frac{1}{2}\int_0^t |\rho_u|^2 du\right), \quad 0 \le t \le T.$$

Moreover, by Girsanov's theorem, the process

$$B^\rho = W + \int \rho\,dt$$

is a Brownian motion under $Q$. The dynamics of $S$ under $Q$ is then written as

$$dS_t = (\mu_t - \sigma_t\rho_t)dt + \sigma_t dB_t^\rho, \quad 0 \le t \le T.$$

Since $S$ is a local martingale under $Q$, we should have

$$\sigma\rho = \mu, \quad \text{on } [0,T] \times \Omega, \ dt \otimes dP \text{ a.e.}$$

By writing $\nu = \rho - \lambda$, and recalling that $\sigma\lambda = \mu$, this shows that $\sigma\nu = 0$, and so $\nu \in K(\sigma)$. Moreover, $Z^\nu = Z$ (which is a martingale), thus $\nu \in K_m(\sigma)$, and so $Q = P^\nu$. $\square$

**Remark 7.2.4** **1.** Part (i) of the previous proof shows that the inclusion $\{P^\nu : \nu \in K_m(\sigma)\} \subset \mathcal{M}_e(S)$ holds always true even without the assumption of Brownian filtration.

**2.** Since $\lambda$ is assumed bounded, we see that the Novikov condition (7.19) is satisfied for any bounded process $\nu$. Actually, this holds also true once $\lambda$ satisfies the Novikov condition $E\left[\exp\left(\frac{1}{2}\int_0^T |\lambda_u|^2 du\right)\right] < \infty$. In particular, the null process $\nu = 0$ lies in $K_m(\sigma)$. The associated martingale probability measure $P^0$ is called minimal martingale measure following the terminology of Föllmer and Schweizer.

**3.** The above remark also shows in particular that once $\lambda$ satisfies the Novikov criterion, $\mathcal{M}_e(S)$ is nonempty, and contains $P^0$. In the case where $Z^0$ is not a martingale, the assumption $\mathcal{M}_e(S) \ne \emptyset$ is not necessarily satisfied, and is equivalent to the existence of some element $\nu$ in $K_m(\sigma)$.

We now give a proof of the optional decomposition theorem in the above framework. Actually, we shall see that in the case of Itô processes and Brownian filtration, the process $C$ in the decomposition is predictable.

**Theorem 7.2.3** *Let $X$ be a nonnegative càd-làg supermartingale under any martingale measure $P^\nu$, $\nu \in K_m(\sigma)$. Then, $X$ admits a decomposition under the form*

$$X = X_0 + \int \alpha dS - C$$

*where $\alpha \in L(S)$ and $C$ is a nondecreasing predictable process, $C_0 = 0$.*

**Proof.** From the Doob-Meyer decomposition theorem applied to the nonnegative supermartingale $X$ under $P^\nu$, for $\nu \in K_m(\sigma)$, we get

$$X_t = X_0 + M_t^\nu - A_t^\nu, \quad 0 \le t \le T,$$

where $M^\nu$ is a local martingale under $P^\nu$, $M_0^\nu = 0$, and $A^\nu$ is a predictable nondecreasing process, integrable (under $P^\nu$), with $A_0^\nu = 0$. By the martingale Itô representation theorem under $P^\nu$, there exists $\psi^\nu \in L^2_{loc}(W^\nu)$ such that

$$X_t = X_0 + \int_0^t \psi_u^\nu . dW_u^\nu - A_t^\nu, \quad 0 \le t \le T. \tag{7.21}$$

Fix some element in $\mathcal{M}_e(S)$, say $P^0$ for simplicity, and compare the decompositions (7.21) of $X$ under $P^\nu$, and $P^0$. By observing that $W^\nu = W^0 + \int \nu dt$, and identifying the (local) martingale and predictable finite variation parts, we obtain a.s.

$$\psi_t^\nu = \psi_t^0, \quad 0 \le t \le T, \tag{7.22}$$

$$A_t^\nu - \int_0^t \nu_u' \psi_u^\nu du = A_t^0, \quad 0 \le t \le T, \tag{7.23}$$

for all $\nu \in K_m(\sigma)$.

Let us define the progressively measurable process $\alpha$ valued in $\mathbb{R}^n$ by

$$\alpha_t = (\sigma_t \sigma_t')^{-1} \sigma_t \psi_t^0, \quad 0 \le t \le T.$$

Observe that $\int_0^T |\alpha_t' \mu_t| dt = \int_0^T |\lambda_t' \psi_t^0| dt < \infty$ and $\int_0^T |\alpha_t' \sigma_t|^2 dt = \int_0^T |\psi_t^0|^2 dt < \infty$ a.s., and so $\alpha \in L(S)$. By writing $\eta_t = \psi_t^0 - \sigma_t' \alpha_t$, we have $\int_0^T |\eta_t|^2 dt < \infty$ a.s., $\sigma \eta = 0$, and so $\eta \in K(\sigma)$. Actually, we wrote the decomposition of $\psi^0$ on $\text{Im}(\sigma')$ and its orthogonal space $K(\sigma)$:

$$\psi_t^0 = \sigma_t' \alpha_t + \eta_t, \quad 0 \le t \le T. \tag{7.24}$$

We now show that $\eta = 0$ by using (7.22)-(7.23). Let us consider, for any $n \in \mathbb{N}$, the process

$$\tilde{\nu}_t = -n \frac{\eta_t}{|\eta_t|} 1_{\eta_t \ne 0}, \quad 0 \le t \le T.$$

Then $\tilde{\nu}$ is bounded, and lies in $K_m(\sigma)$. From (7.22)-(7.23) for $\tilde{\nu}$, and using also (7.24), we get

$$A_T^{\tilde{\nu}} = A_T^0 - n \int_0^T |\eta_t| 1_{\eta_t \ne 0} dt.$$

Since $E^{P^0}[A_T^0] < \infty$ and $E^{P^0}[A_T^{\tilde{\nu}}] \ge 0$, we see by taking expectation under $P^0$, and sending $n$ to infinity into the above relation that

$$\eta = 0, \quad \text{on } [0,T] \times \Omega, \quad dt \otimes dP^0 \text{ a.e.}$$

By recalling the dynamics (7.20) of $S$ under $P^0$, and writing $C = A^0$, the decomposition (7.21) of $X$ under $P^0$ is written as

$$X = X_0 + \int \alpha' \sigma dW^0 - A^0 = X_0 + \int \alpha dS - C,$$

and the proof is complete.                                                    $\square$

## 7.3 Duality for the utility maximization problem

### 7.3.1 Formulation of the portfolio optimization problem

In the context of the financial market model described in Section 7.2.1, we formulate the portfolio utility maximization problem. We are given a function $U(x)$ for the utility of an agent with wealth $x$, and we make the following standard assumptions on the utility function. The function $U : \mathbb{R} \to \mathbb{R} \cup \{-\infty\}$ is continuous on its domain $\mathrm{dom}(U) = \{x \in \mathbb{R} : U(x) > -\infty\}$, differentiable, strictly increasing and strictly concave on the interior of its domain. Without loss of generality, up to a constant to be added, we may assume that $U(\infty) > 0$. Such a function will be called a utility function. We consider the case where

$$\mathrm{int}(\mathrm{dom}(U)) = (0, \infty), \tag{7.25}$$

which means that negative wealth is not allowed.

The utility maximization problem from terminal wealth is then formulated as

$$v(x) = \sup_{\alpha \in \mathcal{A}(S)} E\left[U\left(x + \int_0^T \alpha_t dS_t\right)\right], \quad x > 0. \tag{7.26}$$

Finally, in order to exclude trivial cases, we suppose that the value function is non-degenerate:

$$v(x) < \infty, \quad \text{for some } x > 0. \tag{7.27}$$

Actually, from the increasing and concavity properties of $U$ on its domain, which are transmitted to $v$, this assumption is equivalent to

$$v(x) < \infty, \quad \text{for all } x > 0. \tag{7.28}$$

### 7.3.2 General existence result

In this section, we prove directly the existence of a solution to the utility maximization problem (7.26).

First, observe that since $U(x) = -\infty$ for $x < 0$, it suffices to consider in the supremum of (7.26) the controls $\alpha \in \mathcal{A}(S)$ leading to nonnegative wealth $x + \int_0^T \alpha_t dS_t \geq 0$ a.s. Moreover, by the increasing property of $U$ on $(0, \infty)$, it is clear that

$$v(x) = \sup_{X_T \in \mathcal{C}(x)} E[U(X_T)], \quad x > 0, \tag{7.29}$$

where the set $\mathcal{C}(x)$ was defined in (7.4). It is also clear that if $\hat{X}_T^x \in \mathcal{C}(x)$ is a solution to (7.29), then there exists $\hat{\alpha} \in \mathcal{A}(S)$ such that $\hat{X}_T^x = x + \int_0^T \hat{\alpha}_t dS_t$ and $\hat{\alpha}$ is solution to (7.26).

We show the existence of a solution to (7.29) by means of the dual characterization of $\mathcal{C}(x)$, and actually from its closure property in $L_+^0(\Omega, \mathcal{F}_T, P)$, see Corollary 7.2.1. The idea is to consider a maximizing sequence $(X^n)_{n\geq 1}$ for (7.29), to use a compactness result in $L^0_+(\Omega, \mathcal{F}_T, P)$, which allows us, up to a convex combination, to obtain a limit a.s. $\hat{X}_T$ of $X^n$, and then to pass to the limit in $E[U(X^n)]$. The technical point is to get the uniform integrability of the sequence $U_+(X^n)$. We then make the following assumption:

$$\limsup_{x\to\infty} \frac{v(x)}{x} \leq 0. \tag{7.30}$$

This condition may seem a priori strange and hard to check in practice since it carries on the value function to be determined. Actually, we shall see in the proof below that it is precisely the necessary condition to obtain the convergence of $E[U(X^n)]$ to $E[U(\hat{X}_T)]$. On the other hand, we shall give in the next section some practical conditions carrying directly on the utility function $U$, which ensure (7.30).

**Theorem 7.3.4** *Let $U$ be a utility function satisfying* (7.25), (7.27) *and* (7.30). *Then, for all $x > 0$, there exists a unique solution $\hat{X}_T^x$ to problem $v(x)$ in* (7.29).

**Proof.** Let $x > 0$ and $(X^n)_{n\geq 1}$ be a maximizing sequence in $\mathcal{C}(x)$ for $v(x) < \infty$, i.e.

$$\lim_{n\to\infty} E[U(X^n)] = v(x) < \infty. \tag{7.31}$$

From the compactness theorem A.3.5 in $L^0_+(\Omega, \mathcal{F}_T, P)$, we can find a convex combination $\hat{X}^n \in \text{conv}(X^n, X^{n+1}, \ldots)$, which is still in the convex set $\mathcal{C}(x)$ and such that $\hat{X}^n$ converges a.s. to some nonnegative random variable $\hat{X}_T^x$. Since $\mathcal{C}(x)$ is closed for the convergence in measure, we have $\hat{X}_T^x \in \mathcal{C}(x)$. By concavity of $U$ and from (7.31), we also have

$$\lim_{n\to\infty} E[U(\hat{X}^n)] = v(x) < \infty. \tag{7.32}$$

Denote by $U^+$ and $U^-$ the positive and negative parts of $U$, and observe from (7.32) that: $\sup_n E[U^-(\hat{X}^n)] < \infty$ and $\sup_n E[U^+(\hat{X}^n)] < \infty$. On the other hand, by Fatou's lemma, we have

$$\liminf_{n\to\infty} E[U^-(\hat{X}^n)] \geq E[U^-(\hat{X}_T^x)].$$

The optimality of $\hat{X}_T^x$, i.e. $v(x) = E[U(\hat{X}_T^x)]$, is thus obtained iff we can show that

$$\lim_{n\to\infty} E[U^+(\hat{X}^n)] = E[U^+(\hat{X}_T^x)], \tag{7.33}$$

i.e. the uniform integrability of the sequence $(U^+(X^n))_{n\geq 1}$.

If $U(\infty) \leq 0$, i.e. $U^+ \equiv 0$, there is nothing to check. We recall that $U(\infty) > 0$, and we define

$$x_0 = \inf\{x > 0 : U(x) \geq 0\} < \infty.$$

We argue by contradiction by assuming on the contrary that the sequence $(U^+(X^n))_n$ is not uniformly integrable. Then, there exists $\delta > 0$ such that

$$\lim_{n\to\infty} E[U^+(\hat{X}^n)] = E[U^+(\hat{X}_T^x)] + 2\delta.$$

From Corollary A.1.1, and up to a subsequence still denoted y $(\hat{X}^n)_{n\geq 1}$, we can find disjoint sets $(B^n)_{n\geq 1}$ of $(\Omega, \mathcal{F}_T)$ such that

$$E[U^+(\hat{X}^n)1_{B^n}] \geq \delta, \quad \forall n \geq 1.$$

We thus consider the sequence of random variables in $L^0_+(\Omega, \mathcal{F}_T, P)$

$$H^n = x_0 + \sum_{k=1}^{n} \hat{X}^k 1_{B^k}.$$

For all $Q \in \mathcal{M}_e(S)$, we have

$$E^Q[H^n] \leq x_0 + \sum_{k=1}^{n} E^Q[\hat{X}^k] \leq x_0 + nx,$$

since $\hat{X}^k \in \mathcal{C}(x)$. The characterization (7.16) implies that $H^n \in \mathcal{C}(x_0 + nx)$. Moreover, we get

$$E[U(H^n)] = E\left[U^+\left(x_0 + \sum_{k=1}^{n} \hat{X}^k 1_{B^k}\right)\right]$$

$$\geq E\left[U^+\left(\sum_{k=1}^{n} \hat{X}^k 1_{B^k}\right)\right] = \sum_{k=1}^{n} E[U^+(\hat{X}^k)1_{B^k}] \geq \delta n.$$

We deduce that

$$\limsup_{x\to\infty} \frac{v(x)}{x} \geq \limsup_{n\to\infty} \frac{E[U(H^n)]}{x_0 + nx} \geq \limsup_{n\to\infty} \frac{\delta n}{x_0 + nx} = \delta > 0,$$

which is in contradiction with (7.30). Therefore, (7.33) holds true and $\hat{X}^x_T$ is solution to $v(x)$. The uniqueness follows from the strict concavity of $U$ on $(0, \infty)$. $\qquad\square$

### 7.3.3 Resolution via the dual formulation

The optimization problem $v(x)$ in (7.29) is formulated as a concave maximization problem on infinite dimension in $L^0_+(\Omega, \mathcal{F}_T, P)$ subject to an infinity of linear constraints given by the dual characterization (7.16) of $\mathcal{C}(x)$: for $X_T \in L^0_+(\Omega, \mathcal{F}_T, P)$, we have

$$X_T \in \mathcal{C}(x) \iff E[Z_T X_T] \leq x, \quad \forall Z_T \in \mathcal{M}_e. \tag{7.34}$$

Here and in the sequel, we identify a probability measure $Q \ll P$ with its Radon-Nikodym density $Z_T = dQ/dP$, and we write for simplicity $\mathcal{M}_e = \mathcal{M}_e(S)$.

We can now apply to our context the duality methods for convex optimization problems developed in an abstract framework in the book by Ekeland and Temam [ET74].

We start by outlining the principle of this method in our context, and we then emphasize the arising difficulties and how to overcome them.

Let us introduce the convex conjugate (Fenchel-Legendre transform) of $U$:

$$\tilde{U}(y) = \sup_{x>0} [U(x) - xy], \quad y > 0, \tag{7.35}$$

and define $\mathrm{dom}(\tilde{U}) = \{y > 0 : \tilde{U}(y) < \infty\}$. We require the usual Inada conditions:

$$U'(0) := \lim_{x \downarrow 0} U'(x) = \infty, \quad \text{and} \quad U'(\infty) := \lim_{x \to \infty} U'(x) = 0, \tag{7.36}$$

and we define $I : (0, \infty) \to (0, \infty)$ the inverse function of $U'$ on $(0, \infty)$, which is strictly decreasing, and satisfies $I(0) = \infty$, $I(\infty) = 0$. We recall (see Proposition B.3.5 in Appendix B) that under (7.36), $\mathrm{int}(\mathrm{dom}(\tilde{U})) = (0, \infty)$, $\tilde{U}$ is differentiable, decreasing, strictly convex on $(0, \infty)$ with $\tilde{U}(0) = U(\infty)$ and

$$\tilde{U}' = -(U')^{-1} = - I.$$

Moreover, the supremum in (7.35) is attained at $x = I(y) > 0$, i.e.

$$\tilde{U}(y) = U(I(y)) - yI(y), \quad y > 0, \tag{7.37}$$

and we have the conjugate relation

$$U(x) = \inf_{y>0} \left[ \tilde{U}(x) + xy \right], \quad x > 0,$$

with an infimum attained at $y = U'(x)$.

Typical examples of utility functions satisfying (7.36), and their conjugate functions are

$$U(x) = \ln x, \quad \tilde{U}(y) = -\ln y - 1,$$

$$U(x) = \frac{x^p}{p}, \; p < 1, \; p \neq 0, \quad \tilde{U}(y) = \frac{y^{-q}}{q}, \; q = \frac{p}{1-p}.$$

The starting point of the dual approach is the following. For all $x > 0$, $y > 0$, $X_T \in \mathcal{C}(x)$, $Z_T \in \mathcal{M}_e$, we have by definition of $\tilde{U}$ and the dual characterization (7.34) of $\mathcal{C}(x)$

$$E[U(X_T)] \leq E[\tilde{U}(yZ_T)] + E[yZ_T X_T]$$
$$\leq E[\tilde{U}(yZ_T)] + xy. \tag{7.38}$$

We then introduce the dual problem to $v(x)$:

$$\tilde{v}(y) = \inf_{Z_T \in \mathcal{M}_e} E[\tilde{U}(yZ_T)], \quad y > 0. \tag{7.39}$$

The inequality (7.38) shows that for all $x > 0$

$$v(x) = \sup_{X_T \in \mathcal{C}(x)} E[U(X_T)]$$
$$\leq \inf_{y>0} [\tilde{v}(y) + xy] = \inf_{y>0, Z_T \in \mathcal{M}_e} \left\{ E[\tilde{U}(yZ_T)] + xy \right\}. \tag{7.40}$$

The basic dual resolution method to the primal problem $v(x)$ consists in the following steps: show the existence of a $(\hat{y}, \hat{Z}_T)$ (depending on $x$) solution to the dual problem in the right-hand side of (7.40). Equivalently, we have to show the existence of $\hat{y} > 0$ attaining the minimum of $\tilde{v}(y) + xy$, and to get the existence of a $\hat{Z}_T$ solution to the dual problem $\tilde{v}(\hat{y})$. We then set

$$\hat{X}_T^x = I(\hat{y}\hat{Z}_T), \quad i.e. \quad U'(\hat{X}_T^x) = \hat{y}\hat{Z}_T.$$

From the first-order optimality conditions on $\hat{y}$ and $\hat{Z}_T$, we shall see that this implies

$$\hat{X}_T^x \in \mathcal{C}(x) \quad \text{and} \quad E[\hat{Z}_T\hat{X}_T^x] = x.$$

From (7.37), we then obtain

$$E[U(\hat{X}_T^x)] = E\left[\tilde{U}\left(\hat{y}\hat{Z}_T\right)\right] + x\hat{y},$$

which proves, recalling (7.40), that

$$v(x) = E[U(\hat{X}_T^x)], \quad i.e. \quad \hat{X}_T^x \text{ is solution to } v(x).$$

Moreover, the conjugate duality relations on the primal and dual value functions hold true:

$$v(x) = \inf_{y>0}[\tilde{v}(y) + xy] = \tilde{v}(\hat{y}) + x\hat{y}.$$

Before we mention the difficulties arising in this dual approach, we can already at this step give some sufficient conditions (found in the literature on this topic) ensuring that assumption (7.30) is valid, so that we get the existence of a solution to the primal problem $v(x)$.

**Remark 7.3.5** Suppose that

$$\tilde{v}(y) < \infty, \quad \forall y > 0. \tag{7.41}$$

Then, inequality (7.40) shows immediately that condition (7.30) holds true. The condition (7.41) is obviously satisfied once

$$\forall y > 0, \ \exists Z_T \in \mathcal{M}_e : \ E\left[\tilde{U}\left(yZ_T\right)\right] < \infty. \tag{7.42}$$

A condition carrying directly on $U$ and ensuring (7.42) is: there exists $p \in (0,1)$, positive constants $k_1$, $k_2$ and $Z_T \in \mathcal{M}_e$ such that

$$U^+(x) \le k_1 x^p + k_2, \quad \forall x > 0, \tag{7.43}$$

$$E\left[Z_T^{-q}\right] < \infty, \quad \text{where } q = \frac{p}{1-p} > 0. \tag{7.44}$$

Indeed, in this case, we have

$$\tilde{U}(y) \le \sup_{x>0}[k_1 x^p - xy] + k_2 = (k_1 p)^{\frac{1}{1-p}}\frac{y^{-q}}{q} + k_2, \quad \forall y > 0,$$

and (7.42) is clearly satisfied. We shall see later a weaker condition (actually a minimal condition) on $U$ ensuring (7.41).

Let us turn back to the dual approach formally described above. The delicate point is the existence of a solution to the dual problem $\tilde{v}(y)$, $y > 0$. The set $\mathcal{M}_e$ on which the optimization is achieved is naturally included in $L^1(\Omega, \mathcal{F}_T, P)$, but there is no compactness result in $L^1$. The Komlos theorem states that from any bounded sequence $(Z^n)_n$ in $L^1$, we can find a convex combination converging a.s. to a random variable $\hat{Z} \in L^1$. However, this convergence does not hold true in general in $L^1$. In our problem, the maximizing sequence of probability measures $(Z^n)_{n \geq 1}$ in $\mathcal{M}_e$ (which satisfies $E[Z^n] = 1$) for $\tilde{v}(y)$ does not necessarily converge to a probability measure $\hat{Z}_T$: in general, we have $E[\hat{Z}_T] < 1$. Actually, the space $L^0_+(\Omega, \mathcal{F}_T, P)$ in which the primal variables $X_T \in \mathcal{C}(x)$ vary, is not in suitable duality with $L^1(\Omega, \mathcal{F}_T, P)$. It is more natural to let the dual variables vary also in $L^0_+(\Omega, \mathcal{F}_T, P)$.

We then "enlarge" in $L^0_+(\Omega, \mathcal{F}_T, P)$ the set $\mathcal{M}_e$ as follows. We define $\mathcal{D}$ as the convex, solid and closed envelope of $\mathcal{M}_e$ in $L^0_+(\Omega, \mathcal{F}_T, P)$, i.e. the smallest convex, solid and closed subset in $L^0_+(\Omega, \mathcal{F}_T, P)$ containing $\mathcal{M}_e$. Recall that a subset $\mathcal{S}$ of $L^0_+(\Omega, \mathcal{F}_T, P)$ is said to be solid if: $Y'_T \leq Y_T$ a.s. and $Y_T \in \mathcal{S}$ implies that $Y'_T \in \mathcal{S}$. It is easy to see that $\mathcal{D}$ is written as

$$\mathcal{D} = \left\{ Y_T \in L^0_+(\Omega, \mathcal{F}_T, P) : \exists\, (Z^n)_{n \geq 1} \in \mathcal{M}_e, \ Y_T \leq \lim_{n \to \infty} Z^n \right\},$$

where the limit $\lim_{n \to \infty} Z^n$ should be interpreted in the almost sure convergence. From (7.34) and Fatou's lemma, we deduce that the set $\mathcal{C}(x)$ is also written in duality relation with $\mathcal{D}$: for $X_T \in L^0_+(\Omega, \mathcal{F}_T)$,

$$X_T \in \mathcal{C}(x) \iff E[Y_T X_T] \leq x, \quad \forall Y_T \in \mathcal{D}. \tag{7.45}$$

We then consider the dual problem

$$\tilde{v}(y) = \inf_{Y_T \in \mathcal{D}} E[\tilde{U}(y Y_T)], \quad y > 0. \tag{7.46}$$

We shall see below that this definition is consistent with the one in (7.39), i.e. the infimum in $\tilde{v}(y)$ coincides when it is taken over $\mathcal{M}_e$ or $\mathcal{D}$.

We finally require the so-called condition of reasonable asymptotic elasticity:

$$AE(U) := \limsup_{x \to \infty} \frac{x U'(x)}{U(x)} < 1. \tag{7.47}$$

Typical examples (and counter-examples) of such utility functions are:

- $U(x) = \ln x$, for which $AE(U) = 0$
- $U(x) = \frac{x^p}{p}$, $p < 1$, $p \neq 0$, for which $AE(U) = p$.
- $U(x) = \frac{x}{\ln x}$, for $x$ large enough, for which $AE(U) = 1$.

The following theorem states that under these assumptions on the utility function $U$, the duality theory "works" well in this context. Actually, the condition of reasonable asymptotic elasticity is minimal and cannot be relaxed in the sense that one can find counter-examples of continuous price processes $S$ for which the value function $\tilde{v}(y)$ is not finite for all $y$ and there does not exist a solution to the primal problem $v(x)$, whenever $AE(U) = 1$ (see Kramkov and Schachermayer [KS99]).

**Theorem 7.3.5** *Let $U$ be a utility function satisfying (7.25), (7.27), (7.36) and (7.47). Then, the following assertions hold:*

*(1) The function $v$ is finite, differentiable, strictly concave on $(0, \infty)$, and there exists a unique solution $\hat{X}_T^x \in \mathcal{C}(x)$ to $v(x)$ for all $x > 0$.*

*(2) The function $\tilde{v}$ is finite, differentiable, strictly convex on $(0, \infty)$, and there exists a unique solution $\hat{Y}_T^y \in \mathcal{D}$ to $\tilde{v}(y)$ for all $y > 0$.*

*(3) (i) For all $x > 0$, we have*

$$\hat{X}_T^x = I(\hat{y}\hat{Y}_T), \quad i.e. \quad U'(\hat{X}_T^x) = \hat{y}\hat{Y}_T, \tag{7.48}$$

*where $\hat{Y}_T \in \mathcal{D}$ is the solution to $\tilde{v}(\hat{y})$ with $\hat{y} = v'(x)$ the unique solution to $\mathrm{argmin}_{y>0}[\tilde{v}(y) + xy]$, and satisfying*

$$E[\hat{Y}_T \hat{X}_T^x] = x. \tag{7.49}$$

*(ii) We have the conjugate duality relations*

$$v(x) = \min_{y>0}[\tilde{v}(y) + xy], \quad \forall x > 0,$$

$$\tilde{v}(y) = \max_{x>0}[v(x) - xy], \quad \forall y > 0.$$

*(4) Furthermore, if there exists $y > 0$ such that $\inf_{Z_T \in \mathcal{M}_e} E[\tilde{U}(yZ_T)] < \infty$, then*

$$\tilde{v}(y) = \inf_{Y_T \in \mathcal{D}} E[\tilde{U}(yY_T)] = \inf_{Z_T \in \mathcal{M}_e} E[\tilde{U}(yZ_T)].$$

**Remark 7.3.6** Denote by $\hat{X}_t^x = x + \int_0^t \hat{\alpha}_u dS_u$, $0 \leq t \leq T$, the optimal wealth process associated to problem $v(x)$. (there is no ambiguity of notation at time $T$ since $x + \int_0^T \hat{\alpha}_u dS_u = \hat{X}_T^x = I(\hat{y}\hat{Y}_T)$ is indeed the solution in $\mathcal{C}(x)$ to $v(x)$). The process $\hat{X}^x$ is equal (up to a càd-làg modification) to

$$\hat{X}_t^x = \mathrm{ess} \sup_{Q \in \mathcal{M}_e} E^Q[I(\hat{y}\hat{Y}_T)|\mathcal{F}_t], \quad 0 \leq t \leq T,$$

and the optimal control $\hat{\alpha}$ is determined from the optional decomposition of $\hat{X}^x$. In the case where the dual problem $\tilde{v}(\hat{y})$ admits a solution $\hat{Z}_T$ in $\mathcal{M}_e$ with corresponding probability measure $\hat{Q}$, then the process $\hat{X}^x$ is a nonegative local martingale under $\hat{Q}$, hence a $\hat{Q}$-supermartingale such that $E^{\hat{Q}}[\hat{X}_T^x] = x$ by (7.49). It follows that $\hat{X}^x$ is a $\hat{Q}$-martingale, which is thus written as

$$\hat{X}_t^x = E^{\hat{Q}}[I(\hat{y}\hat{Z}_T)|\mathcal{F}_t], \quad 0 \leq t \leq T.$$

**Remark 7.3.7** The assertion (4) can be proved without assuming that $\inf_{Z_T \in \mathcal{M}_e} E[\tilde{U}(yZ_T)] < \infty$ (see Proposition 3.2 in Kramkov and Schachermayer [KS99]). We give here a simpler proof due to Bouchard and Mazliak [BM03].

The rest of this section is devoted to the proof of Theorem 7.3.5. We split the proof in several propositions and lemmas where we put in evidence the required assumptions for each step.

**Lemma 7.3.1** *Let $U$ be a utility function satisfying (7.25) and (7.36). Then, for all $y > 0$, the family $\{\tilde{U}^-(yY_T), Y_T \in \mathcal{D}\}$ is uniformly integrable.*

**Proof.** Since the function $\tilde{U}$ is strictly decreasing, we consider the case where $\tilde{U}(\infty) = -\infty$ (otherwise there is nothing to prove). Let $\phi$ be the inverse function of $-\tilde{U}$: $\phi$ is a strictly increasing function from $(-\tilde{U}(0), \infty)$ into $(0, \infty)$. Recall that $\tilde{U}(0) = U(\infty) > 0$, and so $\phi$ is well-defined on $[0, \infty)$. For all $y > 0$, we have

$$E[\phi(\tilde{U}^-(yY_T))] \le E[\phi(\tilde{U}(yY_T))] + \phi(0) = yE[Y_T] + \phi(0)$$
$$\le y + \phi(0), \quad \forall Y_T \in \mathcal{D},$$

by (7.45) since $X_T = 1 \in \mathcal{C}(1)$. Moreover, with a trivial change of variable and by the l'Hôpital rule, we have from (7.36)

$$\lim_{x \to \infty} \frac{\phi(x)}{x} = \lim_{y \to \infty} \frac{y}{-\tilde{U}(y)} = \lim_{y \to \infty} \frac{1}{I(y)} = \infty.$$

We conclude with the theorem of la Vallée-Poussin (see Theorem A.1.2).    □

The next result shows that the conjugate duality relations between the value functions of the primal and dual problem hold true.

**Proposition 7.3.2** *(Conjugate duality relations)*
*Let $U$ be a utility function satisfying (7.25), (7.27) and (7.36). Then,*

$$v(x) = \inf_{y>0}[\tilde{v}(y) + xy], \quad \forall x > 0, \tag{7.50}$$

$$\tilde{v}(y) = \sup_{x>0}[v(x) - xy], \quad \forall y > 0. \tag{7.51}$$

**Proof.** By the same argument as for (7.38), we have by using (7.45)

$$\sup_{x>0}[v(x) - xy] \le \tilde{v}(y), \quad \forall y > 0.$$

Fix some $y > 0$. To show (7.51), we can assume w.l.o.g. that $\sup_{x>0}[v(x) - xy] < \infty$. For all $n > 0$, let us consider the set

$$\mathcal{B}_n = \left\{ X_T \in L^0_+(\Omega, \mathcal{F}_T, P) : X_T \le n, \ a.s. \right\}.$$

$\mathcal{B}_n$ is compact in $L^\infty$ for the weak topology $\sigma(L^\infty, L^1)$. It is clear that $\mathcal{D}$ is a convex, closed subset of $L^1(\Omega, \mathcal{F}_T, P)$, and we may apply the min-max theorem B.1.2:

$$\sup_{X_T \in \mathcal{B}_n} \inf_{Y_T \in \mathcal{D}} E[U(X_T) - yX_TY_T] = \inf_{Y_T \in \mathcal{D}} \sup_{X_T \in \mathcal{B}_n} E[U(X_T) - yX_TY_T], \tag{7.52}$$

for all $n$ and $y > 0$. From the duality relation (7.45) between $\mathcal{C}(x)$ and $\mathcal{D}$, we get

$$\lim_{n \to \infty} \sup_{X_T \in \mathcal{B}_n} \inf_{Y_T \in \mathcal{D}} E[U(X_T) - yX_TY_T] = \sup_{x>0} \sup_{X_T \in \mathcal{C}(x)} E[U(X_T) - xy]$$
$$= \sup_{x>0}[v(x) - xy]. \tag{7.53}$$

On the other hand, by defining

$$\tilde{U}_n(y) = \sup_{0 < x \leq n} [U(x) - xy], \quad y > 0,$$

we have

$$\inf_{Y_T \in \mathcal{D}} \sup_{X_T \in \mathcal{B}_n} E[U(X_T) - yX_TY_T] = \inf_{Y_T \in \mathcal{D}} E[\tilde{U}_n(yY_T)] := \tilde{v}_n(y),$$

so that by (7.52) and (7.53)

$$\lim_{n \to \infty} \tilde{v}_n(y) = \sup_{x > 0}[v(x) - xy] < \infty.$$

Thus, to obtain (7.51), we must prove that

$$\lim_{n \to \infty} \tilde{v}_n(y) = \tilde{v}(y). \tag{7.54}$$

Clearly, $\tilde{v}_n(y)$ is an increasing sequence and $\lim_n \tilde{v}_n(y) \leq \tilde{v}(y)$. Let $(Y^n)_{n \geq 1}$ be a minimizing sequence in $\mathcal{D}$ for $\lim_n \tilde{v}_n(y)$:

$$\lim_{n \to \infty} E[\tilde{U}_n(yY_T^n)] = \lim_{n \to \infty} \tilde{v}_n(y) < \infty.$$

From the compactness theorem A.3.5 in $L_+^0(\Omega, \mathcal{F}_T, P)$, we can find a convex combination $\hat{Y}^n \in \text{conv}(Y^n, Y^{n+1}, \ldots)$, which is still lying in the convex set $\mathcal{D}$, and converges a.s. to a nonnegative random variable $Y_T$. Since $\mathcal{D}$ is closed for the convergence in measure, we have $Y_T \in \mathcal{D}$. Notice that $\tilde{U}_n(y) = \tilde{U}(y)$ for $y \geq I(n)$ ($\to 0$ as $n$ goes to infinity). By Fatou's lemma, we first deduce that

$$\liminf_{n \to \infty} E[\tilde{U}_n^+(y\hat{Y}^n)] \geq E[\tilde{U}^+(yY_T)],$$

and on the other hand, by Lemma 7.3.1

$$\lim_{n \to \infty} E[\tilde{U}_n^-(y\hat{Y}^n)] = E[\tilde{U}^-(yY_T)].$$

From the convexity of $\tilde{U}_n$, we then get

$$\lim_{n \to \infty} \tilde{v}_n(y) = \lim_{n \to \infty} E[\tilde{U}_n(yY^n)] \geq \liminf_{n \to \infty} E[\tilde{U}_n(y\hat{Y}^n)]$$

$$\geq E[\tilde{U}(yY_T)] \geq \tilde{v}(y),$$

which proves (7.54), and so (7.51). Under the assumption (7.27), the relation (7.50) follows from the bipolarity property of the Fenchel-Legendre transform for convex functions (see Proposition B.3.5 in appendix B). □

**Remark 7.3.8** From the conjugate duality relation (7.50), we see that assumption (7.27) on the finiteness of $v$ is formulated equivalently in the finiteness of $\tilde{v}$ at some point:

$$\exists \,(\text{or } \forall) \, x > 0, \; v(x) < \infty \Longleftrightarrow \exists y > 0, \; \tilde{v}(y) < \infty, \; i.e. \; \text{dom}(\tilde{v}) \neq \emptyset,$$

where

$$\text{dom}(\tilde{v}) = \{y > 0 : \tilde{v}(y) < \infty\}.$$

We shall see below with the additional assumption (7.47) that $\text{dom}(\tilde{v}) = (0, \infty)$ (and so (7.30) holds true by Remark 7.3.5).

**Proposition 7.3.3** *(Existence of a solution to the dual problem)*
*Let $U$ be a utility function satisfying (7.25), (7.27) and (7.36). Then, for all $y \in dom(\tilde{v})$, there exists a unique solution $\hat{Y}_T^y \in \mathcal{D}$ to $\tilde{v}(y)$. In particular, $\tilde{v}$ is strictly convex on $dom(\tilde{v})$.*

**Proof.** For all $y \in dom(\tilde{v})$, let $(Y^n)_{n \geq 1}$ be a minimizing sequence in $\mathcal{D}$ for $\tilde{v}(y) < \infty$:

$$\lim_{n \to \infty} E[\tilde{U}(yY^n)] = \tilde{v}(y).$$

By the compactness theorem A.3.5 in $L_+^0(\Omega, \mathcal{F}_T, P)$, we can find a convex combination $\hat{Y}^n \in conv(Y^n, Y^{n+1}, \ldots)$ lying in the convex set $\mathcal{D}$, and converging a.s. to a random variable $\hat{Y}_T^y$. Since $\mathcal{D}$ is closed for the convergence in measure, we have $\hat{Y}_T^y \in \mathcal{D}$. As in the proof of Proposition 7.3.2, by convexity of $\tilde{U}$, Fatou's lemma and Lemma 7.3.1, we have

$$\tilde{v}(y) = \lim_{n \to \infty} E[\tilde{U}(yY^n)] \geq \liminf_{n \to \infty} E[\tilde{U}(y\hat{Y}^n)]$$
$$\geq E[\tilde{U}(y\hat{Y}_T^y)] \geq \tilde{v}(y),$$

which proves that $\hat{Y}_T^y$ is a solution to $\tilde{v}(y)$. The uniqueness follows from the strict convexity of $\tilde{U}$, which implies also that $\tilde{v}$ is strictly convex on its domain $dom(\tilde{v})$. $\quad\square$

The next result gives a useful characterization of the reasonable asymptotic elasticity condition in terms of $U$ or $\tilde{U}$.

**Lemma 7.3.2** *Let $U$ be a utility function satisfying (7.25), (7.36). Then, the following assertions are equivalent:*

*(i) $AE(U) < 1$.*

*(ii) There exist $x_0 > 0$ and $\gamma \in \,]0,1[$ such that*

$$xU'(x) < \gamma U(x), \quad \forall x \geq x_0.$$

*(iii) There exist $x_0 > 0$ and $\gamma \in (0,1)$ such that*

$$U(\lambda x) < \lambda^\gamma U(x), \quad \forall \lambda > 1, \, \forall x \geq x_0.$$

*(iv) There exist $y_0 > 0$ and $\gamma \in (0,1)$ such that*

$$\tilde{U}(\mu y) < \mu^{-\frac{\gamma}{1-\gamma}} \tilde{U}(y), \quad \forall 0 < \mu < 1, \, \forall 0 < y \leq y_0.$$

*(v) There exist $y_0 > 0$, $\gamma \in (0,1)$ such that*

$$-y\tilde{U}'(y) < \frac{\gamma}{1-\gamma} \tilde{U}(y), \quad \forall 0 < y \leq y_0.$$

**Proof.** The equivalence $(i) \Leftrightarrow (ii)$ is trivial.

$(ii) \Leftrightarrow (iii)$: Fix some $x \geq x_0$ and consider the functions $F(\lambda) = U(\lambda x)$ and $G(\lambda) = \lambda^\gamma U(x)$ for $\lambda \in [1, \infty)$. $F$ and $G$ are differentiable and we have $F(1) = G(1)$. Observe that $(iii)$ is equivalent to

$$F(\lambda) < G(\lambda), \quad \forall \lambda > 1,$$

for all $x \geq x_0$. Suppose that $(ii)$ holds true. Then, $F'(1) < G'(1)$, and we deduce that there exists $\varepsilon > 0$ such that for all $\lambda \in (1, 1 + \varepsilon]$, $F(\lambda) < G(\lambda)$. We now show that it is indeed valid for all $\lambda > 1$. On the contrary, this would mean that

$$\hat{\lambda} := \inf\{\lambda > 1 : F(\lambda) = G(\lambda)\} < \infty.$$

At this point $\hat{\lambda}$, we should have $F'(\hat{\lambda}) \geq G'(\hat{\lambda})$. But, from $(ii)$:

$$F'(\hat{\lambda}) = xU'(\hat{\lambda}x) < \frac{\gamma}{\hat{\lambda}}U(\hat{\lambda}x) = \frac{\gamma}{\hat{\lambda}}F(\hat{\lambda}) = \frac{\gamma}{\hat{\lambda}}G(\hat{\lambda}) = G'(\hat{\lambda}),$$

which is the required contradiction. Conversely, suppose that $(iii)$ holds true. Then, $F'(1) \leq G'(1)$ and we have

$$U'(x) = \frac{F'(1)}{x} \leq \frac{G'(1)}{x} = \gamma \frac{U(x)}{x},$$

which clearly implies $(ii)$.

$(iv) \Leftrightarrow (v)$: this equivalency is obtained similarly as for $(ii) \Leftrightarrow (iii)$ by fixing $0 < y \leq y_0$ and considering the functions $F(\mu) = \tilde{U}(\mu y)$ and $G(\mu) = \mu^{-\frac{\gamma}{1-\gamma}}\tilde{U}(y)$.

$(ii) \Leftrightarrow (v)$: suppose $(ii)$, and write $y_0 = U'(x_0)$. Then, for all $0 < y \leq y_0$, we have $I(y) \geq I(y_0) = x_0$, and so

$$\tilde{U}(y) = U(I(y)) - yI(y) > \frac{1}{\gamma}I(y)U'(I(y)) - yI(y) = \frac{1 - \gamma}{\gamma}yI(y).$$

Since $\tilde{U}'(y) = -I(y)$, this proves $(v)$. Conversely, suppose $(v)$, and write $x_0 = I(y_0) = -\tilde{U}'(y_0)$. Then, for all $x \geq x_0$, we have $U'(x) \leq U'(x_0) = y_0$, and so

$$U(x) = \tilde{U}(U'(x)) + xU'(x) > -\frac{1 - \gamma}{\gamma}U'(x)\tilde{U}'(U'(x)) + xU'(x) = \frac{1}{\gamma}xU'(x),$$

which is exactly assertion $(ii)$. $\qquad\square$

**Remark 7.3.9 1.** The characterizations $(iv)$ and $(v)$ show that if $AE(U) < 1$, then there exists $y_0 > 0$ such that for all $0 < \mu < 1$

$$yI(\mu y) \leq C\tilde{U}(y), \quad 0 < y \leq y_0, \tag{7.55}$$

where $C$ is a positive constant depending on $\mu$.

**2.** Notice that the characterization (ii) for $AE(U) < 1$ implies the growth condition (7.43) on $U$ mentioned in Remark 7.3.5.

The following result gives a characterization of the solution to the dual problem by deriving the first-order optimality conditions.

**Proposition 7.3.4** *(Characterization of the solution to the dual problem)*
*Let $U$ be a utility function satisfying (7.25), (7.27), (7.36) and (7.47). Then, $\tilde{v}$ is finite, differentiable, strictly convex on $(0, \infty)$ with*

$$-\tilde{v}'(y) = E[\hat{Y}_T^y I(y\hat{Y}_T^y)]$$
$$= \sup_{Y_T \in \mathcal{D}} E[Y_T I(y\hat{Y}_T^y)], \quad y > 0, \tag{7.56}$$

*and thus*

$$I(y\hat{Y}_T^y) \in \mathcal{C}(-\tilde{v}'(y)), \quad y > 0.$$

**Proof. 1.** We have seen in Remark 7.3.8 that assumption (7.27) is equivalent to dom($\tilde{v}$) $\neq \emptyset$, i.e. there exists $y_1 > 0$ such that

$$\tilde{v}(y) < \infty, \quad \forall y \geq y_1, \tag{7.57}$$

by the decreasing feature of $\tilde{U}$, and so of $\tilde{v}$. Since $\tilde{v}(y_1) < \infty$, there exists $Y_T \in \mathcal{D}$ such that $E[\tilde{U}(y_1 Y_T)] < \infty$. Since we also have $\tilde{U}(y_1 Y_T) \geq U(x_0) - x_0 y_1 Y_T$ with $E[Y_T] \leq 1$ and $x_0 > 0$ given, this proves that $\tilde{U}(y_1 Y_T) \in L^1(P)$. Moreover, the characterization $(iv)$ of $AE(U) < 1$ in Lemma 7.3.2 shows that there exists $y_0 > 0$ such that for all $0 < y < y_1$

$$\tilde{U}(yY_T) \leq C(y)\tilde{U}(y_1 Y_T)1_{y_1 Y_T \leq y_0} + \tilde{U}(yY_T)1_{y_1 Y_T > y_0}$$
$$\leq C(y)\left|\tilde{U}(y_1 Y_T)\right| + \left|\tilde{U}\left(\frac{y}{y_1}y_0\right)\right|,$$

for some positive constant $C(y)$. This proves that $\tilde{v}(y) < \infty$ for $y < y_1$ and so dom($\tilde{v}$) $= (0, \infty)$.

**2.** Fix $y > 0$. Then, for all $\delta > 0$, we have by definition of $\tilde{v}$:

$$\frac{\tilde{v}(y + \delta) - \tilde{v}(y)}{\delta} \leq E\left[\frac{\tilde{U}((y + \delta)\hat{Y}_T^y)) - \tilde{U}(y\hat{Y}_T^y)}{\delta}\right]$$
$$\leq E\left[\hat{Y}_T^y \tilde{U}'((y + \delta)\hat{Y}_T^y))\right],$$

by convexity of $\tilde{U}$. Since $\tilde{U}' = -I \geq 0$, we deduce by Fatou's lemma and sending $\delta$ to zero:

$$\limsup_{\delta \downarrow 0} \frac{\tilde{v}(y + \delta) - \tilde{v}(y)}{\delta} \leq -E[\hat{Y}_T^y I(y\hat{Y}_T^y)]. \tag{7.58}$$

Moreover, for any $\delta > 0$ such that $y - \delta > 0$, we have by same arguments as above:

$$\frac{\tilde{v}(y) - \tilde{v}(y - \delta)}{\delta} \geq E\left[\frac{\tilde{U}(y\hat{Y}_T^y)) - \tilde{U}((y - \delta)\hat{Y}_T^y)}{\delta}\right]$$
$$\geq E[\hat{Y}_T^y \tilde{U}'((y - \delta)\hat{Y}_T^y))]. \tag{7.59}$$

Notice (as in step 1) that since $E[\tilde{U}(y\hat{Y}_T^y)] < \infty$, then $\tilde{U}(y\hat{Y}_T^y) \in L^1(P)$. From (7.55) (consequence of $AE(U) < 1$), there exists $y_0 > 0$ such that for all $0 < \delta < y/2$, we get

$$0 \leq -\hat{Y}_T^y \tilde{U}'((y - \delta)\hat{Y}_T^y)) = \hat{Y}_T^y I((y - \delta)\hat{Y}_T^y))$$
$$\leq C(y)\tilde{U}(y\hat{Y}_T^y)1_{y\hat{Y}_T^y \leq y_0} + \hat{Y}_T^y I\left(\frac{y_0}{2}\right)1_{y\hat{Y}_T^y > y_0}$$
$$\leq C(y)\left|\tilde{U}(y\hat{Y}_T^y)\right| + \hat{Y}_T^y I\left(\frac{y_0}{2}\right),$$

where $C(y) < \infty$ and $x_0 > 0$ is arbitrary. The r.h.s. of this last inequality is integrable, and we may then apply dominated convergence theorem to (7.59) by sending $\delta$ to zero:

$$\liminf_{\delta \downarrow 0} \frac{\tilde{v}(y) - \tilde{v}(y - \delta)}{\delta} \geq -E\big[\hat{Y}_T^y I(y\hat{Y}_T^y)\big]. \tag{7.60}$$

By combining with (7.58) and from the convexity of $\tilde{v}$, we deduce that $\tilde{v}$ is differentiable at any $y \in (0, \infty)$ with

$$\tilde{v}'(y) = -E\big[\hat{Y}_T^y I(y\hat{Y}_T^y)\big].$$

**3.** Given an arbitrary element $Y_T \in \mathcal{D}$, we define

$$Y_T^\varepsilon = (1 - \varepsilon)\hat{Y}_T^y + \varepsilon Y_T \in \mathcal{D}, \quad 0 < \varepsilon < 1.$$

Then, by definition of $\tilde{v}(y)$ and convexity of $\tilde{U}$, we have

$$0 \leq E[\tilde{U}(yY_T^\varepsilon)] - E[\tilde{U}(y\hat{Y}_T^y)]$$
$$\leq yE[\tilde{U}'(yY_T^\varepsilon)(Y_T^\varepsilon - \hat{Y}_T^y)] = \varepsilon y E[I(yY_T^\varepsilon)(\hat{Y}_T^y - Y_T)],$$

from which we deduce by the nonincreasing property of $I$

$$E[Y_T I(yY_T^\varepsilon)] \leq E[\hat{Y}_T^y I(y(1 - \varepsilon)\hat{Y}_T^y)].$$

As in step 2, we can apply the dominated convergence theorem (under $AE(U) < 1$) to the r.h.s. and Fatou's lemma to the l.h.s. by sending $\varepsilon$ to zero:

$$E[Y_T I(yY_T)] \leq E[\hat{Y}_T^y I(y\hat{Y}_T^y)],$$

and this holds true for any $Y_T \in \mathcal{D}$. This is relation (7.56). The property $I(y\hat{Y}_T^y) \in \mathcal{C}(-\tilde{v}'(y))$ follows finally from the dual characterization (7.45).    □

**Proof of Theorem 7.3.5**

• The existence of a solution to $v(x)$ for all $x > 0$ follows from the fact that $\text{dom}(\tilde{v}) = (0, \infty)$, which ensures assumption (7.30) (see Remark 7.3.5). The strict convexity of $\tilde{v}$ on $(0, \infty)$ and the conjugate relation (7.50) show (see Proposition B.3.5 in Appendix B) that $v$ is differentiable on $(0, \infty)$. The strict concavity of $v$ on $(0, \infty)$ is a consequence of the strict concavity of $U$ and the uniqueness of a solution to $v(x)$. This implies in turn that $\tilde{v}$ is differentiable on $(0, \infty)$.

• Assertion (2) of the theorem follows from Proposition 7.3.3 and the fact that $\text{dom}(\tilde{v}) = (0, \infty)$.

• Let us check that $\tilde{v}'(\infty) := \lim_{y \to \infty} \tilde{v}'(y) = 0$. Since the function $-\tilde{U}$ is increasing and $-\tilde{U}'(y) = I(y)$ converges to zero as $y$ goes to infinity, then for all $\varepsilon > 0$, there exists $C_\varepsilon > 0$ such that

$$-\tilde{U}(y) \leq C_\varepsilon + \varepsilon y, \quad y > 0.$$

From the l'Hôpital rule, we deduce that

$$0 \leq -\tilde{v}'(\infty) = \lim_{y \to \infty} \frac{-\tilde{v}(y)}{y} = \lim_{y \to \infty} \sup_{Y_T \in \mathcal{D}} E\Big[\frac{-U(yY_T)}{y}\Big]$$

$$\leq \lim_{y \to \infty} \sup_{Y_T \in \mathcal{D}} E\Big[\frac{C_\varepsilon}{y} + \varepsilon Y_T\Big]$$

$$\leq \lim_{y \to \infty} E\Big[\frac{C_\varepsilon}{y} + \varepsilon\Big] = \varepsilon,$$

where we used the relation $E[Y_T] \leq 1$ for all $Y_T \in \mathcal{D}$, by (7.45) with $X_T = 1 \in \mathcal{C}(1)$. This shows that

$$\tilde{v}'(\infty) = 0. \tag{7.61}$$

On the other hand, by (7.56), we get

$$-\tilde{v}'(y) \geq E[Z_T I(y\hat{Y}_T^y)], \quad \forall y > 0, \tag{7.62}$$

where $Z_T > 0$ a.s. is a fixed element in $\mathcal{M}_e$. Notice that since $E[\hat{Y}_T^y] \leq 1$ for all $y > 0$, we have by Fatou's lemma $E[\hat{Y}_T^0] \leq 1$ where $\hat{Y}_T^0 = \liminf_{y \downarrow 0} \hat{Y}_T^y$. In particular, $\hat{Y}_T^0 < \infty$ a.s. By sending $y$ to zero in (7.62), and recalling that $I(0) = \infty$, we obtain by Fatou's lemma

$$\tilde{v}'(0) = \lim_{y \downarrow 0} \tilde{v}'(y) = \infty. \tag{7.63}$$

From (7.61) and (7.63), we deduce that for all $x > 0$, the strictly convex function $y \in (0, \infty) \to \tilde{v}(y) + xy$ admits a unique minimum $\hat{y}$ characterized by $\tilde{v}'(\hat{y}) = -x$ or equivalently $\hat{y} = v'(x)$ since $\tilde{v}' = -(v')^{-1}$ from the conjugate duality relation (7.51).

• Let us prove that $I(\hat{y}\hat{Y}_T)$ is the solution to $v(x)$. From Proposition 7.3.4, we have

$$I(\hat{y}\hat{Y}_T) \in \mathcal{C}(x) \quad \text{and} \quad E[\hat{Y}_T I(\hat{y}\hat{Y}_T)] = x.$$

We then get

$$v(x) \geq E[U(I(\hat{y}\hat{Y}_T))] = E[\tilde{U}(\hat{y}\hat{Y}_T)] + E[\hat{y}\hat{Y}_T I(\hat{y}\hat{Y}_T)]$$
$$= E[\tilde{U}(\hat{y}\hat{Y}_T)] + x\hat{y}$$
$$\geq \tilde{v}(\hat{y}) + x\hat{y}.$$

By combining with the conjugate relation (7.50), this proves that we have equality in the above inequalities, and so $I(\hat{y}\hat{Y}_T)$ is the solution to $v(x)$.

• It remains to prove assertion (4). Under the assumption that $\inf_{Z_T \in \mathcal{M}_e} E[\tilde{U}(yZ_T)] < \infty$, for some given $y > 0$, we can find $Z_T^0 \in \mathcal{M}_e$ such that $\tilde{U}(yZ_T^0) \in L^1(P)$. Consider an arbitrary element $Y_T \in \mathcal{D}$, and let $(Z^n)_{n \geq 1}$ be a sequence in $\mathcal{M}_e$ such that $Y_T \leq \lim_n Z^n$ a.s. For $\varepsilon \in (0,1)$ and $n \geq 1$, we define

$$\bar{Z}^{n,\varepsilon} = (1 - \varepsilon)Z^n + \varepsilon Z_T^0 \in \mathcal{M}_e.$$

Then, by the decreasing property of $\tilde{U}$ and the characterization $(iv)$ of $AE(U) < 1$ in Lemma 7.3.2, there exists $y_0 > 0$ such that

$$\tilde{U}(y\bar{Z}^{n,\varepsilon}) \leq \tilde{U}(y\varepsilon Z_T^0) \leq C_\varepsilon \tilde{U}(yZ_T^0)1_{yZ_T^0 \leq y_0} + \tilde{U}(\varepsilon y_0)1_{yZ_T^0 > y_0},$$

for some positive constant $C_\varepsilon$. We then get

$$\tilde{U}^+(y\bar{Z}^{n,\varepsilon}) \leq C_\varepsilon|\tilde{U}(yZ_T^0)| + |\tilde{U}(\varepsilon y_0)|, \quad \forall n \geq 1,$$

which proves that the sequence $\{\tilde{U}^+(y\bar{Z}^{n,\varepsilon}), n \geq 1\}$ is uniformly integrable. By Fatou's lemma, and the decreasing feature of $\tilde{U}$, we deduce that

$$\inf_{Z_T \in \mathcal{M}_e} E[\tilde{U}(yZ_T)] \leq \limsup_{n\to\infty} E[\tilde{U}(y\bar{Z}^{n,\varepsilon})] \leq E[\tilde{U}(y(1-\varepsilon)Y_T + \varepsilon Z_T^0)]$$
$$\leq E[\tilde{U}(y(1-\varepsilon)Y_T)].$$

By using again the characterization $(iv)$ of $AE(U) < 1$ in Lemma 7.3.2, we get the uniform integrability $\{\tilde{U}^+(y(1-\varepsilon)Y_T), \varepsilon \in (0,1)\}$. By sending $\varepsilon$ to zero in the previous inequality, we obtain

$$\inf_{Z_T \in \mathcal{M}_e} E[\tilde{U}(yZ_T)] \leq E[\tilde{U}(yY_T)],$$

and this holds true for all $Y_T \in \mathcal{D}$. This proves that $\inf_{Z_T \in \mathcal{M}_e} E[\tilde{U}(yZ_T)] \leq v(y)$. The converse inequality is obvious since $\mathcal{M}_e \subset \mathcal{D}$. $\qquad\square$

### 7.3.4 The case of complete markets

In this section, we consider the case where the financial market is complete, i.e. the set of martingale probability measures is reduced to a singleton:

$$\mathcal{M}_e(S) = \{P^0\}.$$

We denote by $Z^0$ the martingale density process of $P^0$. In this context, the dual problem is degenerate:

$$\tilde{v}(y) = E[\tilde{U}(yZ_T^0)], \quad y > 0,$$

and the solution to the dual problem is obviously $Z_T^0$. The solution to the primal problem $v(x)$ is

$$\hat{X}_T^x = I(\hat{y}Z_T^0),$$

where $\hat{y} > 0$ is the solution to

$$E[Z_T^0 I(\hat{y}Z_T^0)] = E^{P^0}[I(\hat{y}Z_T^0)] = x.$$

The wealth process $\hat{X}^x$ and the optimal portfolio $\hat{\alpha}$ are determined by

$$\hat{X}_t^x = x + \int_0^t \hat{\alpha}_u dS_u = E^{P^0}\left[I(\hat{y}Z_T^0)|\mathcal{F}_t\right], \quad 0 \leq t \leq T.$$

**Example: Merton model**

Consider the typical example of the Black-Scholes-Merton model:

$$dS_t = S_t \left( \mu dt + \sigma dW_t \right),$$

where $W$ is a standard Brownian motion on $(\Omega, \mathcal{F}, \mathbb{F} = (\mathcal{F}_t)_{0 \le t \le T}, P)$ with $\mathbb{F}$ the natural filtration of $W$, $\mathcal{F}_0$ trivial and $\mu$, $\sigma > 0$ are constants. Recall (see Section 7.2.4) that the unique martingale measure is given by

$$Z_T^0 = \exp\left( -\lambda W_T - \frac{1}{2}|\lambda|^2 T \right), \quad \text{where } \lambda = \frac{\mu}{\sigma},$$

and the dynamics of $S$ under $P^0$ is

$$dS_t = S_t \sigma dW_t^0,$$

where $W_t^0 = W_t + \lambda t$, $0 \le t \le T$, is a $P^0$ Brownian motion. Take the example of a power utility function:

$$U(x) = \frac{x^p}{p}, \quad p < 1, \ p \ne 0, \ \text{for which} \ I(y) = y^{-r}, \ r = \frac{1}{1-p}.$$

We easily calculate the optimal wealth process for $v(x)$:

$$\hat{X}_t^x = E^{P^0}\left[ (\hat{y} Z_T^0)^{-r} | \mathcal{F}_t \right] = \hat{y}^{-r} E^{P^0}\left[ \exp\left( \lambda r W_T^0 - \frac{1}{2}|\lambda|^2 r T \right) \Big| \mathcal{F}_t \right]$$

$$= \hat{y}^{-r} \exp\left[ \frac{1}{2}(|\lambda r|^2 - |\lambda|^2 r)T \right] \exp\left( \lambda r W_t^0 - \frac{1}{2}|\lambda r|^2 t \right).$$

Since $\hat{y}$ is determined by the equation $\hat{X}_0^x = x$, we obtain

$$\hat{X}_t^x = x \exp\left( \lambda r W_t^0 - \frac{1}{2}|\lambda r|^2 t \right), \quad 0 \le t \le T.$$

In order to determine the optimal control $\hat{\alpha}$, we apply Itô's formula to $\hat{X}^x$:

$$d\hat{X}_t^x = \hat{X}_t^x \lambda r dW_t^0,$$

and we identify with

$$d\hat{X}_t^x = \hat{\alpha}_t dS_t = \hat{\alpha}_t \sigma S_t dW_t^0.$$

This provides the optimal proportion of wealth invested in $S$:

$$\frac{\hat{\alpha}_t S_t}{\hat{X}_t^x} = \frac{\lambda r}{\sigma} = \frac{\mu}{\sigma^2(1-p)}.$$

We find again the same result as in the dynamic programming approach in Section 3.6.1. The computation of the value function $v(x) = E[U(\hat{X}_T^x)]$ is easy, and gives of course the same result as derived in Section 3.6.1:

$$v(x) = \frac{x^p}{p} \exp\left( \frac{1}{2} \frac{\mu^2}{\sigma^2} \frac{p}{1-p} T \right).$$

### 7.3.5 Examples in incomplete markets

In the context of incomplete markets, i.e. $\mathcal{M}_e(S)$ is not reduced to a singleton, and actually of infinite cardinality, we cannot explicitly find the solution to the dual problem. However, some computations may be led more or less explicitly in some particular models. Let us consider the model for Itô price processes as described in Section 7.2.4. With the notation of this section, we consider the set including $\mathcal{M}_e(S)$:

$$\mathcal{M}_{loc} = \{Z_T^\nu : \nu \in K(\sigma)\} \supset \mathcal{M}_e(S) = \{Z_T^\nu : \nu \in K_m(\sigma)\}.$$

It is easy to check by Itô's formula that for any wealth process $X^x = x + \int \alpha dS$, $\alpha \in \mathcal{A}(S)$, and for all $\nu \in K(\sigma)$, the process $Z^\nu X^x$ is a $P$-local martingale. Notice also that for all $\nu \in K(\sigma)$, the bounded process $\nu^n = \nu 1_{|\nu| \le n}$ lies in $K_m(\sigma)$, and $Z_T^{\nu^n}$ converges a.s. to $Z_T^\nu$. Thus, $\mathcal{M}_{loc} \subset \mathcal{D}$ and from the assertion (4) of Theorem 7.3.5 (see also Remark 7.3.7), we get

$$\tilde{v}(y) = \inf_{\nu \in K(\sigma)} E[\tilde{U}(yZ_T^\nu)], \quad y > 0. \tag{7.64}$$

The interest to introduce the set $\mathcal{M}_{loc}$ is that it is explicit (in contrast with $\mathcal{D}$), completely parametrized by the set of controls $\nu \subset K(\sigma)$, and relaxed from the strong constraints of martingale integrability in $K_m(\sigma)$, so that we can hope to find a solution $\hat{\nu}^y$ in $K(\sigma)$ to $\tilde{v}(y)$ in (7.64) by stochastic control methods as in the previous chapters. Actually, if we suppose that the function

$$\xi \in \mathbb{R} \longmapsto \tilde{U}(e^\xi) \quad \text{is convex,}$$

which is satisfied once $x \in (0, \infty) \mapsto xU'(x)$ is increasing (this is the case of power and logarithm utility functions), then it is proved in Karatzas et al. [KLSX91] that for all $y > 0$, the dual problem $\tilde{v}(y)$ admits a solution $Z_T^{\hat{\nu}^y} \in \mathcal{M}_{loc}$. One also shows that for all $\nu \in K(\sigma)$ such that $E[\int_0^T |\nu_t|^2 dt] = \infty$, we have $E[\tilde{U}(yZ_T^\nu)] = \infty$, so that in (7.64), we can restrict to take the infimum over $K_2(\sigma) = \{\nu \in K(\sigma) : E[\int_0^T |\nu_t|^2 dt] < \infty\}$, and thus $\hat{\nu}^y \in K_2(\sigma)$. Notice that in general, this solution $Z_T^{\hat{\nu}^y}$ does not lie in $\mathcal{M}_e(S)$. The solution to the dual problem is given by

$$\hat{X}_T^x = I(\hat{y} Z_T^{\hat{\nu}^y}),$$

where $\hat{y} > 0$ is the solution to $\text{argmin}_{y>0}[\tilde{v}(y) + xy]$ and satisfying

$$E[Z_T^{\hat{\nu}^y} I(\hat{y} Z_T^{\hat{\nu}^y})] = x.$$

For determining the (nonnegative) wealth process $\hat{X}^x$, we observe that the process $Z^{\hat{\nu}^y} \hat{X}^x$ is a nonnegative local martingale, thus a supermartingale, which also satisfies $E[Z_T^{\hat{\nu}^y} \hat{X}_T^x] = x$. Therefore, it is a martingale, and we have

$$\hat{X}_t^x = E\left[\frac{Z_T^{\hat{\nu}^y}}{Z_t^{\hat{\nu}^y}} I(\hat{y} Z_T^{\hat{\nu}^y}) \Big| \mathcal{F}_t\right], \quad 0 \le t \le T.$$

**Logarithm utility function**

Consider the example of the logarithm utility function $U(x) = \ln x$, $x > 0$, for which

$$I(y) = \frac{1}{y} \quad \text{and} \quad \tilde{U}(y) = -\ln y - 1, \quad y > 0.$$

For all $\nu \in K_2(\sigma)$, we have

$$E[\tilde{U}(yZ_T^\nu)] = -\ln y - 1 - \frac{1}{2}E\left[\int_0^T |\lambda_t|^2 + [\nu_t|^2 dt\right], \quad y > 0.$$

This shows that the solution to the dual problem $\tilde{v}(y)$ is attained for $\nu = 0$ (in particular independent of $y$), corresponding to $Z_T^0$ and the optimal wealth process for $v(x)$ is explicitly given by

$$\hat{X}_t^x = \frac{1}{Z_t^0}, \quad 0 \le t \le T.$$

The optimal control is determined by applying Itô's formula to the above expression, and identifying with $d\hat{X}_t^x = \hat{\alpha}_t dS_t$. In the model written under the "geometrical" form

$$dS_t = S_t(\mu_t dt + \sigma_t dW_t),$$

we find the optimal proportion of wealth invested in $S$:

$$\frac{\hat{\alpha}_t S_t}{\hat{X}_t^x} = \frac{\mu_t}{\sigma_t^2}.$$

## Power utility function

Consider the example of the power utility function $U(x) = x^p/p$, $x > 0$, $p < 1$, $p \ne 0$, for which

$$I(y) = y^{\frac{1}{p-1}} \quad \text{and} \quad \tilde{U}(y) = \frac{y^{-q}}{q}, \quad y > 0, \quad q = \frac{p}{1-p}.$$

For all $\nu \in K(\sigma)$, we have

$$E[\tilde{U}(yZ_T^\nu)] = \frac{y^{-q}}{q} E[(Z_T^\nu)^{-q}], \quad y > 0.$$

The solution to the dual problem $\tilde{v}(y)$ does not depend on $y$ and is a solution to the problem

$$\inf_{\nu \in K(\sigma)} E[(Z_T^\nu)^{-q}]. \tag{7.65}$$

This stochastic control problem can be solved in a Markovian framework, typically a stochastic volatility model, by the dynamic programming method. In a more general framework of Itô processes, we can also use methods of BSDE. Denoting by $\hat{\nu}$ the solution to (7.65), the optimal wealth process is given by

$$\hat{X}_t^x = \frac{x}{E[(Z_T^{\hat{\rho}})^{-q}]} E\left[\frac{(Z_T^{\hat{\rho}})^{-q}}{Z_t^{\hat{\rho}}}\middle|\mathcal{F}_t\right], \quad 0 \le t \le T.$$

## 7.4 Quadratic hedging problem

### 7.4.1 Problem formulation

We consider the general framework for continuous semimartingale price process as described in Section 7.2.1. We are given a contingent claim represented by a random variable $H \in L^2(P) = L^2(\Omega, \mathcal{F}_T, P)$, i.e. $H$ $\mathcal{F}_T$-measurable and $E|H|^2 < \infty$. The quadratic hedging criterion consists in minimizing for the $L^2$-norm the difference between the payoff $H$ and the terminal wealth $X_T = x + \int_0^T \alpha_t dS_t$ of a portfolio strategy $\alpha \in L(S)$.

In the utility maximization problem, we considered utility functions on $(0, \infty)$. It was then natural to define admissible portfolio strategies leading to a lower-bounded wealth process. In the quadratic hedging problem, the cost function to be minimized $U(x) = (H-x)^2$ is defined on $\mathbb{R}$, and in general we cannot hope to find a solution $\hat{X} = x + \int \hat{\alpha} dS$, which is lower-bounded. On the other hand, we should exclude doubling strategies. In our context of quadratic optimization, we introduce the following admissibility condition:

$$\mathcal{A}_2 = \left\{ \alpha \in L(S) : \int_0^T \alpha_t dS_t \in L^2(P) \quad \text{and} \right.$$

$$\left. \int \alpha dS \text{ is a } Q-\text{martingale for all } Q \in \mathcal{M}_e^2 \right\},$$

where

$$\mathcal{M}_e^2 = \left\{ Q \in \mathcal{M}_e : \frac{dQ}{dP} \in L^2(P) \right\}$$

is assumed to be nonempty: $\mathcal{M}_e^2 \neq \emptyset$. This admissibility condition permits lower-unbounded wealth while excluding arbitrage opportunuities. Indeed, by fixing some element $Q \in \mathcal{M}_e^2$, we have for all $\alpha \in \mathcal{A}_2$, $E^Q[\int_0^T \alpha_t dS_t] = 0$ and so

$$\not\exists \alpha \in \mathcal{A}_2, \quad \int_0^T \alpha_t dS_t \geq 0, \text{ a.s. } \text{ and } P\left[ \int_0^T \alpha_t dS_t > 0 \right] > 0.$$

Furthermore, with this integrability condition in $\mathcal{A}_2$, we show that the set of stochastic integrals $\{ \int_0^T \alpha_t dS_t : \alpha \in \mathcal{A}_2 \}$ is closed $L^2(P)$, which ensures the existence of a solution to the quadratic minimization problem.

**Proposition 7.4.5** *The set* $\mathcal{G}_T = \{ \int_0^T \alpha_t dS_t : \alpha \in \mathcal{A}_2 \}$ *is closed in* $L^2(P)$.

**Proof.** Let $X^n = \int_0^T \alpha_t^n dS_t$, $n \in \mathbb{N}$, be a sequence in $\mathcal{G}_T$ converging to $X_T$ in $L^2(P)$. Fix some arbitrary element $Q \in \mathcal{M}_e^2$. Then, from the Cauchy-Schwarz inequality, $\int_0^T \alpha_t^n dS_t$ converges to $X_T$ in $L^1(Q)$. By Theorem 1.2.11, we deduce that $X_T = \int_0^T \alpha_t dS_t$ where $\alpha \in L(S)$ is such that $\int \alpha dS$ is a $Q$-martingale. Since $Q$ is arbitrary in $\mathcal{M}_e^2$, we conclude that $\alpha \in \mathcal{A}_2$. $\qquad \square$

The quadratic minimization problem (also called mean-variance hedging problem) of a contingent claim $H \in L^2(P)$ is formulated as

$$v_H(x) = \inf_{\alpha \in \mathcal{A}_2} E\left[ H - x - \int_0^T \alpha_t dS_t \right]^2, \quad x \in \mathbb{R}. \tag{7.66}$$

In other words, we project on the Hilbert space $L^2(P)$, the element $H - x$ on the closed vector subspace $\mathcal{G}_T$. We then already know the existence of a solution to $v_H(x)$. Our goal is to characterize this solution. The resolution method is based on a combination of Kunita-Watanabe projection theorem, convex duality methods, and change of numéraire.

### 7.4.2 The martingale case

In this section, we study the special case where $S$ is a local martingale under $P$, i.e. $P \in \mathcal{M}_e^2$. In this case, the resolution is direct from the Kunita-Watanabe projection theorem. Indeed, by projecting the square-integrable martingale $H_t = E[H|\mathcal{F}_t]$, $0 \le t \le T$, on the continuous local martingale $S$ under $P$, we obtain the decomposition

$$E[H|\mathcal{F}_t] = E[H] + \int_0^t \alpha_u^H dS_u + R_t^H, \quad 0 \le t \le T, \tag{7.67}$$

where $\alpha^H \in L(S)$ satisfies the integrability condition

$$E\left[\int_0^T \alpha_t^H dS_t\right]^2 = E\left[\int_0^T (\alpha_t^H)'d < S >_t \alpha_t^H\right] < \infty, \tag{7.68}$$

and $(R_t^H)_t$ is a square-integrable martingale, orthogonal to $S$, i.e. $< R^H, S > = 0$.

**Theorem 7.4.6** *For any $x \in \mathbb{R}$ and $H \in L^2(P)$, the solution $\alpha^{mv} \in \mathcal{A}_2$ to $v_H(x)$ is equal to $\alpha^H$. Furthermore, we have*

$$v_H(x) = (E[H] - x)^2 + E[R_T^H]^2.$$

**Proof.** Let us check that $\alpha^H \in \mathcal{A}_2$. From condition (7.68), we have $\int_0^T \alpha_t^H dS_t \in L^2(P)$. Moreover, with the Cauchy-Schwarz and Doob inequalities, we have for all $Q \in \mathcal{M}_e^2$

$$E^Q\left[\sup_{0 \le t \le T}\left|\int_0^t \alpha_u^H dS_u\right|\right] \le 2\left(E\left[\frac{dQ}{dP}\right]^2\right)^{\frac{1}{2}}\left(E\left[\int_0^T \alpha_t^H dS_t\right]^2\right)^{\frac{1}{2}} < \infty.$$

This shows that the $Q$-local martingale $\int \alpha^H dS$ is a uniformly integrable $Q$-martingale, and thus $\alpha^H \in \mathcal{A}_2$.

Observe that for all $\alpha \in \mathcal{A}_2$, the stochastic integral $\int \alpha dS$ is a square-integrable $P$-martingale. By the Doob and Cauchy-Schwarz inequalities, we have $E[\sup_t |R_t^H \int_0^t \alpha_u dS_u|]$ $< \infty$. Thus, the orthogonality condition between $R^H$ and $S$ implies that $R^H \int \alpha dS$ is a uniformly integrable $P$-martingale. We then get

$$E\left[R_T^H \int_0^T \alpha_t dS_t\right] = 0, \quad \forall \alpha \in \mathcal{A}_2.$$

By writing from (7.67) that $H = E[H] + \int_0^T \alpha_t^H dS_t + R_T^H$, we derive that for all $\alpha \in \mathcal{A}_2$

$$E\left[H - x - \int_0^T \alpha_t dS_t\right]^2 = (E[H] - x)^2 + E\left[\int_0^T (\alpha_t^H - \alpha_t)dS_t\right]^2 + E[R_T^H]^2,$$

which proves the required result. □

**Example**

Consider the stochastic volatility model

$$dS_t = \sigma(t, S_t, Y_t) S_t dW_t^1$$
$$dY_t = \eta(t, S_t, Y_t) dt + \gamma(t, S_t, Y_t) dW_t^2,$$

where $W^1$ and $W^2$ are standard Brownian motions, supposed uncorrelated for simplicity. We make the standard assumptions on the coefficients $\sigma, \eta, \gamma$ to get the existence and uniqueness of a solution $(S, Y)$ valued in $\mathbb{R}_+ \times \mathbb{R}$ to the above SDE given an initial condition. We consider an option payoff in the form $H = g(S_T)$ where $g$ is a measurable function, and we define the function

$$h(t, s, y) = E[g(S_T)|(S_t, Y_t) = (s, y)], \quad (t, s, y) \in [0, T] \times \mathbb{R}_+ \times \mathbb{R}.$$

Under suitable conditions on $\sigma$, $\eta$, $\gamma$ and $g$, the function $h \in C^{1,2}([0, T) \times \mathbb{R}_+ \times \mathbb{R})$, and is a solution to the Cauchy problem

$$\frac{\partial h}{\partial t} + \eta \frac{\partial h}{\partial y} + \frac{1}{2}\sigma^2 \frac{\partial^2 h}{\partial s^2} + \frac{1}{2}\gamma^2 \frac{\partial^2 h}{\partial y^2} = 0, \quad \text{on} \quad [0, T) \times \mathbb{R}_+ \times \mathbb{R}$$
$$h(T, ., .) = g, \quad \text{on} \quad \mathbb{R}_+ \times \mathbb{R}.$$

The Kunita-Watanabe decomposition (7.67) is simply obtained by Itô's formula applied to the martingale $h(t, S_t, Y_t) = E[g(S_T)|\mathcal{F}_t]$, $0 \le t \le T$:

$$E[g(S_T)|\mathcal{F}_t] = E[g(S_T)] + \int_0^t \frac{\partial h}{\partial s}(u, S_u, Y_u) dS_u + \int_0^t \gamma \frac{\partial h}{\partial y}(u, S_u, Y_u) dW_u^2.$$

The solution to the quadratic minimization problem is then given by

$$\alpha_t^{mv} = \frac{\partial h}{\partial s}(t, S_t, Y_t), \quad 0 \le t \le T,$$

and the value function is

$$v_H(x) = (E[g(S_T)] - x)^2 + E\Big[\int_0^T \Big|\gamma \frac{\partial h}{\partial y}(t, S_t, Y_t)\Big|^2 dt\Big].$$

In the sequel, we consider the general case where $S$ is a continuous semimartingale. The principle of our resolution method is the following: we first choose a suitable wealth process that will be connected to a martingale measure, and derived from a duality relation. We then use this wealth process as the numéraire, and by a method of change of numéraire, we reduce our problem to the martingale case.

### 7.4.3 Variance optimal martingale measure and quadratic hedging numéraire

Let us consider the quadratic minimization problem corresponding to $H = 1$ and $x = 0$:

$$v_1 = \min_{\alpha \in \mathcal{A}_2} E\Big[1 - \int_0^T \alpha_t dS_t\Big]^2. \tag{7.69}$$

Denote by $\alpha^{var} \in \mathcal{A}_2$ the solution to $v_1$ and $X^{var}$ the wealth process

$$X_t^{var} = 1 - \int_0^t \alpha_u^{var} dS_u, \quad 0 \le t \le T.$$

The purpose of this section is to show that $X^{var}$ is related via duality to a martingale probability measure, and is thus strictly positive. We shall use the duality relation between $\mathcal{G}_T$ and $\mathcal{M}_e^2$, and follow a dual approach as for the utility maximization problem, adapted to the $L^2(P)$-context.

Since $E^Q[1 - \int_0^T \alpha_t dS_t] = 1$ for all $\alpha \in \mathcal{A}_2$ and $Q \in \mathcal{M}_e^2$, we get by the Cauchy-Schwarz inequality

$$1 = \left( E\left[ \frac{dQ}{dP} \left( 1 - \int_0^T \alpha_t dS_t \right) \right] \right)^2$$
$$\le E\left[ \frac{dQ}{dP} \right]^2 E\left[ 1 - \int_0^T \alpha_t dS_t \right]^2.$$

We then introduce the dual quadratic problem of $v_1$:

$$\tilde{v}_1 = \inf_{Q \in \mathcal{M}_e^2} E\left[ \frac{dQ}{dP} \right]^2, \tag{7.70}$$

so that

$$1 \le \inf_{Q \in \mathcal{M}_e^2} E\left[ \frac{dQ}{dP} \right]^2 \min_{\alpha \in \mathcal{A}_2} E\left[ 1 - \int_0^T \alpha_t dS_t \right]^2 = \tilde{v}_1 \, v_1.$$

We shall prove the existence of a solution $P^{var}$ to the dual problem $\tilde{v}_1$, and see that it is related to the solution of the primal problem $v_1$ by $\frac{dP^{var}}{dP} = \text{Cte} X_T^{var}$. We then get equality in the previous inequalities: $1 = \tilde{v}_1 \, v_1$.

**Theorem 7.4.7** *There exists a unique solution to $\tilde{v}_1$, denoted by $P^{var}$ and called the variance-optimal martingale measure. The solution to the primal problem $v_1$ is related to the one of dual problem $\tilde{v}_1$ by*

$$X_T^{var} = \frac{\frac{dP^{var}}{dP}}{E\left[ \frac{dP^{var}}{dP} \right]^2} \quad i.e. \quad \frac{dP^{var}}{dP} = \frac{X_T^{var}}{E[X_T^{var}]}. \tag{7.71}$$

*In particular, we have*

$$X_t^{var} > 0, \quad 0 \le t \le T, \ P \ a.s. \tag{7.72}$$

**Remark 7.4.10** Since $X^{var}$ is a $Q$-martingale under any $Q \in \mathcal{M}_e^2$, in particular under $P^{var}$, we have by (7.71), and denoting by $Z^{var}$ the martingale density process of $P^{var}$

$$X_t^{var} = \frac{E^Q[Z_T^{var}|\mathcal{F}_t]}{E[Z_T^{var}]^2} = \frac{E[(Z_T^{var})^2|\mathcal{F}_t]}{Z_t^{var} E[Z_T^{var}]^2}, \quad 0 \le t \le T, \ \forall Q \in \mathcal{M}_e^2. \tag{7.73}$$

The relation (7.71) also shows that

$$E[X_T^{var}]^2 \, E\left[ \frac{dP^{var}}{dP} \right]^2 = 1.$$

For proving Theorem 7.4.7, we cannot apply directly the duality method developed in the previous section for maximization of utility function defined on $(0, \infty)$, since the utility function is here $U(x) = -(H - x)^2$, defined on $\mathbb{R}$, and not increasing on the whole domain $\mathbb{R}$. We shall use the special features of the quadratic utility functions, and the characterizations by projections in Hilbert spaces. We denote by $\mathcal{G}_T^{\perp}$ the orthogonal of $\mathcal{G}_T$ in $L^2(P)$:

$$\mathcal{G}_T^{\perp} = \left\{ Z_T \in L^2(P) : E\left[Z_T \int_0^T \alpha_t dS_t\right] = 0, \ \forall \alpha \in \mathcal{A}_2 \right\}.$$

**Lemma 7.4.3** *The wealth process associated to the solution $\alpha^{var} \in \mathcal{A}_2$ of problem $v_1$ is nonnegative:*

$$X_t^{var} = 1 - \int_0^t \alpha_u^{var} dS_u \geq 0, \quad 0 \leq t \leq T, \ P \ a.s.$$

*Moreover, we have*

$$X_T^{var} \in \mathcal{G}_T^{\perp} \quad and \quad E[X_T^{var}] = E[X_T^{var}]^2 > 0.$$

**Proof.** Consider the stopping time $\tau = \inf\{t \in [0, T] : X_t^{var} \leq 0\}$, with the convention that $\inf \emptyset = \infty$. Since $S$ and so $X^{var}$ is continuous, and $X_0^{var} = 1$, we have

$$X_{\tau \wedge T}^{var} = 0 \text{ on } A := \{\tau \leq T\},$$
$$X^{var} > 0 \text{ on } A^c = \{\tau = \infty\}.$$

Let us define the process $\bar{\alpha}$ by $\bar{\alpha}_t = \alpha_t^{var}$ if $0 \leq t \leq \tau \wedge T$ and 0 otherwise. Since $\alpha^{var} \in \mathcal{A}_2$, it is clear that $\bar{\alpha} \in \mathcal{A}_2$ and

$$1 - \int_0^T \bar{\alpha}_t dS_t = X_{\tau \wedge T}^{var} = X_T^{var} 1_{A^c} \geq 0, \quad a.s.$$

We deduce that $E[1 - \int_0^T \bar{\alpha}_t dS_t]^2 \leq E[X_T^{var}]^2$. Since $\alpha^{var}$ is solution to $v_1$, we must have $X_T^{var} = 1 - \int_0^T \bar{\alpha}_t dS_t \geq 0$. By observing that $X^{var}$ is a martingale under $Q$ arbitrary in $\mathcal{M}_e^2$, we get

$$X_t^{var} = E^Q\left[X_T^{var} | \mathcal{F}_t\right] \geq 0, \quad 0 \leq t \leq T, \ a.s.$$

On the other hand, the solution to problem $v_1$, which is obtained by projection of the element 1 onto the closed vector subspace $\mathcal{G}_T$, is characterized by the property $X_T^{var} \in \mathcal{G}_T^{\perp}$:

$$E\left[X_T^{var} \int_0^T \alpha_t dS_t\right] = 0, \quad \forall \alpha \in \mathcal{A}_2. \tag{7.74}$$

For $\alpha = \alpha^{var}$, this implies in particular that

$$E[X_T^{var}] = E[X_T^{var}]^2 > 0$$

since $X_T^{var}$ cannot be equal to zero $P$ a.s. from the fact that $E^Q[X_T^{var}] = 1$ for $Q \in \mathcal{M}_e^2$. $\qquad \square$

In the sequel, we identify, as usual, an absolutely continuous probability measure with its Radon-Nikodym density.

**Lemma 7.4.4**

$$\mathcal{M}_e^2 = \{Z_T \in L^2(P): \ Z_T > 0, \ a.s., \ E[Z_T] = 1\} \cap \mathcal{G}_T^\perp. \tag{7.75}$$

**Proof.** Denote by $\tilde{\mathcal{M}}_e^2$ the r.h.s. in (7.75). By definition of $\mathcal{A}_2$, it is clear that $\mathcal{M}_e^2 \subset \tilde{\mathcal{M}}_e^2$. Let $Z_T \in \tilde{\mathcal{M}}_e^2$ and $Q \sim P$ the associated probability measure with Radon-Nikodym density $Z_T$. Since $S$ is continuous (hence locally bounded), there exists a sequence of stopping times $\tau_n \uparrow \infty$, such that the stopped process $S^{\tau_n} = (S_{t \wedge \tau_n})_{0 \leq t \leq T}$ is bounded. (for example $\tau_n = \inf\{t \geq 0 : |S_t| \geq n\}$). $\mathcal{A}_2$ contains the simple integrands in the form $\alpha_t = \theta 1_{(s \wedge \tau_n, u \wedge \tau_n]}(t)$ where $0 \leq s \leq u \leq T$ and $\theta$, $\mathcal{F}_{s \wedge \tau_n}$-measurable valued in $\mathbb{R}^n$. Since $Z_T \in \mathcal{G}_T^\perp$, we have

$$0 = E[Z_T \int_0^T \theta 1_{(s \wedge \tau_n, u \wedge \tau_n]}(t) dS_t] = E^Q[\theta.(S_u^{\tau_n} - S_s^{\tau_n})], \tag{7.76}$$

for all $0 \leq s \leq u \leq T$ and $\theta$, $\mathcal{F}_{s \wedge \tau_n}$-measurable. From the characterization of random variables $\mathcal{F}_{s \wedge \tau_n}$-measurables (see Proposition 1.1.2), the relation (7.76) holds also true for all $\theta$, $\mathcal{F}_s$-measurable. This proves that $S^{\tau_n}$ is a $Q$-martingale, i.e. $S$ is a $Q$-local martingale, and thus $Q \in \mathcal{M}_e^2$.  □

We introduce the closure $\mathcal{M}^2$ of $\mathcal{M}_e^2$ in $L^2(P)$, which is given from the previous lemma by

$$\mathcal{M}^2 = \left\{ Q \ll P : \frac{dQ}{dP} \in L^2(P) \ \text{and} \ S \ \text{is a} \ Q - \text{local martingale}\right\}$$
$$= \left\{ Z_T \in L^2(P): \ Z_T \geq 0, \ a.s., \ E[Z_T] = 1\right\} \cap \mathcal{G}_T^\perp.$$

The dual problem is then also written as

$$\tilde{v}_1 = \inf_{Z_T \in \mathcal{M}^2} E[Z_T]^2. \tag{7.77}$$

**Lemma 7.4.5** *There exists a unique solution $Z_T^{var} \in \mathcal{M}^2$ to the dual problem (7.77). This solution is related to problem $v_1$ by*

$$Z_T^{var} = E[Z_T^{var}]^2 X_T^{var}. \tag{7.78}$$

**Proof.** The nonempty set $\mathcal{M}^2$ is clearly convex and closed in $L^2(P)$. The problem (7.77), which is a projection problem in $L^2(P)$ of the zero element onto $\mathcal{M}^2$ then admits a unique solution $Z_T^{var}$. Recall that this solution is characterized by the property that $Z_T^{var} \in \mathcal{M}^2$ and

$$E[Z_T^{var}(Z_T^{var} - Z_T)] \leq 0, \quad \forall Z_T \in \mathcal{M}^2. \tag{7.79}$$

Consider the random variable

$$\bar{Z}_T = \frac{X_T^{var}}{E[X_T^{var}]}. \tag{7.80}$$

From Lemma 7.4.3, $\bar{Z}_T \in \mathcal{M}^2$. Moreover, by definition of $\mathcal{A}_2$ and $\bar{Z}_T$, we have for all $Z_T \in \mathcal{M}_e^2$

$$E[Z_T \bar{Z}_T] = \frac{1}{E[X_T^{var}]} = E[\bar{Z}_T]^2.$$

By density of $\mathcal{M}_e^2$ in $\mathcal{M}^2$, the above relation holds true for all $Z_T \in \mathcal{M}^2$. From the characterization (7.79) of $Z_T^{var}$, this shows that $Z_T^{var} = \bar{Z}_T$, and concludes the proof. $\square$

**Proof of Theorem 7.4.7.**

In view of the previous lemmas, it remains to show that $Z_T^{var} > 0$ a.s., which then defines a probability measure $P^{var} \in \mathcal{M}_e^2$. This is a delicate and technical point, whose proof may be omitted in a first reading.

Let $Z^{var}$ be the nonnegative martingale density process of $P^{var} \in \mathcal{M}^2$: $Z_t^{var} = E[Z_T^{var}|\mathcal{F}_t] = E\left[\frac{dP^{var}}{dP} \big| \mathcal{F}_t\right]$, $0 \le t \le T$. We want to prove that $P$ a.s., $Z_t^{var} > 0$ for all $t \in [0, T]$. Let us consider the stopping time

$$\tau = \inf\{0 \le t \le T : Z_t^{var} = 0\}$$

with the convention $\inf \emptyset = \infty$. Consider also the stopping time

$$\sigma = \inf\{0 \le t \le T : X_t^{var} = 0\}.$$

Fix some element $Q^0 \in \mathcal{M}_2^e$ and denote by $Z^0$ its martingale density process.

On the set $\{\sigma < \tau\} \subset \{\sigma < \infty\}$, we have by the martingale property of the nonnegative process $X^{var}$ under $Q^0$

$$0 = X_\sigma^{var} = E^{Q^0}[X_T^{var}|\mathcal{F}_\sigma].$$

Since $X_T^{var} \ge 0$ by Lemma (7.4.3), this proves that $X_\sigma^{var} = 0$ on $\{\sigma < \tau\}$. With the relation (7.78), we also have $Z_T^{var} = 0$ on $\{\sigma < \tau\}$. By the martingale property of $Z^{var}$, we thus get $Z_\sigma^{var} = 0$ on $\{\sigma < \tau\}$. This is clearly in contradiction with the definition of $\tau$ unless $P\{\sigma < \tau\} = 0$.

On the set $\{\tau < \sigma\} \subset \{\tau < \infty\}$, we have by the martingale property of the nonnegative process $Z^{var}$ under $P$

$$0 = Z_\tau^{var} = E[Z_T^{var}|\mathcal{F}_\tau].$$

Since $Z_T^{var} \ge 0$, this proves that $Z_T^{var} = 0$ on $\{\tau < \sigma\}$ and so by (7.78) that $X_T^{var} = 0$ on $\{\tau < \sigma\}$. By the martingale property of $X^{var}$ under $Q^0$, we have $X_\tau^{var} = 0$ on $\{\tau < \sigma\}$. This is clearly in contradiction with the definition of $\tau$ unless $P\{\tau < \sigma\} = 0$.

We deduce that $\tau = \sigma$ a.s. and by continuity of the nonnegative process $X^{var}$, $\tau$ is a predictable stopping time: there exists a sequence of stopping times $(\tau_n)_{n \ge 1}$, with $\tau_n < \tau$ on $\{\tau > 0\}$, $\tau_n$ converging increasingly to $\tau$ (we say that $\tau$ is announced by $(\tau_n)_n$). Indeed, it suffices to take $\tau_n = \inf\{t \ge 0 : X_t^{var} \le 1/n\} \wedge n$. By the martingale property of the nonnegative process $Z^{var}$ and since $Z_{\tau_n}^{var} > 0$, we have by the Cauchy-Schwarz inequality

$$1 = E\left[\frac{Z_T^{var}}{Z_{\tau_n}^{var}} \Big| \mathcal{F}_{\tau_n}\right] = E\left[\frac{Z_T^{var}}{Z_{\tau_n}^{var}} 1_{Z_\tau^{var} \ne 0} \Big| \mathcal{F}_{\tau_n}\right]$$
$$\le E\left[\left(\frac{Z_T^{var}}{Z_{\tau_n}^{var}}\right)^2 \Big| \mathcal{F}_{\tau_n}\right]^{\frac{1}{2}} E[1_{Z_\tau^{var} \ne 0}|\mathcal{F}_{\tau_n}]^{\frac{1}{2}}. \tag{7.81}$$

Since $E[1_{Z^{var}_\tau \neq 0}|\mathcal{F}_{\tau_n}]$ converges to 0 on the set $\{Z^{var}_\tau = 0\}$, inequality (7.81) proves that

$$E\left[\left(\frac{Z^{var}_T}{Z^{var}_{\tau_n}}\right)^2\Big|\mathcal{F}_{\tau_n}\right] \to \infty, \quad \text{on} \quad \{Z^{var}_\tau = 0\}. \tag{7.82}$$

Since $Z^0$ is a strictly positive $P$-martingale such that $Z^0_T \in L^2(P)$, we get

$$\sup_{0 \leq t \leq T} E\left[\left(\frac{Z^0_T}{Z^0_t}\right)^2\Big|\mathcal{F}_t\right] < \infty, \quad a.s.$$

Suppose that $P[Z^{var}_\tau = 0] > 0$. From (7.82), we see that for $n$ large enough, the set

$$A_n = \left\{E\left[\left(\frac{Z^0_T}{Z_{\tau_n}}\right)^2\Big|\mathcal{F}_{\tau_n}\right] < E\left[\left(\frac{Z^{var}_T}{Z^{var}_{\tau_n}}\right)^2\Big|\mathcal{F}_{\tau_n}\right]\right\}$$

is nonempty in $\mathcal{F}_{\tau_n}$. Let us then define the martingale

$$Z_t = \begin{cases} Z^{var}_t & t < \tau_n \\ Z^0_t \frac{Z^{var}_{\tau_n}}{Z^0_{\tau_n}} & t \geq \tau_n \quad \text{on } A_n \\ Z^{var}_t & t \geq \tau_n \quad \text{outside } A_n. \end{cases}$$

We easily check that $ZS$ inherits the local martingale property of $Z^{var}S$ and $Z^0S$, and so $Z_T \in \mathcal{M}^2$. Morever, by construction, we get

$$E[Z_T]^2 < E[Z^{var}_T]^2,$$

which is in contradiction with the definition of $Z^{var}_T$. We then conclude that $P[Z^{var}_\tau = 0] = 0$ and thus $Z^{var}_t > 0$, for all $t \in [0, T]$, $P$ a.s. Finally, the strict positivity of $X^{var}$ follows from the strict positivity of $X^{var}_T$ and the martingale property of $X^{var}$ under $Q^0 \sim P$.    $\square$

The strictly positive wealth process $X^{var}$ is called quadratic hedging numéraire, and shall be used in the next section for the resolution of the mean-variance hedging problem.

### 7.4.4 Problem resolution by change of numéraire

We use $X^{var}$ as numéraire: we consider the discounted price process $S^{var}$ valued in $\mathbb{R}^{n+1}$ by

$$S^{var,0} = \frac{1}{X^{var}} \quad \text{and} \quad S^{var,i} = \frac{S^i}{X^{var}}, \; i = 1, \ldots, n.$$

Recall that for any $Q \in \mathcal{M}^2_e$, the process $X^{var}$ is a $Q$-martingale with initial value 1. We then define the set of probability measures with martingale density $X^{var}$ with respect to a probability measure $Q$ in $\mathcal{M}^2_e$:

$$\mathcal{M}^{2,var}_e = \Big\{Q^{var} \text{ probability on } (\Omega, \mathcal{F}_T) : \exists Q \in \mathcal{M}^2_e$$
$$\frac{dQ^{var}}{dQ}\Big|_{\mathcal{F}_t} = X^{var}_t, \; 0 \leq t \leq T.\Big\}.$$

Since by definition, $\mathcal{M}_e^2$ is the set of probability measures $Q \sim P$ with square-integrable Radon-Nikodym density, under which $S$ is a local martingale, we easily deduce from Bayes formula that $\mathcal{M}_e^{2,var}$ is also written as

$$\mathcal{M}_e^{2,var} = \left\{ Q^{var} \sim P : \frac{1}{X_T^{var}} \frac{dQ^{var}}{dP} \in L^2(P) \text{ and} \right.$$

$$\left. S^{var} \text{ is a } Q^{var} - \text{local martingale} \right\}. \tag{7.83}$$

We then introduce the set of admissible integrands with respect to $S^{var}$:

$$\Phi_2^{var} = \left\{ \phi \in L(S^{var}) : X_T^{var} \int_0^T \phi_t dS_t^{var} \in L^2(P) \quad \text{and} \right.$$

$$\left. \int \phi dS^{var} \text{ is a } Q^{var} - \text{martingale under any } Q^{var} \in \mathcal{M}_e^{2,var} \right\}.$$

We first state a general invariance result for stochastic integrals by change of numéraire.

**Proposition 7.4.6** *For all $x \in \mathbb{R}$, we have*

$$\left\{ x + \int_0^T \alpha_t dS_t : \alpha \in \mathcal{A}_2 \right\} = \left\{ X_T^{var} \left( x + \int_0^T \phi_t dS_t^{var} \right) : \phi \in \Phi_2^{var} \right\}. \tag{7.84}$$

*Furthermore, the correspondence relation between $\alpha = (\alpha^1, \ldots, \alpha^n) \in \mathcal{A}_2$ and $\phi = (\phi^0, \ldots, \phi^n) \in \Phi_2^{var}$ is given by $\phi = F_x^{var}(\alpha)$ where $F_x^{var} : \mathcal{A}_2 \to \Phi_2^{var}$ is defined by*

$$\phi^0 = x + \int \alpha dS - \alpha.S \quad \text{and} \quad \phi^i = \alpha^i, \quad i = 1, \ldots, n, \tag{7.85}$$

*and $\alpha = F_x^{-1,var}(\phi)$, with $F_x^{-1,var} : \Phi_2^{var} \to \mathcal{A}_2$ determined by*

$$\alpha^i = \phi^i - \alpha^{var,i} \left( x + \int \phi dS^{var} - \phi.S^{var} \right), \quad i = 1, \ldots, n. \tag{7.86}$$

**Proof.** The proof is essentially based on Itô's product rule, the technical point concerning the integrability questions on the integrands.

(1) By Itô's product, we have

$$d\left( \frac{S}{X^{var}} \right) = S d\left( \frac{1}{X^{var}} \right) + \frac{1}{X^{var}} dS + d < S, \frac{1}{X^{var}} > . \tag{7.87}$$

Let $\alpha \in \mathcal{A}_2$ and consider the truncated bounded integrand $\alpha^{(n)} = \alpha 1_{|\alpha| \leq n}$, which is integrable with respect to $S/X^{var}$, $1/X^{var}$ and $< S, \frac{1}{X^{var}} >$. We then get

$$\int \alpha^{(n)} d\left( \frac{S}{X^{var}} \right) = \int \alpha^{(n)} S d\left( \frac{1}{X^{var}} \right) + \int \alpha^{(n)} \frac{1}{X^{var}} dS$$

$$+ \int \alpha^{(n)} d < S, \frac{1}{X^{var}} > . \tag{7.88}$$

Denote by $X^{x,\alpha^{(n)}} = x + \int \alpha^{(n)} dS$. Then, by Itô's formula and (7.88), we have

$$d\left(\frac{X^{x,\alpha^{(n)}}}{X^{var}}\right) = X^{x,\alpha^{(n)}} d\left(\frac{1}{X^{var}}\right) + \frac{1}{X^{var}}\alpha^{(n)} dS + \alpha^{(n)} d < S, \frac{1}{X^{var}} >$$
$$= (X^{x,\alpha^{(n)}} - \alpha^{(n)}.S) d\left(\frac{1}{X^{var}}\right) + \alpha^{(n)} d\left(\frac{S}{X^{var}}\right)$$
$$= \phi^{(n)} dS^{var}, \tag{7.89}$$

with $\phi^{(n)} = F_x^{var}(\alpha^{(n)}) \in L(S^{var})$. The relation (7.89) shows that

$$x + \int \alpha^{(n)} dS = X^{var}\left(x + \int \phi^{(n)} dS^{var}\right). \tag{7.90}$$

Since $\alpha$ is $S$-integrable, i.e. $\alpha \in L(S)$, we know that $\int \alpha^{(n)} dS$ converges to $\int \alpha dS$ for the semimartingale topology as $n$ goes to infinity. This implies that $X^{x,\alpha^{(n)}}/X^{var}$ converges also for the semimartingale topology. From (7.90), we deduce that $\int \phi^{(n)} dS^{var}$ converges to $\int \psi d\tilde{X}$ for the semimartingale topology with $\psi \in L(S^{var})$, since the space $\{\int \psi dS^{var} : \psi \in L(S^{var})\}$ is closed for the semimartingale topology. Since $\phi^{(n)}$ converges a.s. to $\phi = F_x^{var}(\alpha)$, we get $\psi = \phi$. We then obtain by sending $n$ to infinity in (7.90)

$$x + \int \alpha dS = X^{var}\left(x + \int \phi dS^{var}\right). \tag{7.91}$$

Since $X_T^{var}$ and $\int_0^T \alpha_t dS_t \in L^2(P)$, we get by (7.91) $X_T^{var} \int_0^T \phi_t dS_t^{var} \in L^2(P)$. Since $\int \alpha dS$ is a $Q$-martingale for any $Q \in \mathcal{M}_e^2$, it follows by definition of $\mathcal{M}_e^{2,var}$, by (7.91) and the Bayes formula that $\int \phi dS^{var}$ is a $Q^{var}$-martingale for all $Q^{var} \in \mathcal{M}_e^{2,var}$. Thus, $\phi \in \Phi_2^{var}$, and the inclusion $\subseteq$ in (7.84) is proved.

(2) The proof of the converse is similar. By Itô's product, we have

$$d(X^{var}X) = X^{var}dS^{var} + S^{var}dX^{var} + d < X^{var}, S^{var} > . \tag{7.92}$$

Let $\phi \in \Phi_2^{var}$ and consider the truncated bounded integrand $\phi^{(n)} = \phi 1_{|\phi|\leq n}$. Then, from (7.92) and the definitions of $S^{var}$ and $X^{var}$, we get

$$d\left(X^{var}\left(x + \int \phi^{(n)} dS^{var}\right)\right)$$
$$= \left(x + \int \phi^{(n)} dS^{var}\right) dX^{var} + X^{var}\phi^{(n)} dS^{var} + \phi^{(n)} d < X^{var}, S^{var} >$$
$$= \left(x + \int \phi^{(n)} dS^{var}\right) dX^{var} + \phi^{(n)} d(X^{var} S^{var}) - \phi^{(n)}.S^{var} dX^{var}$$
$$= \alpha^{(n)} dS,$$

with $\alpha^{(n)} = F_x^{-1,var}(\phi^{(n)}) \in L(S)$. By the same arguments as in point (1), we obtain by sending $n$ to infinity

$$X^{var}\left(x + \int \phi dS^{var}\right) = x + \int \alpha dS, \tag{7.93}$$

with $\alpha = F_x^{-1,var}(\phi) \in L(S)$. We also check as in (1) that $\alpha \in \mathcal{A}_2$ since $\phi \in \Phi_2^{var}$. The inclusion $\supseteq$ in (7.84) is proved, and the proof is complete. $\qquad\square$

Notice that in the proof of Proposition 7.4.6, we used only the strict positivity of the process $X^{var} = 1 - \int \alpha^{var} dS$. The previous invariance result for the space of stochastic integrals by change of numéraire is actually valid for any choice of numéraire $X^{num} = 1 - \int \alpha^{num} dS_t$, $\alpha^{num} \in \mathcal{A}_2$, with $X_t^{num} > 0$, $0 \le t \le T$. The particular choice of $X^{num} = X^{var}$, the solution to problem (7.69), is now used in a crucial way for the resolution of the quadratic minimization problem. We shall prove, by means of this suitable numéraire, how one can reduce the original mean-variance problem to the martingale case of Section 7.4.2.

To the variance-optimal martingale measure $P^{var} \in \mathcal{M}_e^2$, we associate $P^{2var} \in \mathcal{M}_e^{2,var}$ defined by

$$\frac{dP^{2var}}{dP^{var}} = X_T^{var}. \tag{7.94}$$

From the duality relation (7.71), the Radon-Nikodym density of $P^{2var}$ with respect to $P$ is

$$\frac{dP^{2var}}{dP} = E\left[\frac{dP^{var}}{dP}\right]^2 (X_T^{var})^2. \tag{7.95}$$

Since $H \in L^2(P)$, the relation (7.95) implies in particular that the discounted payoff $H^{var} = H/X_T^{var} \in L^2(P^{2var})$. Recall also that $S^{var}$ is a (continuous) local martingale under $P^{2var}$ by the characterization (7.83) of $\mathcal{M}_e^{2,var}$. We can then apply the Kunita-Watanabe projection theorem for the square-integrable $P^{2var}$-martingale $H_t^{var} = E^{P^{2var}}[H^{var}|\mathcal{F}_t]$, $0 \le t \le T$, onto $S^{var}$, and we derive

$$E^{P^{2var}}\left[\frac{H}{X_T^{var}}\Big|\mathcal{F}_t\right] = E^{P^{2var}}\left[\frac{H}{X_T^{var}}\right] + \int_0^t \phi_u^H dS_u^{var} + R_t^{var,H}, 0 \le t \le T, \tag{7.96}$$

where $\phi^H \in L(S^{var})$ satisfies

$$E^{P^{2var}}\left[\int_0^T \phi_t^H dS_t^{var}\right]^2 = E^{P^{2var}}\left[\int_0^T (\phi_t^H)' d < S^{var} >_t \phi_t^H\right] < \infty, \tag{7.97}$$

and $R^{var,H}$ is a square-integrable $P^{2var}$-martingale, orthogonal to $S^{var}$.

**Theorem 7.4.8** *For all $x \in \mathbb{R}$ and $H \in L^2(P)$, the solution $\alpha^{mv}$ to $v_H(x)$ is given by*

$$\alpha^{mv} = F_x^{-1,var}(\phi^H), \tag{7.98}$$

*where $\phi^H$ defined in (7.96) lies in $\Phi_2^{var}$ and $F_x^{-1,var}$ is defined in Proposition 7.4.6. Moreover, we have*

$$v_H(x) = \frac{\left(E^{P^{var}}[H] - x\right)^2}{E\left[\frac{dP^{var}}{dP}\right]^2} + E\left[X_T^{var} R_T^{var,H}\right]^2. \tag{7.99}$$

**Proof.** By similar arguments as in the proof of Theorem 7.4.6 (by using the Doob and Cauchy-Schwarz inequalities), we see that the integrability condition (7.97) on $\phi^H$ implies that (actually is equivalent to) $\phi^H \in \Phi_2^{var}$. From Proposition 7.4.6 on the invariance by change of numéraire and relation (7.95), we have

$$v_H(x) := \inf_{\alpha \in \mathcal{A}_2} E\Big[H - x - \int_0^T \alpha_t dS_t\Big]^2 \tag{7.100}$$

$$= \inf_{\phi \in \Phi_2^{var}} E\Big[H - X_T^{var}\Big(x + \int_0^T \phi_t dS_t^{var}\Big)\Big]^2$$

$$= \frac{1}{E\Big[\frac{dP^{var}}{dP}\Big]^2} \inf_{\phi \in \Phi_2^{var}} E^{P^{2var}}\Big[\frac{H}{X_T^{var}} - x - \int_0^T \phi_t dS_t^{var}\Big]^2, \tag{7.101}$$

and the solutions to (7.100) and (7.101) are related via the correspondence function $F_x^{-1,var}$. Now, problem (7.101) is a quadratic minimization problem as in the martingale case, whose solution is determined by the Kunita-Watanabe decomposition (7.96). This proves (7.98) and also that

$$v_H(x) = \frac{1}{E\Big[\frac{dP^{var}}{dP}\Big]^2}\Big\{\Big(E^{P^{2var}}\Big[\frac{H}{X_T^{var}}\Big] - x\Big)^2 + E^{P^{2var}}[R_T^{var}]^2\Big\}.$$

We finally get the expression (7.99) of $v_H(x)$ with (7.94) and (7.95). □

**Remark 7.4.11** The solution $x_{mv}(H)$ to problem $\inf_{x \in \mathbb{R}} v_H(x)$, called quadratic approximation price of $H$, is given from (7.99) by

$$x_{mv}(J) = E^{P^{var}}[H].$$

The above theorem shows that the quadratic hedging problem may be solved in the following three steps:

(1) Determine the solution to problem $v_1$ that defines the quadratic numéraire $X^{var}$ or equivalently the solution to the dual problem $\tilde{v}_1$ defining the variance-optimal martingale measure. Of course, if $S$ is already a martingale under $P$, the solution is trivial: $X^{var} = 1$ and $P^{var} = P$. We give in the next section some other examples of models where the computations of $X^{var}$ and $P^{var}$ are explicit.

(2) Change of numéraire by discounting the price process $S$, the option payoff $H$ and the variance-optimal martingale measure by $X^{var}$. We then define the price process $S^{var} = (1/X^{var}, S/X^{var})$, the payoff $H^{var} = H/X_T^{var}$ and the probability $P^{2var}$ with Radon-Nikodym density with respect to $P^{var}$: $X_T^{var}$. We then project according to the Kunita-Watanabe decomposition, the $P^{2var}$-martingale $E^{P^{2var}}[H^{var}|\mathcal{F}_t]$ onto $S^{var}$. In a Markovian framework, for example a diffusion, this decomposition is derived in the smooth case by Itô's formula. In the more general case, the integrand of $S^{var}$ in the decomposition can be expressed by means of the Clark-Ocone formula or Malliavin derivative.

(3) The solution to the mean-variance hedging problem is finally given by the correspondence relation between the space of stochastic integrals with respect to $S$ and the space of stochastic integrals with respect to $S^{var}$.

### 7.4.5 Example

We consider the model and the notation of Section 7.2.4. Recall that we have an explicit description of $\mathcal{M}_e^2$:

$$\mathcal{M}_e^2 = \Big\{ P^\nu : \frac{dP^\nu}{dP} = Z_T^\nu, \ \nu \in K_m^2(\sigma) \Big\}, \tag{7.102}$$

where

$$Z_t^\nu = \exp\Big( - \int_0^t (\lambda_u + \nu_u).dW_u - \frac{1}{2} \int_0^t |\lambda_u|^2 + |\nu_u|^2 du \Big), \quad 0 \le t \le T.$$

and $K_m^2(\sigma)$ is the set of elements $\nu$ in $K(\sigma)$ such that $Z^\nu$ is a square-integrable martingale.

We assume in this example that the quantity

$$\hat{K}_T = \int_0^T |\lambda_t|^2 dt,$$

called the mean-variance ratio, is deterministic. This is a generalization of the case where $S$ is a local martingale under $P$ for which $\hat{K}_T = 0$.

Consider for any $\nu \in K_m^2(\sigma)$, the Doléans-Dade exponential local martingale

$$\xi_t^\nu = \exp\Big( - 2 \int_0^t (\lambda_u + \nu_u).dW_u - 2 \int_0^t |\lambda_u|^2 + |\nu_u|^2 du \Big), \quad 0 \le t \le T.$$

It is clear that $|\xi_t^\nu| \le |Z_t^\nu|^2$. Since $Z^\nu$ is a square-integrable martingale, we have: $E[\sup_{0 \le t \le T} |Z_t^\nu|^2] < \infty$. It follows that $\xi^\nu$ is uniformly integrable, and so is a martingale. We can then define a probability measure $Q^\nu$ equivalent to $P$ with martingale density process $\xi^\nu$. We have for any $\nu \in K_m^2(\sigma)$:

$$E\Big[\frac{dP^\nu}{dP}\Big]^2 = E\Big[ \exp\Big( - 2 \int_0^T (\lambda_u + \nu_u).dW_u - \int_0^T |\lambda_u|^2 + |\nu_u|^2 du \Big)\Big]$$

$$= E^{Q^\nu}\Big[ \exp\Big( \int_0^T |\lambda_u|^2 + |\nu_u|^2 du \Big)\Big]$$

$$= \exp(\hat{K}_T) E^{Q^\nu}\Big[ \exp\Big( \int_0^T |\nu_u|^2 du \Big)\Big]$$

$$\ge \exp(\hat{K}_T), \tag{7.103}$$

where the third equality follows from the fact that $\hat{K}_T$ is deterministic. Notice that the equality in (7.103) holds for $\nu = 0$, which proves that the solution to the problem defining the variance-optimal martingale measure, given from (7.102) by

$$\tilde{v}_1 = \inf_{\nu \in K_m^2(\sigma)} E\Big[\frac{dP^\nu}{dP}\Big]^2 \tag{7.104}$$

is attained for $\nu = 0$. We then get

$$P^{var} = P^0 \text{ and } \tilde{v}_1 = E\Big[\frac{dP^0}{dP}\Big]^2 = \exp(\hat{K}_T).$$

We calculate the quadratic hedging numéraire by means of the expression (7.73):

$$X_t^{var} = \frac{1}{E[Z_T^0]^2} E^{P^0}[Z_T^0|\mathcal{F}_t]$$

$$= E^{P^0}\Big[ \exp\Big( - \int_0^T \lambda_u.dW_u^0 - \frac{1}{2} \int_0^T |\lambda_u|^2 du \Big)\Big|\mathcal{F}_T\Big],$$

where $W^0 = W + \int \lambda dt$ is a $P^0$ Brownian motion. Notice that the Novikov condition $E^{P^0}[\exp(\frac{1}{2}\int_0^T |\lambda_t|^2 dt)] = \exp(\hat{K}_T/2) < \infty$ is satisfied so that

$$X_t^{var} = \exp\left(-\int_0^t \lambda_u.dW_u^0 - \frac{1}{2}\int_0^t |\lambda_u|^2 du\right), \quad 0 \le t \le T.$$

Since $dX_t^{var} = -(\alpha_t^{var})'\sigma_t dW_t^0$, we deduce by identification and recalling the definition $\lambda = \sigma'(\sigma\sigma')^{-1}\mu$:

$$\alpha^{var} = (\sigma\sigma')^{-1}\mu X^{var}.$$

In the general case where $\hat{K}_T$ is random, the dual problem (7.104) defining the variance-optimal martingale measure is a stochastic control problem, which can be studied by the dynamic programming methods or BSDEs. We give some references in the last section of this chapter.

## 7.5 Bibliographical remarks

The optional decomposition theorem for supermartingales was originally proved in the context of Itô processes by El Karoui and Quenez [ElkQ95]. It was then extended for locally bounded semimartingales by Kramkov [Kr96]. The more general version for semimartingales is due to Föllmer and Kabanov [FoK98].

The dual approach to the utility maximization problem was initially formulated in a complete market by Pliska [Pli86], Karatzas, Lehoczky and Shreve [KLS87] and Cox and Huang [CH89]. It was then extended to the incomplete markets case for Itô processes, independently by Karatzas et al. [KLSX91] and He and Pearson [HeP91]. The general study for semimartingale price processes, and under the minimal assumption of reasonable asymptotic elasticity, is due to Kramkov and Schachermayer [KS99], [KS01]. Our presentation follows the main ideas of their works. The case of utility defined on the whole domain $\mathbb{R}$, typically the exponential utility, was studied in Delbaen et al. [DGRSSS02], Bellini and Frittelli [BF02] and Schachermayer [Scha01]. We also mention the recent book by Frittelli, Biagini and Scandolo [FBS09] for a detailed treatment of duality methods in finance.

The quadratic hedging criterion was introduced by Föllmer and Sondermann [FoS86] in the martingale case. The resolution method in the general case of continuous semimartingale price process, detailed in Section 7.4, is due to Gouriéroux, Laurent and Pham [GLP98]. Rheinländer and Schweizer [RhS97] proposed an alternative approach based on the so-called Föllmer-Schweizer decomposition. We also mention the recent work by Cerny and Kallsen [CeKal07] for another probabilistic approach to the mean-variance hedging problem. The existence result for the variance-optimal martingale measure (notion introduced by Schweizer [Schw96]) equivalent to the initial probability is proved in Delbaen and Schachermayer [DS96]. The example in Section 7.4.5 is inspired by Pham, Rheinländer and Schweizer [PRS98]. Other explicit computations of the variance-optimal martingale measure and the quadratic hedging numéraire in stochastic volatility models are developed in Laurent and Pham [LP99] and Biagini, Guasoni and Pratelli [BGP00].

# A

# Complements of integration

We are given a probability space $(\Omega, \mathcal{F}, P)$ and $L^1 = L^1(\Omega, \mathcal{F}, P)$ is the set of integrable random variables.

## A.1 Uniform integrability

**Definition A.1.1** *(Uniformly integrable random variables)*
*Let $(f_i)_{i \in I}$ be a family of random variables in $L^1$. We say that $(f_i)_{i \in I}$ is uniformly integrable if*

$$\lim_{x \to \infty} \sup_{i \in I} E[|f_i| 1_{|f_i| \geq x}] = 0.$$

Notice that any family of random variables, bounded by a fixed integrable random variable (in particular any finite family of random variables in $L^1$) is uniformly integrable.

The following result extends the dominated convergence theorem.

**Theorem A.1.1** *Let $(f_n)_{n \geq 1}$ be a sequence of random variables in $L^1$ converging a.s to a random variable $f$. Then $f$ is integrable and the convergence of $(f_n)$ to $f$ holds in $L^1$ if and only if the sequence $(f_n)_{n \geq 1}$ is uniformly integrable. When the random variables $f_n$ are nonnegative, this is equivalent to*

$$\lim_{n \to \infty} E[f_n] = E[f].$$

The following corollary is used in the proof of Theorem 7.3.4.

**Corollary A.1.1** *Let $(f_n)_{n \geq 1}$ be a sequence of nonnegative random variables bounded in $L^1$, i.e. $\sup_n E[f_n] < \infty$, converging a.s to a nonnegative random variable $f$ and such that $\lim_{n \to \infty} E[f_n] = E[f] + \delta$ with $\delta > 0$. Then, there exists a subsequence $(f_{n_k})_{k \geq 1}$ of $(f_n)_{n \geq 1}$ and a disjoint sequence $(A_k)_{k \geq 1}$ of $(\Omega, \mathcal{F})$ such that*

$$E[f_{n_k} 1_{A_k}] \geq \frac{\delta}{2}, \quad \forall k \geq 1.$$

H. Pham, *Continuous-time Stochastic Control and Optimization with Financial Applications*, Stochastic Modelling and Applied Probability 61, DOI 10.1007/978-3-540-89500-8, © Springer-Verlag Berlin Heidelberg 2009

**Proof.** We define $B_n = \{f_n \geq (f + \delta) \vee 1/\delta\}$. The sequence $(f_n 1_{\Omega \setminus B_n})_{n \geq 1}$ is uniformly integrable and converges a.s to $f$. This implies that $E[f_n 1_{\Omega \setminus B_n}]$ converges to $E[f]$, and so $E[f_n 1_{B_n}]$ converges to $\delta$. Thus, there exists $N = N(\delta) \geq 1$ such that

$$E[f_n 1_{B_n}] \geq \frac{3\delta}{4}, \quad \forall n \geq N.$$

We set $n_1 = N$. the sequence $(f_{n_1} 1_{B_m})_{m \geq 1}$ is uniformly integrable and converges a.s to $0$. Thus, there exists $n_2 \geq n_1 + 1$ such that

$$E[f_{n_1} 1_{B_{n_2}}] \leq \frac{\delta}{4}.$$

We then write $A_1 = B_{n_1} \setminus B_{n_2}$ so that

$$E[f_{n_1} 1_{A_1}] \geq E[f_{n_1} 1_{B_{n_1}}] - E[f_{n_1} 1_{B_{n_2}}] \geq \frac{\delta}{2}.$$

The sequence $(f_{n_2} 1_{B_{n_1}} 1_{B_m})_{m \geq 1}$ is uniformly integrable and converges a.s to $0$. Thus, there exists $n_3 \geq n_2 + 1$ such that

$$E[f_{n_2} 1_{B_{n_1} \cup B_{n_3}}] \leq \frac{\delta}{4}.$$

Define $A_2 = B_{n_2} \setminus (B_{n_1} \cup B_{n_3})$ so that $A_2$ is disjoint from $A_1$ and

$$E[f_{n_2} 1_{A_2}] \geq \frac{\delta}{2}.$$

We repeat this procedure: at step $k$, the sequence $(f_{n_k} 1_{\cup_{i=1}^{k-1} B_{n_i}} 1_{B_m})_{m \geq 1}$ is uniformly integrable and converges a.s to $0$. Thus, there exists $n_{k+1} \geq n_k + 1$ such that

$$E[f_{n_k} 1_{\cup_{i=1}^{k-1} B_{n_i} \cup B_{n_{k+1}}}] \leq \frac{\delta}{4}.$$

We then define $A_k = B_{n_k} \setminus (\cup_{i=1}^{k-1} B_{n_i} \cup B_{n_{k+1}})$ so that $A_k$ is disjoint from $A_i$, $i \leq k-1$, and

$$E[f_{n_k} 1_{A_k}] \geq \frac{\delta}{2}.$$

$\square$

The following result, due to la Vallée-Poussin, gives a practical condition for proving the uniform integrability.

**Theorem A.1.2** *(la Vallée-Poussin)*
*Let $(f_i)_{i \in I}$ be a family of random variables. We have equivalence between:*

*(1) $(f_i)_{i \in I}$ is uniformly integrable*

*(2) there exists a nonnegative function $\varphi$ defined on $\mathbb{R}_+$, $\lim_{x \to \infty} \varphi(x)/x = \infty$, such that*

$$\sup_{i \in I} E[\varphi(|f_i|)] < \infty.$$

In practice, we often use the implication $(2) \Longrightarrow (1)$. For example, by taking $\varphi(x) = x^2$, we see that any family of random variables bounded in $L^2$ is uniformly integrable. One can find the proofs of Theorem A.1.1 and A.1.2 in the book by [Do94].

## A.2 Essential supremum of a family of random variables

**Definition A.2.2** *(Essential supremum)*
*Let $(f_i)_{i \in I}$ be a family of real-valued random variables. The essential supremum of this family, denoted by $\operatorname{ess\,sup}_{i \in I} f_i$ is a random variable $\hat{f}$ such that*

*(a) $f_i \leq \hat{f}$ a.s., for all $i \in I$*

*(b) if $g$ is a random variable satisfying $f_i \leq g$ a.s., for all $i \in I$, then $\hat{f} \leq g$ a.s.*

The next result is proved in Neveu [Nev75].

**Theorem A.2.3** *Let $(f_i)_{i \in I}$ be a family of real-valued random variables. Then, $\hat{f} = \operatorname{ess\,sup}_{i \in I} f_i$ exists and is unique. Moreover, if the family $(f_i)_{i \in I}$ is stable by supremum, i.e. for all $i, j$ in $I$, there exists $k$ in $I$ such that $f_i \vee f_j = f_k$, then there exists an increasing sequence $(f_{i_n})_{n \geq 1}$ in $(f_i)_{i \in I}$ satisfying*

$$\hat{f} = \lim_{n \to \infty} \uparrow f_{i_n} \quad a.s$$

We define the essential infimum of a family of real-valued random variables $(f_i)_{i \in I}$ by: $\operatorname{ess\,inf}_{i \in I} f_i = -\operatorname{ess\,sup}_{i \in I}(-f_i)$.

## A.3 Some compactness theorems in probability

This first compactness result is well-known, and due to Komlos [Kom67].

**Theorem A.3.4** *(Komlos)*
*Let $(f_n)_{n \geq 1}$ be a sequence of random variables bounded in $L^1$. Then, there exists a subsequence $(f_{n_k})_{k \geq 1}$ of $(f_n)_{n \in \mathbb{N}}$ and a random variable $f$ in $L^1$ such that*

$$\frac{1}{k} \sum_{j=1}^{k} f_{n_j} \to f \quad a.s \text{ when } k \text{ goes to infinity.}$$

The following compactness theorem in $L^0_+(\Omega, \mathcal{F}, P)$ is very useful for deriving existence results in optimization problems in finance. It is proved in the appendix of Delbaen and Schachermayer [DS94].

**Theorem A.3.5** *Let $(f_n)_{n \geq 1}$ be a sequence of random variables in $L^0_+(\Omega, \mathcal{F}, P)$. Then, there exists a sequence $g_n \in \operatorname{conv}(f_n, f_{n+1}, \ldots)$, i.e. $g_n = \sum_{k=n}^{N_n} \lambda_k f_k$, $\lambda_k \in [0,1]$ and $\sum_{k=n}^{N_n} \lambda_k = 1$, such that the sequence $(g_n)_{n \geq 1}$ converges a.s. to a random variable $g$ valued in $[0, \infty]$.*

# B

## Convex analysis considerations

Standard references for convex analysis are the books by Rockafellar [Ro70] and Ekeland and Temam [ET74]. For the purpose of our book, we mainly focus (unless specified) to the case in $\mathbb{R}^d$. We define $\bar{\mathbb{R}} = \mathbb{R} \cup \{-\infty, \infty\}$.

## B.1 Semicontinuous, convex functions

Given a function $f$ from $\mathcal{O}$ open set of $\mathbb{R}^d$ into $\bar{\mathbb{R}}$, we define the functions $f_*$ and $f^* : \mathcal{O} \to \bar{\mathbb{R}}$ by

$$f_*(x) = \liminf_{y \to x} f(y) \;:=\; \lim_{\varepsilon \to 0} \inf\{f(y) : y \in \mathcal{O}, |y - x| \leq \varepsilon\}$$
$$f^*(x) = \limsup_{y \to x} f(y) \;:=\; \lim_{\varepsilon \to 0} \sup\{f(y) : y \in \mathcal{O}, |y - x| \leq \varepsilon\}.$$

**Definition B.1.1** *(Semicontinuity)*
*Let $f$ be a function from $\mathcal{O}$ open set in $\mathbb{R}^d$ into $\bar{\mathbb{R}}$. We say that $f$ is lower-semicontinuous (l.s.c.) if one of the following equivalent conditions is satisfied:*

*(i)* $\forall\, x \in \mathcal{O}, f(x) \leq \liminf_{n \to \infty} f(x_n)$, *for any sequence $(x_n)_{n \geq 1}$ converging to $x$.*

*(ii)* $\forall\, x \in \mathcal{O}, f(x) = f_*(x)$.

*(iii)* $\{x \in \mathcal{O} : f(x) \leq \lambda\}$ *is closed for all $\lambda \in \mathbb{R}$.*

*We say that $f$ is upper-semicontinuous (u.s.c.) if $-f$ is lower-semicontinuous.*

Notice that $f$ is continuous on $\mathcal{O}$ if and only if $f$ is lower and upper-semicontinuous. The function $f_*$ is called a lower-semicontinuous envelope of $f$: it is the largest l.s.c. function below $f$. The function $f^*$ is called a upper-semicontinuous envelope of $f$: it is the smallest u.s.c. function s.c.s. above $f$.

**Theorem B.1.1** *A l.s.c. (resp. u.s.c.) function attains its minimum (resp. maximum) on any compact.*

Given a convex subset $C$ of $E$ vector space, we recall that a function $f$ from $C$ into $\bar{\mathbb{R}}$ is convex if for all $x, y \in C$, $\lambda \in [0,1]$, $f(\lambda x + (1-\lambda)y) \leq \lambda f(x) + (1-\lambda)f(y)$. We

H. Pham, *Continuous-time Stochastic Control and Optimization with Financial Applications*, Stochastic Modelling and Applied Probability 61,
DOI 10.1007/978-3-540-89500-8, © Springer-Verlag Berlin Heidelberg 2009

say that $f$ is strictly convex on $C$ if for all $x, y \in C$, $x \neq y$, $\lambda \in (0,1)$, $f(\lambda x + (1 - \lambda)y)$ $< \lambda f(x) + (1 - \lambda)f(y)$. We say that $f$ is (strictly) concave if $-f$ is (strictly) convex.

The following min-max theorem is proved in Strasser [Str85], Theorem 45.8.

**Theorem B.1.2** *(Min-max)*
*Let $\mathcal{X}$ be a convex subset of a normed vector space $E$, compact for the weak topology $\sigma(E, E')$, and $\mathcal{Y}$ a convex subset of a vector space. Let $f : \mathcal{X} \times \mathcal{Y} \to \mathbb{R}$ be a function satisfying:*

*(1) $x \to f(x, y)$ is continuous and concave on $\mathcal{X}$ for all $y \in \mathcal{Y}$*

*(2) $y \to f(x, y)$ is convex on $\mathcal{Y}$ for all $x \in \mathcal{Y}$.*

*Then, we have*

$$\sup_{x \in \mathcal{X}} \inf_{y \in \mathcal{Y}} f(x, y) = \inf_{y \in \mathcal{Y}} \sup_{x \in \mathcal{X}} f(x, y).$$

In the sequel, we shall restrict ourselves to the case $E = \mathbb{R}^d$. Given a convex function $f$ from $\mathbb{R}^d$ into $\bar{\mathbb{R}}$, we define its domain by

$$\mathrm{dom}(f) = \left\{ x \in \mathbb{R}^d : f(x) < \infty \right\},$$

which is a convex set of $\mathbb{R}^d$. We say that a convex function $f$ from $\mathbb{R}^d$ into $\bar{\mathbb{R}}$ is *proper* if it never takes the value $-\infty$ and if $\mathrm{dom}(f) \neq \emptyset$.

We have the following continuity result for convex functions.

**Proposition B.1.1** *A proper convex function from $\mathbb{R}^d$ into $\bar{\mathbb{R}}$ is continuous on the interior of its domain.*

We focus on the differentiability of convex functions.

**Definition B.1.2** *(Subdifferential)*
*Given a convex function $f$ from $\mathbb{R}^d$ into $\bar{\mathbb{R}}$, we define the subdifferential of $f$ in $x \in \mathbb{R}^d$, denoted by $\partial f(x)$, as the set of points $y$ in $\mathbb{R}^d$ such that*

$$f(x) + y.(z - x) \leq f(z), \quad \forall z \in \mathbb{R}^d.$$

**Proposition B.1.2** *Let $f$ be a convex function from $\mathbb{R}^d$ into $\bar{\mathbb{R}}$.*
*(1) If $f$ is finite and continuous at $x \in \mathbb{R}^d$, then $\partial f(x) \neq \emptyset$.*
*(2) $f$ is finite and differentiable at $x \in \mathbb{R}^d$ with gradient $Df(x)$ if and only if $\partial f(x)$ is reduced to a singleton and in this case $\partial f(x) = \{Df(x)\}$.*

## B.2 Fenchel-Legendre transform

**Definition B.2.3** *(Polar functions)*
*Given a function $f$ from $\mathbb{R}^d$ into $\bar{\mathbb{R}}$, we define the polar function (or conjugate) of $f$ as the function $\tilde{f}$ from $\mathbb{R}^d$ into $\bar{\mathbb{R}}$ where*

$$\tilde{f}(y) = \sup_{x \in \mathbb{R}^d} [x.y - f(x)], \quad y \in \mathbb{R}^d.$$

When $f$ is convex, we also say that $\tilde{f}$ is the Fenchel-Legendre transform of $f$. It is clear that in the definition of $\tilde{f}$, we may restrict in the supremum to the points $x$ lying in the domain of $f$. The polar function $\tilde{f}$ is defined as the pointwise supremum of the affine functions $y \to x.y - f(x)$. Thus, it is a convex function on $\mathbb{R}^d$.

We may also define the polar function of a polar function. We have the following bipolarity result.

**Theorem B.2.3** *(Fenchel-Moreau)*
*Let $f$ be a proper, convex l.s.c. function from $\mathbb{R}^d$ into $\bar{\mathbb{R}}$ and $\tilde{f}$ its Fenchel-Legendre transform. Then,*

$$f(x) = \sup_{y \in \mathbb{R}^d} \left[ x.y - \tilde{f}(y) \right], \quad x \in \mathbb{R}^d.$$

We state the connection between differentiability and polar functions.

**Proposition B.2.3** *Let $f$ be a proper, convex l.s.c. function from $\mathbb{R}^d$ into $\bar{\mathbb{R}}$ and $\tilde{f}$ its Fenchel-Legendre transform. Then, for all $x, y \in \mathbb{R}^d$, we have equivalence between*

$$y \in \partial f(x) \iff x \in \partial \tilde{f}(y) \iff f(x) = x.y - \tilde{f}(y).$$

**Proposition B.2.4** *Let $f$ be a proper, convex l.s.c. function from $\mathbb{R}^d$ into $\bar{\mathbb{R}}$, strictly convex on $\mathrm{int}(\mathrm{dom}(f))$. Then its Fenchel-Legendre transform $\tilde{f}$ is differentiable on $\mathrm{int}(\mathrm{dom}(\tilde{f}))$. Furthermore, if $f$ is differentiable on $\mathrm{int}(\mathrm{dom}(f))$, then the gradient of $f$, $Df$, is one-to-one from $\mathrm{int}(\mathrm{dom}(f))$ into $\mathrm{int}(\mathrm{dom}(f))$ with $Df = (D\tilde{f})^{-1}$ and $\tilde{f}$ is strictly convex on $\mathrm{int}(\mathrm{dom}(\tilde{f}))$.*

# B.3 Example in $\mathbb{R}$

In Chapter 7, Section 7.3, we often meet the following situation. We have a function $u : (0, \infty) \to \mathbb{R}$, increasing, concave on $(0, \infty)$ and we consider the function $\tilde{u} : (0, \infty) \to R \cup \{\infty\}$ defined by

$$\tilde{u}(y) = \sup_{x>0}[u(x) - xy], \quad y > 0.$$

$\tilde{u}$ is a decreasing function, convex on $(0, \infty)$, and we define $\mathrm{dom}(\tilde{u}) = \{y > 0 : \tilde{u}(y) < \infty\}$. The next proposition collects some results used in Section 7.3.

**Proposition B.3.5** *We have the conjugate relation*

$$u(x) = \inf_{y>0}[\tilde{u}(y) + xy], \quad x > 0, \tag{B.1}$$

*and*

$$\tilde{u}(0) := \lim_{y \downarrow 0} \tilde{u}(y) = u(\infty) := \lim_{x \to \infty} u(x).$$

*Suppose that $u$ is strictly concave on $(0, \infty)$. Then, $\tilde{u}$ is differentiable on $\mathrm{int}(\mathrm{dom}(\tilde{u}))$. Furthermore, if one of the two following equivalent conditions:*

*(i) u is differentiable on $(0, \infty)$*

*(ii) $\tilde{u}$ is strictly convex on $int(dom(\tilde{u}))$*

*is satisfied, then the derivative $u'$ is one-to-one from $(0, \infty)$ into $int(dom(\tilde{u})) \neq \emptyset$ with $I := (u')^{-1} = -\tilde{u}'$, and we have*

$$\tilde{u}(y) = u(I(y)) - yI(y), \quad \forall y \in int(dom(\tilde{u})).$$

*Finally, under the additional conditions*

$$u'(0) = \infty \quad and \quad u'(\infty) = 0, \tag{B.2}$$

*we have $int(dom(\tilde{u})) = dom(\tilde{u}) = (0, \infty)$.*

**Proof.** Since the function $u$ is concave, and finite on $(0, \infty)$, it satisfies a linear growth condition. It follows that $dom(\tilde{u}) \neq \emptyset$ and its interior is under the form

$$int(dom(\tilde{u})) = (y_0, \infty),$$

where $y_0 = \inf\{y > 0 : \tilde{u}(y) < \infty\}$.

Notice that $u(\infty)$ exists in $\bar{\mathbb{R}}$ by the increasing property of $u$ on $(0, \infty)$. Similarly, $\tilde{u}(0)$ exists in $\bar{\mathbb{R}}$. From the definition of $\tilde{u}$, we have $\tilde{u}(y) \geq u(x) - xy$ for all $x, y > 0$, and so $\tilde{u}(0) \geq u(\infty)$. Moreover, we have for all $y > 0$, $\tilde{u}(y) \leq \sup_{x>0} u(x) = u(\infty)$ by the increasing property $u$. This proves that $\tilde{u}(0) = u(\infty)$.

Let us consider the function $f$ from $\mathbb{R}$ into $\bar{\mathbb{R}}$ defined by

$$f(x) = \begin{cases} -u(x), & x \geq 0 \\ \infty, & x < 0. \end{cases}$$

$f$ is a proper, convex l.s.c. function on $\mathbb{R}$ and $int(dom(f)) = (0, \infty)$. Its Fenchel-Legendre transform is given by

$$\tilde{f}(y) = \sup_{x \in \mathbb{R}}[xy - f(x)] = \sup_{x>0}[xy + u(x)], \quad y \in \mathbb{R}.$$

When $y < 0$, we have by definition of $\tilde{u}$, $\tilde{f}(y) = \tilde{u}(-y)$. When $y > 0$, we have by the increasing property of $u$, $\tilde{f}(y) \geq \lambda x_0 y + u(x_0)$ for all $\lambda > 1$ and $x_0 > 0$ fixed. This proves that $\tilde{f}(y) = \infty$ for $y > 0$. For $y = 0$, we have $\tilde{f}(0) = \sup_{x>0} u(x) = u(\infty) = \tilde{u}(0)$. We thus get

$$\tilde{f}(y) = \begin{cases} \tilde{u}(-y), & y \leq 0 \\ \infty, & y > 0, \end{cases}$$

and $int(dom(\tilde{f})) = -int(dom(\tilde{u})) = (-\infty, -y_0)$. From the bipolarity theorem B.2.3, we have for all $x \in \mathbb{R}$

$$f(x) = \sup_{y \in \mathbb{R}}[xy - \tilde{f}(y)] = \sup_{y<0}[xy - \tilde{u}(-y)]$$

$$= \sup_{y>0}[-xy - \tilde{u}(y)] = -\inf_{y>0}[xy + \tilde{u}(y)].$$

In particular, we deduce the relation (B.1) for $x > 0$.

Moreover, if $u$ is strictly concave on $(0, \infty)$, then $f$ is strictly convex on $\mathrm{int}(\mathrm{dom}(f))$. From Proposition B.2.4, $\tilde{f}$ is then differentiable on $\mathrm{int}(\mathrm{dom}(\tilde{f}))$, i.e. $\tilde{u}$ is differentiable on $\mathrm{int}(\mathrm{dom}(\tilde{u}))$.

The equivalence between conditions $(i)$ and $(ii)$ of the proposition follows from the equivalence between:

(i') $f$ is differentiable on $\mathrm{int}(\mathrm{dom}(f))$

(ii') $\tilde{f}$ is strictly convex on $\mathrm{int}(\mathrm{dom}(\tilde{f}))$.

This is indeed a consequence of Proposition B.2.4 applied on one hand to $f$ and on the other hand to $\tilde{f}$, which is also proper, convex and l.s.c. on $\mathbb{R}$. Under one of these conditions, we deduce that $f'$ is one-to-one from $\mathrm{int}(\mathrm{dom}(f))$ into $\mathrm{int}(\mathrm{dom}(\tilde{f}))$ with $(f')^{-1} = \tilde{f}'$ and by Proposition B.2.3, we have for all $y \in \mathrm{int}(\mathrm{dom}(\tilde{f}))$

$$\tilde{f}(y) = xy - f(x) \quad \text{where } x = \tilde{f}'(y), \text{ i.e. } y = f'(x).$$

This proves the required relations on $u$ and $\tilde{u}$.

Finally, under the conditions (B.2), the function $f'$ maps $(0, \infty)$ into $(-\infty, 0) = \mathrm{int}(\mathrm{dom}(\tilde{f}))$, which means that $\mathrm{int}(\mathrm{dom}(\tilde{u})) = (0, \infty) = \mathrm{dom}(\tilde{u})$.    $\square$

# References

[ADEH99] Artzner P., F. Delbaen, J.M. Eber and D. Heath (1999): "Coherent measures of risk", *Mathematical Finance*, 9, 203-228.

[ALP95] Avellaneda M., A. Levy and A. Paras (1995): "Pricing and hedging derivative securities in markets with uncertain volatilities", *Applied Mathematical Finance*, 2, 73-88.

[Ba95] Barles G. (1995): Solutions de viscosité des équations d'Hamilton-Jacobi, Springer-Verlag, Mathématiques and Applications.

[BElk04] Barrieu P. and N. El Karoui (2004): "Optimal design of derivatives under dynamic risk measures", Proceedings of the AMS, Spring.

[BF02] Bellini F. and M. Frittelli (2002): "On the existence of minimax martingale measures", *Mathematical Finance*, 12, 1-21.

[Be57] Bellman R. (1957): Dynamic programming, Princeton University Press.

[Ben92] Bensoussan A. (1992): Stochastic control of partially observable systems, Cambridge University Press.

[BL78] Bensoussan A. and J.L. Lions (1978): Applications des inéquations variationnelles en contrôle stochastique, Dunod.

[BL82] Bensoussan A. and J.L. Lions (1982): Contrôle impulsionnel et inéquations quasi-variationnelles contrôle stochastique, Dunod.

[BN91] Bensoussan A. and H. Nagai (1991): "An ergodic control problem arising from the principal eigenfunction of an elliptic operator", *J. Math. Soc. Japan*, 43, 49-65.

[BeBo07] Bentahar I. and B. Bouchard (2007): "Explicit characterization of the super-replication strategy in financial markets with partial transaction costs", *Stochastic Processes and Their Applications*, 117, 655-672.

[BeSh78] Bertsekas D. and S. Shreve (1978): Stochastic optimal control; the discrete-time case, Math. in Sci. and Eng., Academic Press.

[BGP00] Biagini F., P. Guasoni and M. Pratelli (2000): "Mean-variance hedging for stochastic volatility models", *Mathematical Finance*, 10, 109-123.

[BP99] Bielecki T. and S. Pliska (1999): "Risk-sensitive dynamic asset management", *Applied Math. Optim.*, 39, 337-360.

[Bis76] Bismut, J.M. (1976): "Théorie probabiliste du contrôle des diffusions", *Mem. Amer. Math. Soc.*, 4, n° 167.

[Bis78] Bismut J.M. (1978): "Contrôle des systèmes linéaires quadratiques: applications de l'intégrale stochastique", Sem. Prob. XII, Lect. Notes in Math., 649, 180-264.

[BElJM08] Blanchet-Scalliet C., N. El Karoui, M. Jeanblanc and L. Martellini (2008): "Optimal investment decisions when time horizon is uncertain", *Journal of Mathematical Economics*, 44, 1100-1113.

224    References

[Bor89] Borkar V. (1989): Optimal control of diffusion processes, Pitman Research Notes in Math., 203.

[BM03] Bouchard B. and L. Mazliak (2003): "A multidimensional bipolar theorem in $L^0(\mathbb{R}^d, \Omega, \mathcal{F}, P)$", Stoch. Proc. Applic., 107, 213-231.

[BO94] Brekke K. and B. Oksendal (1994): "Optimal switching in an economic activity under uncertainty", SIAM J. Cont. Optim., 32, 1021-1036.

[BuH98] Buckdahn R. and Y. Hu (1998): "Hedging contingent claims for a large investor in an incomplete market", Advances in Applied Probability, 30, 239-255.

[CaDa06] Carlier G. and R.A. Dana (2006): "Law invariant concave utility functions and optimization problems with monotonicity and comonotonicity constraints", Statistics and Decisions, 24, 127-152.

[CeKal07] Cerny A. and J. Kallsen (2007): "On the structure of general mean-variance hedging strategies", Annals of Probability, 35, 1479-1531.

[CS93] Chatelain M. and C. Stricker (1995): "Componentwise and vector stochastic integration with respect to certain multi-dimensional continuous local martingales", Seminar on Stochastic Analysis, Random fields and Applications, Ascona 1993, Prog. Proba., 36, 319-325.

[CST05] Cheridito P., M. Soner and N. Touzi (2005): "The multi-dimensional super-replication problem under Gamma constraints", Annales Inst. H. Poincaré, Anal. non linéaire, 22, 633-666.

[CTZ03] Choulli T., M. Taksar and X.Y. Zhou (2003): "A diffusion model for optimal dividend distribution with constraints on risk control", SIAM J. Cont. Optim., 41, 1946-1979.

[CH89] Cox J. and C.F. Huang (1989): "Optimal consumption and portfolio policies when asset prices follow a diffusion process", Journal of Economic Theory, 49, 33-83.

[CIL92] Crandall M., H. Ishii and P.L. Lions (1992): "User's guide to viscosity solutions of second order partial differential equations", Bull. Amer. Math. Soc., 27, 1-67.

[CKS98] Cvitanic J., I. Karatzas and M. Soner (1998): "Backwards stochastic differential equations with constraints on the gains process", Annals of Probability, 26, 1522-1551.

[CPT99a] Cvitanic J., H. Pham and N. Touzi (1999): "Superreplication in stochastic volatility models under portfolio constraints", Journal of Appplied Probability, 36, 523-545.

[CPT99b] Cvitanic J., H. Pham and N. Touzi (1999): "A closed form solution for the super-replication problem under transaction costs", Finance and Stochastics, 3, 35-54.

[Da77] Davis M. (1977): Linear estimation and stochastic control, Chapman and Hall.

[DN90] Davis M. and A. Norman (1990): "Portfolio selection with transaction costs", Math. of Oper. Research, 15, 676-713.

[DGRSSS02] Delbaen F., P. Grandits, T. Rheinländer, D. Samperi, M. Schweizer and C. Stricker (2002): "Exponential hedging and entropic penalties", Mathematical Finance, 12, 99-123.

[DS94] Delbaen F. and W. Schachermayer (1994): "A general version of the fundamental theorem of asset pricing", Math. Annalen, 300, 463-520.

[DS96] Delbaen F. and W. Schachermayer (1996): "The variance-optimal martingale measure for continuous processes", Bernoulli, Vol 2, 1, 81-105.

[DM75] Dellacherie C. and P.A. Meyer (1975): Probabilités and potentiel, ch. I à IV, Théorie des martingales, Hermann.

[DM80] Dellacherie C. and P.A. Meyer (1980): Probabilités and potentiel, ch. V à VIII, Théorie des martingales, Hermann.

[DeMa06] Denis L. and C. Martini (2006): "A theoretical framework for the pricing of contingent claims in the presence of model uncertainty", Annals of Applied Probability, 16, 827-852.

[DP94] Dixit A. and R. Pindick (1994): Investment under uncertainty, Princeton University Press.

[Do94]  Doob J.L. (1994): Measure theory, Springer-Verlag.

[DZ00]  Duckworth K. and M. Zervos (2000): "An investment model with entry and exit decisions", *J. Applied Prob.*, 37, 547-559.

[DZ01]  Duckworth K. and M. Zervos (2001): "A model for investment decisions with switching costs", *Annals of Applied Probability*, 11, 239-250.

[Dyn63]  Dynkin E. (1963): "The optimal choice of the instant for stopping a Markov process", *Dolk. Acad. Nauk USSR*, 150, 238-240.

[ET74]  Ekeland I. and R. Temam (1974): Analyse convexe and problèmes variationnels, Dunod.

[Elk81]  El Karoui N. (1981): Les aspects probabilistes du contrôle stochastique, Lect. Notes in Math., 816, Springer-Verlag.

[ElkNJ87]  El Karoui N., D. Huu Nguyen and M. Jeanblanc-Picqué (1987): "Compactification methods in the control of degenerate diffusions: existence of optimal controls", *Stochastics and Stochastics Reports*, 20, 169-219.

[EKPPQ97]  El Karoui N., C. Kapoudjian, E. Pardoux, S. Peng and M.C. Quenez (1997): "Reflected solutions of backward SDEs, and related obstacle problems for PDEs", *Annals of Probability*, 25, 702-737.

[ElkM97]  El Karoui N. and L. Mazliak (editors) (1997): Backward stochastic differential equations, Pitman Research Notes in Mathematics Series.

[ElkPQ97]  El Karoui N., S. Peng and M.C. Quenez (1997): "Backward stochastic differential equations in finance", *Mathematical Finance*, 7, 1-71.

[ElkQ95]  El Karoui N. and M.C. Quenez (1995): "Dynamic programming and pricing contingent claims in incomplete markets", *SIAM J. Cont. Optim.*, 33, 29-66.

[ElkR00]  El Karoui N. and R. Rouge (2000): "Pricing via utility maximization and entropy", *Mathematical Finance*, 7, 1-71.

[Em79]  Emery M. (1979): "Une topologie sur l'espace des semimartingales", *Sem. de Prob.*, XIII, vol. 721, 152-160, Lect. Notes in Math., Springer-Verlag.

[FM95]  Fleming W. and W. McEneaney (1995): "Risk-sensitive control on an infinite horizon", *SIAM J. Cont. and Optim.*, 33, 1881-1915.

[FR75]  Fleming W. and R. Rishel (1975): Deterministic and stochastic optimal control, Springer-Verlag.

[FP05]  Fleming W. and T. Pang (2005): "A stochastic control model of investment, production and consumption", *Quarterly of Applied Mathematics*, 63, 71-87.

[FS00]  Fleming W. and S. Sheu (2000): "Risk sensitive control and an optimal investment model", *Math. Finance*, 10, 197-213.

[FSo93]  Fleming W. and M. Soner (1993): Controlled Markov processes and viscosity solutions, Springer-Verlag.

[FoK98]  Föllmer H. and Y. Kabanov (1998): "Optional decomposition and Lagrange multipliers", *Finance and Stochastics*, 1, 69-81.

[FoS02]  Föllmer H. and A. Schied (2002): Stochastic finance. An introduction in discrete-time., Berlin de Gruyter Studies in Mathematics.

[FoS86]  Föllmer H. and D. Sondermann (1986): "Hedging of non-redundant contingent claims", *Contributions to Mathematical Economics*, eds. A. Mas-Colell and W. Hildenbrand, North-Holland, 205-223.

[Fr75]  Friedman A. (1975): Stochastic differential equations and applications, Vol. 1, Academic Press.

[FBS09]  Frittelli M, S. Biagini and G. Scandolo (2009): Duality in mathematical finance, Springer, forthcoming.

[FG04]  Frittelli M. and M. Rosazza Gianin (2004): "Dynamic convex risk measures", New risk measures in the 21th century, G. Szego ed., Wiley.

[Ga76] Galtchouk L. (1976): "Représentation des martingales engendrées par un processus à accroissement indépendants", *Ann. Inst. H. Poincaré*, 12, 199-211.

[GS72] Gihman I. and A. Skorohod (1972): Stochastic differential equations, Springer-Verlag.

[GT85] Gilbarg D. and N. Trudinger (1985): Elliptic differential equations of second order, Springer-Verlag.

[GS89] Gilboa I. and D. Schmeidler (1989): "Maxmin expected utility with non-unique prior", *J. Math. Econ.*, 18, 141-153.

[GLP98] Gouriéroux C., J.P. Laurent and H. Pham (1998): "Mean-variance hedging and numéraire", *Mathematical Finance*, 8, 179-200.

[GV02] Gozzi F. and T. Vargiolu (2002): "Superreplication of European multiasset derivatives with bounded stochastic volatility", *Mathematical Methods of Operations Research*, 55(1), 69-91.

[Gu05] Gundel A. (2005): "Robust utility maximization for complete and incomplete market models", *Finance and Stochastics*, 9, 151-176.

[GP05] Guo X. and H. Pham (2005): "Optimal partially reversible investment with entry decision and general production function", *Stoc. Proc. Applic.*, 115, 705-736.

[GS01] Guo X. and L. Shepp (2001): "Some optimal stopping problems with non-trivial boundaries for pricing exotic options," *Journal of Applied Probability*, 38, 647-658.

[HJ07] Hamadène S. and M. Jeanblanc (2007): "On the stopping and starting problem : application to reversible investment", *Mathematics of Operations Research*, 32, 182-192.

[HL95] Hamadène S. and J.-P. Lepeltier (1995): "Backward equations, stochastic control and zero-sum stochastic differential games", *Stochastics and Stochastic Reports*, 54, 221-231.

[HP81] Harrison M. and S. Pliska (1981): "Martingales and stochastic integrals in the theory of continuous trading", *Stoch. Proc. Applic.*, 11, 215-260.

[HaNaSh08] Hata H., H. Nagai and S.J. Sheu (2008): "Asymptotics of the probability minimizing down-side risk", preprint, Osaka University.

[HeP91] He H. and N.D. Pearson (1991): "Consumption and portfolio policies with Incomplete markets and short sale constraints", *Journal of Economic Theory*, 54, 259-305.

[Hi80] Hida T. (1980): Brownian motion, Springer-Verlag.

[HIM05] Hu Y., P. Imkeller and M. Müller (2005): "Utility maximization in incomplete markets", *Ann. Appl. Probability*, 15, 1691-1712.

[IW81] Ikeda N. and S. Watanabe (1981): Stochastic differential equations and diffusion processes, North-Holland.

[Ish89] Ishii H. (1989): "On uniqueness and existence of viscosity solutions of fully nonlineqr second order elliptic PDE's", *Comm. Pure. Appl. Math.*, 42, 15-45.

[Ja93] Jacka S. (1993): "Local times, optimal stopping and semimartingales", *Annals of Probability*, 21, 329-339.

[Jac79] Jacod J. (1979): Calcul stochastique et problèmes de martingales, Lect. Notes in Math., 714, Springer-Verlag.

[JLL90] Jaillet P., D. Lamberton and B. Lapeyre (1990): "Variational inequalities and the pricing of American options", *Acta Appl. Math.*, 21, 263-289.

[JS95] Jeanblanc-Picqué M. and A. Shiryaev (1995): "Optimization of the flow of dividends", *Russian Math. Surveys*, 50, 257–277.

[Je88] Jensen R. (1988): "The maximum principle for viscosity solutions of second order fully nonlinear partial differential equations", *Arch. Rat. Mech. Anal.*, 101, 1-27.

[JiZh08] Jin H. and X.Y. Zhou (2008): "Behavorial portfolio selection in continuous time", *Mathematical Finance*, 18, 385-426.

[KS81] Kamien M. and N. Schwartz (1981): Dynamic optimization, North Holland.

[Kar80] Karatzas I. (1980): "On a stochastic representation for the principal eigenvalue of a second order differential equation", *Stochastics and Stochastics Reports*, 3, 305-321.

[KLS87] Karatzas I., J. Lehoczky and S. Shreve (1987): "Optimal portfolio and consumption decisions for a small investor on a finite horizon", *SIAM J. Cont. Optim.*, 25, 297-323.

[KLSX91] Karatzas I., J. Lehoczky, S. Shreve and G. Xu (1991): "Martingale and duality methods for utility maximization in incomplete market", *SIAM J. Cont. Optim.*, 29, 702-730.

[KaSh88] Karatzas I. and S. Shreve (1988): Brownian motion and stochastic calculus, Springer-Verlag.

[KaSh98] Karatzas I. and S. Shreve (1998): Methods of mathematical finance, Springer-Verlag.

[KMZ98] Knudsen T., B. Meister and M. Zervos (1998): "Valuation of investments in real assets with implications for the stock prices", *SIAM Journal on Control and Optimization*, 36, 2082-2102.

[Ko00] Kobylanski M. (2000): "Backward stochastic differential equations and partial differential equations with quadratic growth", *Annals of Probability*, 28, 558-602.

[KT02] Kohlmann M. and S. Tang (2002): "Global adapted solution of one-dimensional backward stochastic differential Riccati equations, with application to the mean-variance hedging", *Stochastic Process. Appl.*, 97, 255–288.

[KZ00] Kohlmann M. and X.Y. Zhou (2000): "Relationship between backward stochastic differential equations and stochastic controls: a linear-quadratic approach", *SIAM Journal on Control and Optimization*, 38, 1392-1407.

[Kom67] Komlos J. (1967): "A generalisation of a theorem of Steinhaus", *Acta Math. Acad. Sci. Hung.*, 18, 217-229.

[Kor97] Korn R. (1997): Stochastic models for optimal Investment and risk management in continuous time, World Scientific.

[Kr96] Kramkov D. (1996): "Optional decomposition of supermartingales and hedging contingent claims in incomplete security markets", *Prob. Theor. Rel. Fields*, 145, 459-480.

[KS99] Kramkov D. and W. Schachermayer (1999): "The asymptotic elasticity of utility functions and optimal investment in incomplete markets", *Annals of Applied Probability*, 9, 904-950.

[KS01] Kramkov D. and W. Schachermayer (2001): "Necessary and sufficient conditions in the problem of optimal investment in incomplete markets", *Annals of Applied Probability*, 13, 1504-1516.

[Kry80] Krylov N. (1980): Controlled diffusion processes, Springer-Verlag.

[Kry87] Krylov N. (1987): Nonlinear elliptic and parabolic equations of second order, Boston, D. Reidel.

[KW67] Kunita H. and S. Watanabe (1967): "On square integrable martingales", *Nagoya Math. J.*, 30, 209-245.

[Ku75] Kushner H. (1975): "Existence results for optimal stochastic controls", *J. Optim. Theory and Appl.*, 15, 347-359.

[Lam98] Lamberton D. (1998): "American options", *Statistics and Finance*, D. Hand, S. Jacka eds. Arnold.

[Las74] Lasry J.M. (1974): Thèse détat, Université Paris Dauphine.

[LP99] Laurent J.P. and H. Pham (1999): "Dynamic programming and mean-variance hedging", *Finance and stochastics*, 23, 83-110.

[LeG89] Le Gall J.F. (1989): "Introduction au mouvement Brownien", *Gazette des mathématiciens*, 40, 43-64. Soc. Math. France.

[LMX05] Lepeltier J.P., A. Matoussi and M. Xu (2005): "Reflected backward stochastic differential equations under monotonicity and general increasing growth conditions", *Advanced Applied Probability*, 37, 1-26.

[Lio83] Lions P.L. (1983): "Optimal control of diffusion processes and Hamilton-Jacobi-Bellman equations", *Comm. P.D.E.*, 8, Part I, 1101-1134, Part II, 1229-1276.

[LS84] Lions P.L. and A. Snitzman (1984): "Stochastic differential equations with reflecting boundary conditions", *Comm. Pure. Appl. Math.*, 37, 511-537.

[LP07] Ly Vath V. and H. Pham (2007): "Explicit solution to an optimal switching problem in the two-regime case", *SIAM Journal on Control and Optimization*, 395-426.

[MY00] Ma J. and J. Yong (2000): Forward-backward stochastic differential equations and their applications, Lect. Notes in Math., 1702.

[Mac65] McKean H.P. (1965): "A free boundary problem for the heat equation arising from a problem in mathematical economics", appendix to a paper by R. Samuelson, *Indus. Man. Rev.*, 6, 32-39.

[Ma03] Mania M. (2003): "A semimartingale backward equation and the variance-optimal martingale measure under general information flow", *SIAM J. Control Optim.*, 42, 1703-1726.

[Ma52] Markowitz H. (1952): "Portfolio selection", *J. of Finance*, 7, 77-91.

[Mer69] Merton R. (1969): "Lifetime portfolio selection under uncertainty: the continuous time case", *Rev. Econ. Stat.*, 51, 239-265.

[Mer73] Merton R. (1973): "Optimum consumption and portfolio rules in a continuous-time model", *J. Econ. Theory*, 3, 373-413.

[MuZa07] Musiela M. and T. Zariphopoulou (2007): "Portfolio choice under dynamic investment performance criteria", to appear in *Quantitative Finance*.

[Na03] Nagai H. (2003): "Optimal strategies for risk-sensitive portfolio optimization problems for general factor models", *SIAM J. Cont. Optim.*, 41, 1779-1800.

[Nev75] Neveu J. (1975): Martingales à temps discret, Masson.

[Nis81] Nisio M. (1981): Lectures on stochastic control theory, ISI Lect. Notes, 9, Kaigai Publ. Osaka.

[Oks00] Øksendal B. (2000): Stochastic differential equations: an introduction with applications, 6th edition, Springer-Verlag.

[OR98] Øksendal B. and K. Reikvam (1998): "Viscosity solutions of optimal stopping problems", *Stoc. and Stoc. Reports*, 62, 285-301.

[OS02] Øksendal B. and A. Sulem (2002): "Optimal consumption and portfolio with both fixed and proportional transaction costs: a combined stochastic control and impulse control model", *SIAM J. Control and Optim.*, 40, 1765-1790.

[OS04] Øksendal B. and A. Sulem (2004): Applied stochastic control of jump diffusion, Springer-Verlag.

[Pa98] Pardoux E. (1998): "Backward stochastic differential equations and viscosity solutions of systems of semilinear parabolic and elliptic PDEs of second order", Stochastic analysis and related topics, VI (Geilo, 1996), 79–127, Progr. Probab., 42, Birkhäuser, Boston, MA, 1998.

[PaPe90] Pardoux E. and S. Peng (1990): "Adapted solutions of a backward stochastic differential equation", *Systems and Control Letters*, 14, 55-61.

[Pe99] Peng S. (1999): "Monotonic limit theory of BSDE and nonlinear decomposition theorem of Doob-Meyer's type", *Prob. Theory and Rel. Fields*, 113, 473-499.

[PeSh06] Peskir G. and A. Shiryaev (2006): Optimal stopping and free boundary problems, Lect. in Mathematics, ETH Zürich, Birkhäuser.

[Pha98] Pham H. (1998): "Optimal stopping of controlled jump diffusion processes: a viscosity solution approach", *J. Math. Sys. Est. Cont.*, 8, 1-27 (electronic).

[Pha03a] Pham H. (2003a): "A large deviations approach to optimal long term investment'", *Finance and Stochastics*, 7, 169-195.

[Pha03b] Pham H. (2003b): "A risk-sensitive control dual approach to a large deviations control problem", *Systems and Control Letters*, 49, 295-309.

[Pha07] Pham H. (2007): "On the smooth-fit property for one-dimensional optimal switching problem", *Séminaire de Probabilités*, XL, 187-201.

[PLZ07] Pham H., V. Ly Vath and X.Y. Zhou (2007): "Optimal switching over multiple regimes", Preprint University Paris 7.

[PRS98] Pham H., T. Rheinländer and M. Schweizer (1998): "Mean-variance hedging for continuous processes : new proofs and examples", *Finance and Stochastics*, 2, 173-198.

[Pli86] Pliska S. (1986): "A stochastic calculus model of continuous trading: optimal portfolios", *Math. Oper. Res.*, 11, 371-382.

[Pro90] Protter P. (1990): Stochastic integration and differential equations, Springer-Verlag.

[Rev94] Revuz D. (1994): Mesures and intégration, Hermann.

[Rev97] Revuz D. (1997): Probabilités, Hermann.

[ReY91] Revuz D. and M. Yor (1991): Continuous martingale and Brownian motion, Springer-Verlag.

[RhS97] Rheinländer T. and M. Schweizer (1997): "On $L^2$-projections on a space of stochastic integrals", *Annals of Probability*, 25, 1810-1831.

[Ro70] Rockafellar R. (1970): Convex analysis, Princeton University Press.

[Scha01] Schachermayer W. (2001): "Optimal investment in incomplete markets when wealth may become negative", *Annals of Applied Probability*, 11, 694-734.

[Schi05] Schied A. (2005): "Optimal investment for robust utility functionals in complete markets", *Mathematics of Operations Research*, 30, 750-764.

[Schm08] Schmidli H. (2008): Stochastic control in insurance, Springer.

[Schw96] Schweizer M. (1996): "Approximation pricing and the variance-optimal martingale measure", *Annals of Probability*, 24, 206-236.

[SS87] Seierstad A. and K. Sydsaeter (1987): Optimal control theory with economic applications, North Holland.

[Se06] Sekine J. (2006): "Exponential hedging by solving a backward stochastic differential equation: an alternative approach", *Applied Mathematics and Optimization*, 54, 131-158.

[SZ94] Sethi S.P. and Q. Zhang (1994): Hierarchical decision making in stochastic manufacturing systems, in series Systems and Control: Foundations and Applications, Birkhauser Boston, Cambridge.

[Sh78] Shiryayev A. (1978): Optimal stopping rules, Springer-Verlag.

[ShSo94] Shreve S. and M. Soner (1994): "Optimal investment and consumption with transaction costs", *Annals of Applied Probability*, 4, 609-692.

[ST00] Soner M. and N. Touzi (2000): "Super replication under gamma constraints", *SIAM Journal on Control and Optimization*, 39 (1), 73-96.

[ST02] Soner M. and N. Touzi (2002): "Stochastic target problems, dynamic programming and viscosity solutions", *SIAM Journal on Control and Optimization*, 41, 404-424.

[Str85] Strasser H. (1985): Mathematical theory of statistics: statistical experiments and asymptotic decision theory, W. de Gruyter.

[TY93] Tang S. and J. Yong (1993): "Finite horizon stochastic optimal switching and impulse controls with a viscosity solution approach", *Stoc. and Stoc. Reports*, 45, 145-176.

[T04] Touzi N. (2004): "Stochastic control problem, viscosity solutions and applications to finance", *Scuola Normale Superiore Pisa*, Quaderni.

[Van76] Van Moerbecke P. (1976): "On optimal stopping and free boundary problems", *Arch. Rat. Mech.*

[YZ00] Yong J. and X.Y. Zhou (2000): Stochastic controls, Hamiltonian systems and HJB equations, Springer-Verlag.

[Yo78] Yor M. (1978): "Sous-espaces denses dans $L^1$ ou $H^1$ and représentation des martingales", Séminaire de Probabilités XII, Lect. Notes in Math., 649, Springer-Verlag, 265-309.

[Zar88] Zariphopoulou T. (1988): Optimal investment-consumption models with constraints, PhD thesis, Brown University.

[ZL00] Zhou X.Y. and D. Li (2000): "Continuous-time mean-variance portfolio selection: a stochastic LQ framework", *Applied Mathematics and Optimization*, 42, 19-33.

# Index

H. Pham, *Continuous-time Stochastic Control and Optimization with Financial
Applications*, Stochastic Modelling and Applied Probability 61,
DOI 10.1007/978-3-540-89500-8, © Springer-Verlag Berlin Heidelberg 2009

# Stochastic Modelling and Applied Probability
## formerly: Applications of Mathematics

# Stochastic Modelling and Applied Probability
formerly: Applications of Mathematics